Ethnobotany

This book is dedicated to the memory of Diana Burlingham

Photographs courtesy of Cath Cotton

Ethnobotany
Principles and Applications

C.M. Cotton
School of Life Sciences,
Roehampton Institute London, UK

JOHN WILEY & SONS
Chichester • New York • Brisbane • Toronto • Singapore

Other Wiley Editorial Offices

John Wiley & Sons, Inc., 605 Third Avenue,
New York, NY 10158-0012, USA

Jacaranda Wiley Ltd, 33 Park Road, Milton,
Queensland 4064, Australia

John Wiley & Sons (Canada) Ltd, 22 Worcester Road,
Rexdale, Ontario M9W 1L1, Canada

John Wiley & Sons (Asia) Pte Ltd, 2 Clementi Loop #02-01
Jin Xing Distripark, Singapore 0512

Library of Congress Cataloging-in-Publication Data
Cotton, C. M.
    Ethnobotany : principles and applications / C. M. Cotton.
        p.   cm.
    Includes bibliographical references and index.
    ISBN 0 471 95537 X (alk. paper)
    1. Ethnobotany   I. Title.
    GN476.73.C67   1996
    581.6'1—dc20   95-5721
                                                          CIP

British Library Cataloguing in Publication Data

A catalogue record for this book is available from the British Library

ISBN 0 471 95537 X

Typeset in 10/12pt Century Schoolbook by Servis Filmsetting Ltd
Printed and bound by Antony Rowe Ltd, Eastbourne

Contents

Preface and Acknowlededgments   vii

1.   **Introduction to Ethnobotany**   1

2.   **Plant Structures, Functions and Applications**   19

3.   **Traditional Botanical Knowledge**   59

4.   **Methods in Ethnobotanical Study**   90

5.   **Traditional Botanical Knowledge and Subsistence: Wild Plant Resources**   127

6.   **Traditional Botanical Knowledge and Subsistence: Domesticated Plants and Traditional Agriculture**   159

7.   **Plants in Material Culture**   190

8.   **Traditional Phytochemistry**   216

9.   **Understanding Traditional Plant Use and Management: Indigenous Perceptions of the Natural World**   245

10.   **The History of Plant–Human Interaction: Palaeoethnobotanical Evidence**   278

11.   **Applied Ethnobotany: Commercialisation and Conservation**   313

12.   **Applying Ethnobotany in Sustainable Development: Practical Considerations**   347

Contents

Bibliography	**375**
Postscript	**401**
Index	**402**

Preface and Acknowledgements

The intention behind this text is that it will enable students of ethnobotany to get to grips with the main principles and potential applications of this diverse and fascinating discipline. In addition, while it is aimed primarily at an undergraduate audience, I hope that the book will also prove useful to established anthropologists (as an introduction to plant biology) and to botanists (as an introduction to anthropology). Whether these aims are achieved however, remains to be seen.

The book itself has been organised such that it may be read in two ways: a general reading of the main text gives a fairly brief overview of the subject under discussion, while reference to the many tables, figures and boxes will allow the reader to gain much more detailed information on his or her particular areas of interest. Each of the tables and boxes has therefore been presented as completely self contained units, which may be consulted directly by interested parties.

In addition, while limitations on both time and space have prohibited the inclusion of a glossary, any specialist terms in the text have been italicised, and reference to standard botanical or anthropological dictionaries should prove enlightening. For example, Allaby's *Concise Oxford Dictionary of Botany* and Seymour-Smith's *Macmillan Dictionary of Anthropology* are both referenced in the bibliography.

A large number of people have been crucial to the production of this text. Many have been supportive—if not long-suffering—others have kindly given me both their time and the benefit of their considerable expertise. I would sincerely like to thank all of these people, many of whom I have not met, as without their help the book would not have been written.

For their help in providing specialist information and encouragement I would like to thank Dan Austin (Florida Atlantic University, US),

Mike Balick (New York Botanical Garden, US), Wendy Beck (University of New England, Australia), Carole Biggam (University of Glasgow, UK), Vaughn Bryant (Texas A&M University, US), Robert Bye (Universidad Nacional Autónoma de México, Mexico), Bob Carling (Chapman & Hall, UK), Brian Coppins (Royal Botanic Garden Edinburgh, UK), Paul Cox (Brigham Young University, US), BJ Cundy (Australian Institute of Aboriginal and Torres Strait Islander Studies, Australia), Brian Egloff (University of Canberra, Australia), Roy Ellen (University of Kent, UK), Trish Flaster (Shaman Pharmaceuticals, US), Gabrielle Hatford (Norwich, UK), Toby Hodgkin (International Plant Genetic Resources Institute, Italy), Peter Houghton (King's College London, UK), Johanne Kawell (University of California, US), Peter Lapinskas (Scotia Pharmaceuticals Ltd, UK), Edelmira Linares (Universidad Nacional Autónoma de México, Mexico), Bitte Linder (Swedish University of Agricultural Sciences, Sweden), Lynda Macfarlane (Griffith University, Australia), Russell McGregor (James Cook University of North Queensland, Australia), Brian Meilleur (Missouri Botanical Garden, US), William Milliken (Royal Botanic Garden Kew, UK), Sue Minter (Chelsea Physic Garden, UK), Sharon Moroschan (University of Alberta, Canada), Brian Morris (University of London, UK), John Murray (Australian Catholic University, Australia), Mike Norton-Griffiths (Center for Social and Economic Research on the Global Environment), Ríonach uí Ógáin (University College Belfield, Dublin), Oliver Phillips (Missouri Botanical Garden, US), Darrell Posey (Oxford University, UK), Tony Rains (Winchester, UK), Amala Raman (King's College London, UK), Daniela Soleri (Centre for People, Food and Environment, US), Roy Vickery (The Natural History Museum, UK), GW von Liebestein (Centre for International Research and Advisory Networks, Netherlands), Fiona Watson (Survival International, UK), Fiona Wilson (Scotia Pharmaceuticals Ltd, UK), and Peter Winkler (National Geographic Society, US).

For their help and patience during the preparation and production of the text I would like to thank Mandy Collison, Mike Davis, Elaine Hutton and Hilary Rowe (all of John Wiley & Sons, Chichester).

For covering many of my duties at work during the presentation of the manuscript I would like to thank (enormously) Clive Bullock, Ann Maclarnon, Claire Ozanne, Graham Reed, Nigel Reeve, Chris Roger and Caroline Ross (all of the Roehampton Institute London). I would also like to thank Pete Wesson both for granting me study leave and for his moral support.

Special thanks are also due to Imogen Palmer who produced the graphs, and Jim Brown who produced the line drawings in their original form.

Finally, I must thank a number of people who have given me both moral support and help or advice on the text: Jackie Bailey, Charlie Errock, Joan King, Claire Ozanne, Tony and Mary Rains, Nick Swann and of course, John and Wendy Cotton.

C.M. Cotton

Chapter 1

Introduction to Ethnobotany

The best ethnobotanist would be a member of an ethnic minority who, trained in both botany and anthropology, would study . . . the traditional knowledge, cultural significance, and the management and uses of the flora. And it would be even better—for him and his people—if his study could result in economic and cultural benefits for his own community.

A Barrera (cited in Martin 1995).

INTRODUCTION

Since its conception in 1895 (Harshberger 1896), '*ethnobotany*' has proved a rather difficult term to define. Harshberger himself regarded it as simply 'the use of plants by aboriginal peoples', yet during the century which has intervened, considerable attention has focused not only on how plants are used, but also on how they are perceived and managed, and on the reciprocal relationships between human societies and the plants on which they depend. As a result, ethnobotany has been repeatedly redefined and even now no definitive agreement in its interpretation has been reached (Yen 1993). However, for the purposes of this text, ethnobotany is considered to encompass all studies which concern the mutual relationships between plants and *traditional peoples*.

HISTORY AND DEVELOPMENT OF ETHNOBOTANICAL STUDY

Much of the controversy surrounding the definition of ethnobotany has stemmed from differences in the interests of workers involved in its study, which for many years has included students from several disciplines. For while early botanists concentrated on the economic potential of plants used by aboriginal societies, anthropologists argued the

Table 1.1   Changes in the interpretation of 'ethnobotany'. Prior to the introduction of the term ethnobotany in 1895, the study of traditional botanical knowledge focused almost entirely on the applications and economic potential of plants used by native peoples; during the first half of this century, however, anthropological and ecological aspects became increasingly important. Later, as ethnobotanical studies escalated during the 1980s, a whole range of interpretations appeared as more and more disciplines became involved, culminating with Martin's interpretation in late 1995

Date	Interpretation of ethnobotany	Source
1873	*Aboriginal botany**—the study of all forms of vegetation which aborigines used for commodities such as medicine, food, textiles and ornaments.	Powers 1873 (in Castetter 1944)
1895	*Ethno-botany**—the use of plants by aboriginal peoples.	Harshberger (1896)
1916	Not just a record of plant use, but the traditional impressions of the total environment as revealed through custom and ritual.	Robbins *et al.* (1916) (in Castetter 1944)
1932	Not only tribal economic botany, but the whole range of traditional knowledge of plants and plant life.	Gilmour (1932)
1941	The study of the relations which exist between humans and their ambient vegetation.	Schultes (1941) (in Castetter 1944)
1941	The study of the interrelations between 'primitive' humans and plants.	Jones (1941) (in Castetter 1944)
1981	The study of the *direct* relationships between humans and plants.	Ford (1978)
1990	The study of useful plants prior to commercialisation and eventual domestication.	Wickens (1990)
1993	The recording and evaluation of environmental knowledge that different cultures have accumulated throughout millennia.	FEB (1993)
1994	All studies (concerning plants) which describe local people's interaction with the natural environment.	Martin (1995)

* Terms in italics introduced at this time.

need to understand how different perceptions of the natural world could influence subsistence decisions. Similarly, as some concentrated solely on the study of non-literate societies, others broadened their studies to include *folk knowledge* held by 'middle class Americans' (Ford 1978). Some of the major changes in the interpretation of ethnobotany are summarised in Table 1.1

Ethnobotanical studies are now in progress throughout the world: in India and other parts of Asia, many projects are aimed at documenting knowledge of traditional medicinal plants; in Africa, traditional agricultural knowledge is increasingly incorporated into rural development

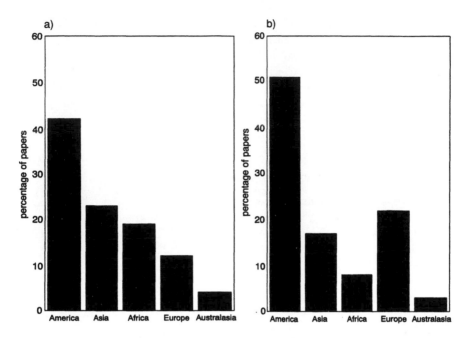

Figure 1.1   Current studies in ethnobotany. While ethnobotanical studies are currently under way in many parts of the world, a large proportion of this research is carried out throughout the American continent. (a) illustrates where much of the current ethnobotanical research is undertaken; (b) illustrates where workers who are publishing reports on ethnobotanical research are based. (Data taken from the international journals *Economic Botany* and *Journal of Ethnopharmacology* between 1992 and 1993)

programmes; in Australia, traditional methods of vegetation management are receiving considerable attention from the ecological community. However, a large proportion of the current research in ethnobotany remains focused on the American continent, where up to 41 per cent of studies are carried out (Figure 1.1). This is perhaps not surprising considering the continent's wealth of biological and cultural diversity, its rich archaeological record, and the European fascination for the New World since its Portuguese discovery just over 500 years ago.

Ethnobotany in the New World

Historically, the field of ethnobotany has belonged to the explorers and adventurers of Europe who observed and documented the uses of plants by the aboriginal peoples they encountered on their travels. For example, Christopher Columbus, discovered tobacco (*Nicotiana* spp) in

Cuba during his infamous voyage of 1492, while almost 350 years later, British explorer Richard Spruce first noted the psychoactive properties of the South American vine *Banisteriopsis caapi*; both resulted from observations of plants used by local peoples (Simpson & Conner-Ogorzaly 1986, Schultes 1983). These early botanical discoveries in the New World marked the beginnings of a long tradition of ethnobotanical study in the American continent—a tradition which culminated in the formalisation of ethnobotany as a field of academic study.

From Columbus to the Pilgrim Fathers (1492–1620)

On 12 October 1492, after more than 2 months at sea, the Genoese mariner Christopher Columbus made shore on the Bahaman island which he named San Salvador (Williamson 1992). Convinced that he had navigated a new westward sea passage to the mainland of Asia, and keen to assess the economic potential of his discovery he cruised the other islands of the Bahamas in search of gold. Yet when he finally arrived in what is now Cuba he found not gold but tobacco—a plant previously unknown to Europeans—which was rolled up and smoked by the native Cuban peoples. From this and his subsequent voyages Columbus finally brought back not only tobacco but also corn (*Zea mays*), allspice (*Pimenta dioica*) and cotton (*Gossypium* spp) all of which were collected on the basis of observations of local use (Hobhouse 1992; Simpson & Conner-Ogorzaly 1986).

During the centuries which followed, more native uses of plants were recorded by immigrants arriving in the New World from both mainland Europe and the British Isles: conquistadors and missionaries from Spain carefully documented the use of plants as foods and medicines by the Aztec, Maya and Inca peoples (Ortiz de Montellano 1990); Spanish settlers in South America were quick to modify native uses of rubber (*Hevea brasiliensis*) to waterproof their hats and cloaks (Simpson & Conner-Ogorzaly 1986); the Pilgrim Fathers—with some expertise in printing and shopkeeping, but little practical knowledge of farming or hunting—survived in their new home only by learning how to plant corn from members of the local Wampanoag Indians (Bryson 1994).

Most of these early observations documented the use of plants as foods and medicines, including potatoes (*Solanum tuberosum*), tobacco and cocoa (*Theobroma cacao*), many of which were adopted by the settlers. Soon, several of these new plants had been introduced to Europe, and by 1556, tobacco cultivation had begun in France, while corn—initially the mainstay of the ancient Mesoamerican and Andean civilisations—soon spread to many parts of the world and now constitutes one

Table 1.2 **Estimated market values for early ethnobotanical discover-ies—New World.** Many of the world's most economically important plants were originally discovered during the Middle Ages as a result of observations of their use by indigenous peoples in the New World—most of which were rapidly introduced into Europe following their discovery during the sixteenth century. The market values shown here are the global values estimated on the basis of data provided by the USDA (1994a, b; 1995a, b); Anon (1994); PT Data Consult Inc. (1994); ICCO (1995)

Commodity		Market value (US$ *per annum*)
Tobacco	(*Nicotiana* spp)	22 billion
Corn	(*Zea mays*)	36.5 billion
Rubber	(*Hevea brasiliensis*)	1.7 billion
Cocoa	(*Theobroma cacao*)	3.1 billion

of the most important grain crops produced world-wide. Together with other useful New World plants such as rubber, these early ethnobotanical discoveries now account for total sales worth tens of billions of dollars each year, thus illustrating the great importance of ethnobotany in Europe's economic history (Table 1.2).

An Era of Scientific Exploration (1663–1870)

Throughout the early period of colonial interest in the economic potential of the New World, the study of its ethnobotany was based largely on casual observation and anecdotal evidence. However, this type of information was increasingly supplemented by a number of more scientific reports as naturalists from Europe and elsewhere organised expeditions into these new territories. For example, almost half a century after the arrival of the Pilgrims, John Josselyn arrived in New England with the express intention 'to discover all along the Natural, physical and Chirurgicall Rarities of this New-found World' (Griggs 1981). There, he spent 8 years observing the use of herbs by native Indians, which he published in 1672 in his *New England's Rarities Discovered* (Griggs 1981).

By 1831, naturalists such as Charles Darwin were beginning to flood the museums and gardens of London with their collections of exotic corpses and plants. And although Darwin himself admitted to knowing 'no more about the plants which I had collected than the Man in the Moon' (Desmond & Moore 1992), other nineteenth century naturalists proved to be keen botanists. One of these was the Briton, Richard Spruce, a scientist and early explorer of the northwest Amazon and northern Andes, who has been described as one of the greatest naturalists ever (Schultes 1983). Between 1851 and 1854, he explored much of the Rio Negro and its tributaries, describing the ritual use of a number of psychoactive plants, as other botanists continued to document the useful plants of New England (see Ford 1978).

However, for centuries, this wealth of ethnobotanical information remained scattered in the chronicles of sixteenth century Spanish missionaries, the diaries of European adventurers, and the many works on Native American herbal medicine. Only in 1870, were these disparate data drawn together in a systematic treatise, as American botanist Edward Palmer published his *Food Products of the North American Indians* (see Castetter 1944). By this time, almost a century after the Declaration of American Independence, much of the research into indigenous plant use was carried out by settled Americans, and the formal emphasis shifted from Europe to America, heralding the beginnings of the new American tradition in the study of the science soon to become known as ethnobotany (Table 1.3).

A New Discipline Emerges (1873–1980s)

During the second half of the nineteenth century, the study of plants used by Native Americans had become more rigorous than previously, as accurate taxonomic descriptions of the continent's flora became available, and Palmer and another American botanist Stephen Powers, brought scientific exactness to the field (Ford 1978). Their careful work soon precipitated the publication of many systematic treatments of the subject and in 1873 Powers introduced the term '*aboriginal botany*' to describe the botanical investigation of native plant use, a term which was readily accepted by the academic community over the next 25 years.

However, as the nineteenth century drew to a close, interest in aboriginal botany began to broaden, particularly during preparations for the 1893 World's Fair which involved both anthropologists and archaeologists in the collection of traditionally useful plant products (Ford 1978). Significantly, this exhibition included the Hazzard collection, a range of preserved plant products used by the ancestors of the Pueblo Indians in Mancos Cañon in Colorado, and which was later sent to the University of Pennsylvania for analysis. There, botanist John Harshberger examined the collection, and in December 1895 he finally delivered a lecture in which he described items of food, dress, household utensils and agricultural tools of plant origin preserved in the Hazzard collection; it was during this lecture that the term 'ethnobotany' was first used (Harshberger 1896).

In the decades which followed, the study of ethnobotany entered a

Table 1.3   Summary of the study of New World ethnobotany (see text for details)

Date	Events
1492	The discovery of the New World initiates the identification of several plants of considerable economic value based on observations of native people.
1663	John Josselyn begins his study of the natural history of New England, later publishing his text on native herbal medicine, *New England's Rarities Discovered*.
1871–78	Seminal works by botanists Palmer and Powers herald 25 years during which the field is largely dominated by economic botanists.
1893	World Fair preparations stimulate anthropological interest in aboriginal botany, leading to increasing emphasis on the cultural significance of plants.
1895	Colville publishes *Directions for Collecting Specimens and Information Illustrating Aboriginal Uses of Plants*; Harshberger introduces the term ethnobotany.
1896	Fewkes introduces ethnobotany to the anthropological literature.
1898	The Department of Ethnology in the US National Museum proposes to document all plants used, for whatever purpose, by North American Indians.
1900	David Barrows is awarded the first doctoral dissertation in ethnobotany, while universities show an increasing interest in the subject.
1919	Gilmore emphasises the active modification of the plant world by traditional peoples, initiating increasing interest in traditional resource management.
1930–50	Castetter establishes a masters programme in ethnobotany at the University of New Mexico.
1950–70	Ethnobotany becomes increasingly equated with linguistic concepts and classification, while Conklin also highlights the practical significance of understanding folk classification systems. Meanwhile, interest in palaeoethnobotany increases as archaeobotanical techniques improve.
1980s	The Society of Ethnobiology publishes the first issue of its *Journal of Ethnobiology* in 1981.
1990s	Both postgraduate and undergraduate programmes in ethnobotany become increasingly available, while many research projects focus on the practical applications of traditional plant knowledge.

phase of rapid expansion and change. Only a year after his historical lecture, anthropologist Walter Fewkes introduced Harshberger's term to the anthropological literature, where he emphasised Hopi Indian plant names and their etymology; in 1900 the first doctoral dissertation in ethnobotany—*The ethno-botany of the Coahuilla Indians of Southern California*—was awarded to David Barrows by the University of

Chicago; by 1916 'ethnobotany' had expanded to include not only how plants were used by indigenous peoples, but also how they were perceived and understood within different cultures (Robbins *et al.* 1916 cited in Castetter 1944). This last point was later expanded by American ethnologist Melvin Gilmore who argued both the need to interpret ethnobotanical data within its cultural context, and the important role of linguistics in ethnobotanical study (Gilmore 1932).

As ethnobotany was redefined, the relevant data accumulated at a rapid pace: ethnologies specifically devoted to uses of plants included general studies of plants used currently in daily and ceremonial life by extant tribes (Gilmore 1919; Densmore 1928) as well as those used in the past (Harshberger 1896); the uses of particular plant species by a range of groups were compared (Bell & Castetter 1937); investigations of traditional agriculture and wild plant foods increased (Barrows 1931; Castetter & Bell 1951; Conklin 1954a). Soon, the study of traditional plant knowledge began to play a significant part in the development of anthropological theory, and while the study of Tsembaga horticulture in New Guinea contributed to early ideas in cultural ecology (Rappaport 1968), the analysis of plant names and systems of folk classification provided an increasingly popular basis for the exploration of human cognition (Conklin 1954b; Berlin *et al.* 1973).

By the mid-1980s ethnobotany had become widely recognised in the USA, not only in academic circles, but also in the public eye as articles on ethnobotany and pollen analysis appeared in publications including the *Southwest Airline's Flight Magazine* of May 1985, and *Forbes* magazine in August of the same year (Bohrer 1986). At about this time too, the American-based Society of Ethnobiology was formed with the first issue of its *Journal of Ethnobiology* published in 1981. This in turn was soon followed by the formation of the International Society of Ethnobiology, while the Society of American Archaeology organised its first general session on ethnobotany in 1983. Hence, as the final decade of the twentieth century approached, with increasing public and academic recognition, ethnobotany finally emerged from relative obscurity and stood poised to enter a new phase in its development.

Outside America

Although the discovery of the American continent certainly marks a significant point in the evolution of ethnobotanical study, it by no means marks the beginning of interest in indigenous plant use. Indeed it was in part the lure of exotic Eastern spices such as cinnamon (*Cinnamomum zeylandica*) and black pepper (*Piper nigrum*) which precipitated

Table 1.4 Estimated market values for early ethnobotanical discoveries—Old World. Even before the discovery of the American continent, several economically important plants had been discovered as a result of ethnobotanical observations in the Old World. The market values shown here are the global values estimated on the basis of data provided by McIntosh Baring (1993); USDA (1994c, 1995c)

Commodity		Market value (US$ *per annum*)
Coffee	(*Coffea arabica*)	3.2 billion
Tea	(*Camellia sinensis*)	3.8 billion
Sugarcane	(*Saccharum officinarum*)	2.7 billion

Columbus' voyages across the Atlantic. Like tobacco and rubber from the New World, a number of plants originally used by traditional societies in the Old World have also played a crucial role in world economic history, many retaining their value in modern times. For example, sugarcane (*Saccharum officinarum*) from the Far East, tea (*Camellia sinensis*) from China, and coffee from Ethiopia (*Coffea arabica*) had all been known in Europe since the Middle Ages, and by the eighteenth century sugar alone represented the single most important commodity traded in the world (Hobhouse 1992). Even today, sugar generates a worldwide revenue of almost US$3 billion *per annum* (Table 1.4).

Similarly, while the new-found lands of America preoccupied many Europeans for centuries, their discovery did not diminish explorations elsewhere, and the global search for gold and other riches continued. These explorations finally culminated in Captain Cook's famous voyage in 1770, facilitating the first detailed observations of the Australian Aborigines and their use of plants. Unlike the New World, Australia apparently offered little in the way of plants with economic potential (see Fitzgerald 1982), yet since that time, ethnobotanical discoveries in Australia have included the antibacterial properties of the tea tree (*Melaleuca* spp), which now represents an important ingredient in a wide range of toiletry and cleaning products throughout the world (Thursday Plantation nd).

Developments in Europe

Despite the American dominance in ethnobotany, European ethnobotanists have continued to make invaluable contributions to the field. for example, in recent years workers such as Brian Morris and Roy Ellen in the UK have contributed much to the debate on ethnotaxonomy

(Morris 1976; Ellen & Fukui 1994), while the potential for using traditional plant management in sustainable dryland agriculture has been assessed in Sicily (Barbera *et al.* 1992). However, traditionally speaking, two of the most important contributions made by European ethnobotanists lie in the fields of *ethnopharmacology*—the scientific evaluation of traditional medicines, and *palynology*—the study of fossilised pollen.

Long before naturalists such as John Josselyn began to compile herbals in the New World, Europe had enjoyed a long tradition in the production of herbal pharmacopoeia—a tradition which dates back to the herbals of ancient Greece and which reached its zenith in the medieval publications of English physicians John Gerard and Nicholas Culpeper (Griggs 1981). Such herbals, which documented popular folk remedies along with botanical descriptions of medicinal plants, provided a starting point in the much cited work of William Withering, another English physician, whose research into 'a family recipe for the cure of dropsy' marked the beginning of the empirical study of folk medicine.

Dropsy was the medieval name for oedema—the swelling of parts of the body due to accumulation of fluids—which often results from long-term effects of heart disease (Mann 1994). With no effective treatment for this condition available in eighteenth century England, Withering became intrigued when asked for his opinion on a family cure for the disease—a secret remedy kept by an old woman in Shropshire. Presented with the formula, which contained 20 or more different herbs, Withering soon determined that the active constituent was the purple foxglove (*Digitalis purpurea*), which gave the crude drug its name of digitalis. While attempting to standardise a suitable dose, he further discovered not only that the leaves contained the active principle, but that their level of activity varied throughout the year (Griggs 1981). Following a further 10 years of research, Withering published his conclusions about dosage and the likelihood of adverse reactions and made some suggestions for further research, including the possibility of experimenting on 'insects and quadrupeds'—an innovative approach at that time. Yet despite Withering's careful experiments, there remained little confidence in the efficacy of his digitalis preparation, and, following his death in 1799, its use declined.

This lack of confidence in Withering's findings, however, was largely due to the fact that many of his contemporaries had tried to use digitalis for ailments ranging from bronchitis to psychiatric disorders, with little regard for its specific biological activity (Mann 1994). Fortunately, research into the drug continued regardless, and throughout the sub-

Table 1.5   European contributions to the development of ethnobotany—ethnopharmacology. The pharmacological verification of Withering's early work resulted from a number of important developments in organic chemistry and pharmacology, many of which took place in Europe. In 1803, the successful isolation of the morphine alkaloid from the opium poppy (*Papaver somniferum*) reflected the development of increasingly effective methodologies for phytochemical analysis, which facilitated the successful isolation of an active principle from the foxglove in 1841. Later, during the first decades of the twentieth century, physician James MacKenzie found digitalis effective in treating the condition of atrial fibrillation, while Windaus and his coworkers reported the structures of two active isolates—the *cardiac glycosides* digitoxin and digitalin (Windaus *et al.* 1928 cited in Mann 1994). Around the same time, another British worker, Sydney Smith isolated a related compound, digoxin, from the foxglove's woolly relative, *Digitalis lanata*. More potent than anything from *Digitalis purpurea*, digoxin was marketed as Lanoxin which is still widely used in the treatment of congestive heart failure (Mann 1994)

Date	Events	Source
1597	John Gerard publishes his *Herbal* or *General Historie of Plantes* (UK)	Griggs (1981)
1651	Nicholas Culpeper publishes the *English Physician* (UK)	Griggs (1981)
1785	William Withering publishes *An account of the foxglove and some of its medical uses: with practical remarks on dropsy and other diseases* (UK)	Griggs (1981)
1803	Friedrich Wilhelm Serturner isolates morphine crystals from crude opium (Germany)	Griggs (1981)
1841	Homolle and Quevenne isolate an active principle from *Digitalis purpurea* (France)	Mann (1994)
1905	James MacKenzie recognises the condition of atrial fibrillation (UK)	Mann (1994)
1928	Windaus reports the structures of digitoxin and digitalin	Mann (1994)
1930	Sydney Smith isolates digoxin from *Digitalis lanata* (UK)	Mann (1994)

sequent centuries, a number of developments in organic chemistry and pharmacology have verified the validity of Withering's claims (Table 1.5). His careful analysis, therefore, represents a milestone not only in the development of modern pharmacognosy, but also in the science of ethnopharmacology, and illustrates the enormous potential of ethnopharmacological investigations.

Europe's second major contribution to the development of ethnobotanical study lies in its crucial contributions to palynology, a branch of *archaeobotany* which has contributed much to *palaeoethnobotany*—the study of past relationships between humans and plants. As early as 1830, Christian Ehrenberg first reported his observations of

Table 1.6 European contributions to the development of ethnobotany—palaeoethnobotany. Although Ehrenberg's initial reports of preserved fossil pollen received little attention at the time, the development of palynology in Europe proceeded rapidly from the beginning of the twentieth century. Pollen analysis now represents a powerful tool used by palaeoethnobotanists throughout the world

Date	Events	Source
1803	Christian Ehrenberg reports fossil pollen preserved in sedimentary rocks	Bryant (1989)
1916	Lennart von Post presents a basic theory of quantitative pollen analysis as a tool for assessing changes in past vegetation (Norway)	Bryant and Holloway (1983)
1928–37	Fritz Zetzsche and his co-workers identify sporopollenin (Switzerland)	Bryant (1989)
1941	Johannes Iversen applies pollen analysis specifically to archaeological questions concerning prehistoric subsistence (Denmark)	Iversen (1941)
1963	Dimbleby presents a new technique for quantifying pollen samples (UK)	Dimbleby (1963)
1969	First major international seminar on plant and animal domestication held at the Institute of Archaeology in London, UK	Ucko and Dimbleby (1969)
1983	Symposium *Recent Advances in the Understanding of Plant Domestication and Early Agriculture* held as part of the World Archaeological Congress in Southampton, UK	Harris and Hillman (1989a)

fossil pollen preserved in sedimentary rocks—he even suggested its potential role as an indicator of past environmental change. However, it was not until the early 1890s that scientists began to look at preserved plant materials in earnest (Bryant 1989), and even then much of the initial interest centred on the recovery and identification of macro-fossils such as seeds, and fragments of leaves and wood. Yet a few of these early palaeobotanists did begin to investigate the preserved pollen found in the deposits, and by 1916 the Norwegian geologist Lennart von Post, proposed a novel technique for the quantitative analysis of pollen, which could be used as a tool to study past changes in vegetation (Table 1.6).

By 1937, a Swiss chemist, Fritz Zetzsche discovered sporopollenin, a polymeric compound found in spores and pollen grains, whose durability explains the excellent preservation of these microfossils often over millions of years; 4 years later a Dane—geologist Johannes Iversen—was the first to apply pollen analysis to archaeological studies, determining the nature and timing of the transformation from hunting and gathering to agriculture in the Barkaer site of northern Denmark (Iversen 1941). Soon, other palynologists in both Europe and North America began to use pollen analysis to help in both the interpretation of archaeological data and the exploration of ancient anthropogenic effects on vegetation (Dimbleby 1963). Since then, palynological methods have been continually refined and pollen analysis now constitutes an extremely powerful tool in palaeoethnobotanical study.

Ethnobotany in Asia, Australasia and Africa

Outside Europe and America, academic research into indigenous plant knowledge has become widespread: by 1878 in Australia, Joseph Bancroft presented an ethnopharmacological paper on the chemical properties and Aboriginal uses of *pituri* (*Duboisia* spp) to the Queensland Philosophical Society (Bancroft 1878); in India the publication of ethnobotanical data intensified during the 1920s as traditional herbal medicine received increasing attention (see Jain 1989 and references therein). Since these early reports, ethnobotanical research in both Asia and Australasia has expanded to include studies on aspects ranging from the representation of plants in art and myth to the role of indigenous practice in vegetation management, many of which may be found in the publications of the Society of Ethnobotanists in India (see Jain 1981, 1987, 1989) and the Australian Institute of Aboriginal and Torres Strait Islander Studies in Canberra, South Australia.

Equally, in Africa, traditional rangeland management and the uses of wild and cultivated plants have received academic scrutiny (e.g. Wickens *et al.* 1985; Abbiw 1990; Homewood & Rodgers 1991; Smith AB 1992), although perhaps some of the most influential work has been encapsulated in Paul Richards' *Indigenous Agricultural Revolution*. In this seminal text, Richards highlights the value of traditional farmers' knowledge in West Africa, and demonstrates that even after the 'Green Revolution' of Western science, many of the most successful innovations in food-crop production have been based on indigenous knowledge (Richards 1985).

Ethnobotany in the 1990s

By the beginning of this decade, the study of traditional plant-lore had gathered considerable momentum, and between 1990 and 1994, academic publications in ethnobotany almost doubled compared with the previous 5-year period (Figure 1.2). This same period also saw a number of other important developments in ethnobotany, including the establish-

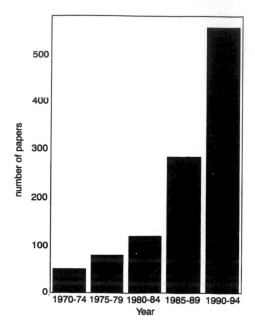

Figure 1.2   Increased publication of ethnobotanical research (1970–94).
These data (taken from **BIOSYS** biological abstracts database) illustrate how
numbers of published works on ethnobotany have increased during the last
25 years

ment of the People and Plants collaboration between WWF international, UNESCO and the Royal Botanic Gardens, Kew (Hamilton 1994),
the launch of the Oxford-based Foundation for Ethnobiology (FEB
1993), and the publication of the first issue of CIRAN's *Indigenous
Knowledge and Development Monitor* (Ciran 1993).

   This recent intensification in research activity, has been reflected to
some extent by an increase in the teaching of ethnobotany, particularly
in the USA, where the first masters degree programme in ethnobotany
was established at the University of New Mexico (Ford 1978). Graduate
programmes are now available at several universities both in America
and the UK, while undergraduate studies in economic botany and
ethnobotany are becoming increasingly widespread (Jain *et al.* 1986;
Flaster 1994). However, interest in ethnobotany has not been confined
simply to academic circles and since the late 1980s articles have
appeared in journals ranging from *Scientific American* (Cox & Balick
1994) to *The Economist* and *Newsweek* (Anon 1988; Begley 1988), while
popular books such as Anna Lewington's *Plants for People* and Richard

Rudgley's *Alchemy of Culture* have proliferated in recent years (Lewington 1990; Rudgley 1993).

The growing interest in ethnobotany is due, at least in part, to changing attitudes towards traditional peoples. For during the middle of the twentieth century, when it seemed that the world's indigenous peoples were about to disappear, traditional societies and their knowledge attracted widespread scholarly attention, primarily as part of an anthropological rescue operation (Burch & Ellanna 1994). Since then, however, many scientists have begun to realise the practical and academic value of ethnobotanical data, and are beginning to acknowledge that traditional peoples have much to teach Western science.

Current Scope and Potential Applications

Since the early ethnobotanical studies in aboriginal plant use, the scope of the subject has expanded enormously, encompassing the botanical aspects of a number of ethnoscientific fields including ethnomedicine, ethnotaxonomy and ethnoecology as well as the anthropological and botanical study of material culture and subsistence mode (Figure 1.3). For the sake of clarity, six major fields of investigation are distinguished here: ethnoecology, traditional agriculture, cognitive ethnobotany, material culture, traditional phytochemistry and palaeoethnobotany, the main points of which are summarised in Table 1.7. Of necessity, each of these areas of ethnobotanical study draws from the theory and techniques of a range of established disciplines, several of which may be pertinent to any given project. For example, an ethnopharmacological study of traditional herbal pharmacopoeia might require anthropological and medical assessments of the ethnomedical system, the taxonomic skills of botanists, the linguistic techniques of ethnotaxonomists and etymologists, and the analytical expertise of phytochemists and pharmacognosists.

Much of the expansion and diversification of ethnobotany has occurred in the last 25 years, during which it has developed into a truly multidisciplinary field of natural science, combining the talents of anthropologists and archaeologists with those of molecular biologists and ecologists. But what is the driving force behind this increasing research activity? For some, stimulation comes from a desire to identify patterns in systems of knowledge which may in turn assist in the understanding of human cognition; others are intent on elucidating details of human evolution through the analysis of prehistoric relationships between humans and plants. More recently, however, there has been an increasing awareness of the considerable practical and social value of

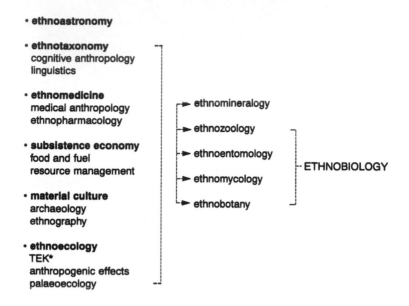

Figure 1.3   The nature of ethnoscientific study. Ethnobiological studies involve the examination of the reciprocal relationships which occur between traditional societies and the natural world, in extant cultures and reflected in the archaeological record. They encompass both the analysis of traditional biological knowledge, and the assessment of human influence on the biological environment. Specifically, ethnobotany includes any such studies which relate to plants, including how they are classified and named, how they are used and managed, and how their exploitation has influenced their evolution. *Traditional environmental knowledge (TEK) includes topics such as ethnopedology, traditional climatology and knowledge of biological components of local environments

traditional knowledge, and many workers are becoming involved in *applied ethnobotany*—the practical application of ethnobotanical data in areas such as biodiversity prospecting and conservation biology (Table 1.8). For example, ethnobotanical investigations by the San Francisco-based company, Shaman Pharmaceuticals Inc. have already led to the development of two antiviral products which are now in clinical trials (King & Tempesta 1994), while indigenous management practices have been shown to have a profound influence on factors such as plant genetic diversity and habitat conservation (Soleri & Cleveland 1993; Rajasekaran & Warren 1994).

Table 1.7 Main areas of modern ethnobotanical investigation. Ethnobotany now constitutes a diverse field of study which examines all aspects of the reciprocal relationships between plants and traditional peoples. It is, by necessity, multidisciplinary in its approach and draws from a broad range of subject areas

Field	Main areas of investigation
Ethnoecology	Traditional knowledge of plant phenology, adaptations and interactions with other organisms
	Nature and environmental impact of traditional vegetation management
Traditional agriculture	Traditional knowledge of crop varieties and agricultural resources
	Nature and environmental impact of crop selection and crop management
Cognitive ethnobotany	Traditional perceptions of the natural world (through the analysis of symbolism in ritual and myth) and their ecological consequences
	Organisation of knowledge systems (through ethnotaxonomic study)
Material culture	Traditional knowledge and use of plants and plant products in art and technology
Traditional phytochemistry	Traditional knowledge and use of plants for plant chemicals, for example in pest control and traditional medicine
Palaeoethnobotany	Past interactions of human populations and plants based on the interpretation of archaeobotanical remains

About this Text

Throughout recent decades, the scope of ethnobotany has become extremely broad and its studentship diverse. Pertinent publications may appear, not only in those journals which are specifically dedicated to ethnobiological research (*Journal of Ethnobiology*, *Journal of Ethnopharmacology*, *Ethnobotany*), but also in a whole range of publications from different academic disciplines, including anthropology, botany, archaeology, palaeobotany, phytochemistry and conservation biology. The present text draws from many of these sources in an attempt to synthesise these disparate influences, and to present an introduction to the underlying principles and practical applications of modern ethnobotany.

For the sake of clarity, the available material has been organised according to the categories defined in Table 1.7, with additional sections covering background information and applied ethnobotany: Chapters 2

Table 1.8 Potential applications of ethnobotanical inquiry. In recent decades the traditional applications of ethnobotanical data have expanded considerably. In addition to its traditional roles in economic botany and the exploration of human cognition, ethnobotanical research has been applied to practical areas such as biodiversity prospecting and vegetation management. It is hoped that in the future, ethnobotany may play an increasingly important role in sustainable development and biodiversity conservation (Rajasekaran & Warren 1994)

Application	Examples
Economic botany	
Agriculture	Identification of novel species for foods, fibres and other commodities; conservation of traditional germplasm with qualities such as drought tolerance and pest resistance
Arts and crafts	Development of alternative sources of income for sustainable development
Pharmaceuticals	Identification of new drugs based on traditional medicinal plants
Ecology	
Vegetation management	Identification of practices which may facilitate the sustainable use of biological resources, particularly in marginal areas
Biodiversity	Conservation of practices which both promote and conserve biological and genetic diversity
Human ecology	Assessment of past and present anthropogenic affects on the plant environment

and 3 introduce the plants and people forming the basis of ethnobotanical study, while Chapter 4 discusses the range of relevant methodologies. Then Chapters 5–10 each discuss one of the main fields of study, often using case studies to illustrate a particular point. In Chapters 11 and 12, the potential benefits and practical problems of applied ethnobotany are discussed.

Chapter 2

Plant Structures, Functions and Applications

During the many millennia of interaction between people and plants, humans have learnt to exploit plants in diverse ways for myriad purposes. . . .

DR Harris, GC Hillman 1989b in *Foraging and Farming*

INTRODUCTION

Plants are fundamental to almost all life on earth, providing protection and sustenance for organisms ranging from bacteria to large mammals. With their unique capacity for *photosynthesis*, they form the basis of the biological food web, meanwhile producing oxygen and mopping up excess levels of the greenhouse gas carbon dioxide. Plants also perform a number of other important environmental services, recycling essential nutrients, stabilising soils, protecting water catchment areas, and helping to control rainfall via the process of *transpiration.*

Most of the world's plants today are *vascular plants*, the earliest of which invaded the terrestrial habitat more than 400 million years ago. At that time they were rather primitive organisms lacking any obvious differentiation into leaves, roots or complex reproductive structures (Gifford & Foster 1989); since then, these ancestral forms have evolved to produce the vast array of terrestrial plants we see today. Now different plant species exhibit a range of specific *adaptations*—physical structures and chemical characteristics favoured by a particular ecological niche—many of which have also played a fundamental part in the evolution of the human culture, providing not only food and fuel, but also a whole range of materials used in shelter, clothing and medicine.

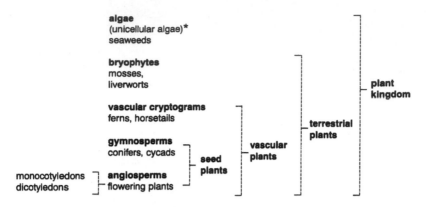

Figure 2.1   Major groups within the plant kingdom. The main groups of plants are largely differentiated on the basis of their reproductive behaviour, only the gymnosperms and angiosperms producing embryos which are protected by the formation of a seed. In turn, while the seeds of gymnosperms remain naked, those produced in angiosperms are protected further by their enclosure within the fruit, which may be fleshy as in the case of apples (*Malus pumila*) or grapes (*Vitus vinifera*), or dry as in nuts and cereal grains (Fahn 1990).
  * Unicellular algae may also be included within the kingdom, protista

BASIC BIOLOGY OF HIGHER PLANTS

It is estimated that at least 300 000 species of plant exist throughout the world, the majority of which are mainly tropical in distribution (Walters & Hamilton 1993). According to differences in their reproductive behaviour they are divided into four main groups (Figure 2.1), the largest of which—and that of greatest significance to humans—contains the *angiosperms* or flowering plants (Table 2.1), upon which much of this text will focus.

The Nature of Flowering Plants

The flowering plants of the world occur in many different physical forms, and exhibit a wide range of behavioural patterns, all of which can affect both their ecological role and their applications in human society. For example, they may be *monocotyledons* or *dicotyledons*; *woody* or *herbaceous*; *annual* or *perennial* (Table 2.2). Nevertheless, while they may differ in detail, all angiosperms hold many general features in common, including their mode of reproduction, their basic organs and tissues and their underlying biochemical processes.

Table 2.1 Estimated numbers of species occurring within main plant groups. The flowering plants or angiosperms form by far the largest group of plants on the earth today, conservative estimates suggesting that there are about 250 000–300 000 extant species (Farnsworth 1990; Walters & Hamilton 1993); yet while many of these are of great current and/or potential value to humans, only a small proportion have so far been studied scientifically. Unfortunately, the majority of plant species exist within endangered habitats such as the world's rapidly diminishing tropical rainforests, coral reefs and coastal wetlands (Wilson 1988), where species are becoming extinct at a rate faster than that at which they can be studied

Plant group	Estimated numbers of species
Flowering plants	250 000+
Conifers and cycads	700
Ferns and horsetails	12 000
Mosses and liverworts	22 000
Algae	2 500

Table 2.2 Physical and behavioural nature of angiosperms. Angiosperms exhibit a wide range of physical and behavioural characteristics which affect both their ecological functions and their roles in human society. For example, many successful weed species are herbacious and ephemeral, while large woody perennials may be particularly valuable in stabilising soils and protecting watersheds

Term	Meaning
Monocotyledon	A plant whose developing embryo typically has one *cotyledon* (Figure 2.2) e.g. grasses and palms
Dicotyledon	A plant whose developing embryo typically has two cotyledons, e.g. oak and chrysanthemum
Herbaceous	Non-woody plants which do not undergo *secondary growth*
Woody	Trees and shrubs—produce wood by virtue of secondary growth from the *vascular cambium* (typically gymnosperms and dicotyledenous angiosperms)
Ephemeral	A plant which completes its life cycle very rapidly, in some cases germinating, blooming, and setting seed several times in 1 year
Annual	A plant which lives for 1 year, e.g. wheat (*Triticum aestivum*)
Biennial	A plant which lives for 2 years, e.g. onion (*Allium cepa*)
Perennial	A plant which normally lives for more than 2 years, and which generally reproduces annually following maturity, e.g. apple (*Malus pumila*)

The Flowering Plant Body

The body of a plant is made up of millions of cells, all of which are genetically identical and begin life in specialised tissues known as *meristems*. While all meristematic tissues are characterised as being sites of active cell division, three distinct types may be distinguished: *apical meristems*, which occur terminally in the developing tips of roots and shoots; *intercalary meristems*, which are found between mature tissues (such as those occurring in the *internodes* of grasses); and *lateral meristems*; which are situated parallel to the circumference and are responsible for the thickening of tissues (Fahn 1990). Although initially very similar, the new cells produced here soon differentiate into one of a wide range of specialised cell types which together make up the different tissues and organs of a plant.

For the sake of convenience, the mature plant body can be divided into the roots, stems, leaves and reproductive parts, each of which have their own specific structures and functions. The roots provide anchorage in the soil, are responsible for absorbing the water and nutrients required for plant growth, and in root vegetables such as carrots (*Daucus carota*) they may play an important part in storing energy. From early in their development, root systems are branched and often produce millions of *root hairs*—tiny outgrowths which can increase the absorbing surface by severalfold (Salisbury & Ross 1978). These hairs are essentially elongations of individual cells in the root *epidermis*, and although they are relatively short-lived, new ones are constantly produced near the growing root tip.

Plant stems generally provide a support for both leaves and reproductive structures, and are responsible for the long-distance transport of water and nutrients between different parts of the plant. However, in a few species stems are more specialised and carry out specific functions. For example, in potatoes (*Solanum tuberosum*) and onions (*Allium cepa*), their tubers and bulbs are actually underground stems capable of storing energy, while the succulent stems of leafless cacti store water and carry out photosynthesis—the process during which plants capture light energy from the sun to fuel the synthesis of carbohydrate.

In most species, however, the major sites of photosynthesis are the leaves which are distributed along the stem in characteristic patterns according to species. In order to photosynthesise, leaves must take up carbon dioxide from the atmosphere, and this is achieved through small pores on the leaf surface known as *stomata*. These stomatal apertures may be opened or closed as environmental conditions fluctuate, allowing the plant to take up carbon dioxide while conditions are favourable, yet restricting water loss when water availability is low.

The reproductive structures begin to develop when the vegetative plant reaches maturity. In some short-lived annuals this may occur after only a few weeks; in others such as the perennial bamboos (Bambusoidae), this may take up to 50 years (Salisbury & Ross 1978). Flowers, like the cones of gymnosperms, consist of a group of specialised leaves which protect the egg and pollen cells as they develop. Following *pollination* and *fertilisation*, tissues around the developing embryo differentiate into the new seed, which, in angiosperms, is further contained within the developing fruit (Figure 2.2).

Cells and Tissues

The main components of a generalised plant cell are the *cell wall*, and the *protoplast* (Figure 2.3). The protoplast essentially consists of a chemically complex viscous fluid (the *cytosol*), which is delimited by a membrane known as the *plasmalemma* (plasma membrane), and this in turn is surrounded by the cell wall. It contains the cell's *organelles*—small, often membrane-bound bodies with specialised functions—as well as the *vacuole* and precipitated materials, such as starch grains (Table 2.3). The presence of the cell wall distinguishes plant cells from those of animals, and provides mechanical support for the delicate protoplast. Composed primarily of cellulose, the cell walls of adjoining cells are cemented together by a mixture of pectins at the *middle lamella*, while cytoplasmic continuity between cells is maintained by small pores known as *plasmodesmata*. Although these basic cellular components are common to almost all types of plant cell, mature cells vary enormously as during plant development specialised cells form a range of tissues with specific properties and functions (Table 2.4).

A large proportion of the plant body is composed of *parenchyma* (Figure 2.4) and while the constituent cells generally show relatively little specialisation their exact structure varies according to a range of different physiological functions in which they may be involved: those involved in photosynthesis contain chloroplasts and are known as *chlorenchyma*; others serve a storage function accumulating reserve materials or—in the case of succulent plants—water; still others, develop thick secondary walls impregnated with a strong polymeric material known as *lignin*, and provide some support in vascular tissues. However, much of the support in plants is provided by two further tissue

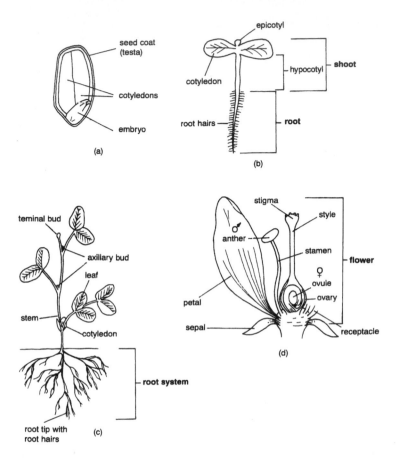

Figure 2.2   Major stages in the life cycle of angiosperms. The life cycle of most flowering plants begins with the production of seed which germinates to produce the mature plant body: (a) the seed consists of the embryo and either one or two *cotyledons* contained within a water-repellent seed coat or *testa*—the cotyledons are essentially storage tissues, which provide nutrients and energy to the developing embryo; (b) as the embryo develops, differentiation into distinct shoot and root systems occurs and finally the seedling shoot emerges into the light, allowing the cotyledons to begin *photosynthesis* and the root system to absorb water and nutrients; (c) from this stage, the young plant soon begins to produce new leaves, while the root system branches as it extends into the soil; (d) finally, as the vegetative plant matures the reproductive structures develop. Pollen, which contains the male *gametes* is produced in the anthers, and is transfered to the *stigma* of another individual during *pollination*. The pollen grain then germinates forming a *pollen tube* which extends down through the *style* until it reaches the egg-containing *ovule*. At this stage, the male gamete migrates down the pollen tube and is able to fuse with the *egg* during *fertilisation*. As the resulting seed develops, the surrounding ovule develops into the *testa*, while the *ovary* develops into the surrounding *fruit*. In the case of fleshy fruits the *receptacle* is also involved in fruit formation

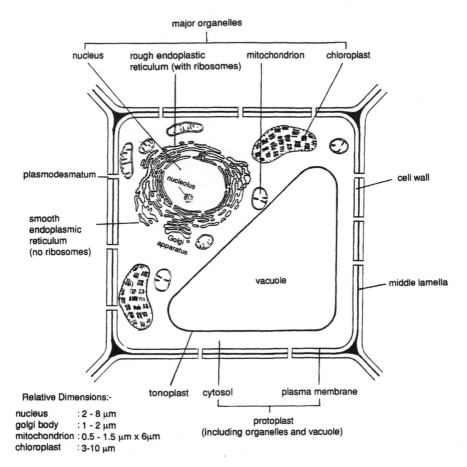

major organelles

nucleus   rough endoplastic   mitochondrion   chloroplast
reticulum (with ribosomes)

plasmodesmatum

nucleolus

cell wall

smooth
endoplasmic
reticulum
(no ribosomes)

Golgi
apparatus

vacuole

middle lamella

tonoplast   cytosol   plasma membrane

Relative Dimensions:-

nucleus	: 2 - 8 μm
golgi body	: 1 - 2 μm
mitochondrion	: 0.5 - 1.5 μm x 6μm
chloroplast	: 3-10 μm

protoplast
(including organelles and vacuole)

Figure 2.3 Generalised structure of the plant cell. Although this generalised plant cell contains all of the major plant organelles, specialised cells may differ considerably in their internal structure. For example, meristematic and glandular cells tend to have no large vacuole, while the vacuole of parenchymatous cells may occupy about 90 per cent of the total cell volume; similarly, while most plant cells contain a nucleus, mature phloem sieve elements do not, their nuclear functions being carried out in their companion cells (Figure 2.4). In addition, the size of both cells and organelles may vary considerably. For example, while parenchymatous cells vary between 10 and 100 μm, fibre cells can extend up to several cm in length (Fahn 1990); equally, the plant nucleus varies between 2 and 8 μm, while chloroplasts range between 3 and 10 μm (Goodwin & Mercer 1983). The Golgi body ranges between 1 and 2 μm and the mitochondrion is 0.5–1.5 μm × 6 μm

Table 2.3 Major components of a generalised plant cell. Plant cells contain a number of organelles which carry out specific functions in the plant cell. These include synthetic, storage and transport functions which are essential to normal cellular activity and hence to the growth and maintenance of the whole organism (Fahn 1990; Goodwin & Mercer 1990)

Component	Nature and function
Vacuole	Often occupying up to 90% of the cell volume, the vacuole is a compartment bound by the *tonoplast* membrane and filled with an aqueous solution. It functions in the maintenance of cellular turgidity, and the storage of waste products, toxins and food reserves.
Nucleus	The largest of the organelles, the nucleus is bound by a double membrane and contains the *chromosomes*, which contain most of the cell's genetic information.
Mitochondrion	The mitochondrion has two membranes and contains the enzymes required for cellular respiration. Here, carbohydrate is broken down to release free energy required for growth and reproduction.
Chloroplast	Unique to plants, the chloroplasts—the sites of photosynthesis—occur in all photosynthetic tissues. They consist of a membranous *thylakoid* system surrounded by a double membrane, and all contain the green pigment *chlorophyll*, which is essential to the photosynthetic process.
Membrane systems	Most cells contain two related membrane systems the *endoplasmic reticulum* (ER) and *Golgi apparatus*. The ER membrane is continuous with the nuclear envelope and can be involved in the synthesis and intracellular transport of proteins and lipophilic substances. The Golgi apparatus is also involved in secretion.
Ribosomes	Ribosomes are composed of ribonucleic acid (RNA) and protein and play a crucial part in protein synthesis. They occur in mitochondria and chloroplasts, as well as the cytosol where they are often attached to the ER.
Reserve materials	Reserve materials stored in plant cells include starch grains, protein bodies, calcium oxalate crystals and silica bodies, occurring in various cell compartments.
Cell wall	In many cells the first cell wall—the *primary cell wall* remains the only cell wall; in others this wall undergoes *secondary thickening* as subsequent layers are laid down forming the *secondary cell wall*, which may become impregnated with water-repellent substances such as *lignin* or *cutin*.

Table 2.4 Structure and function of plant tissues. The main tissues of the plant organs are generally made up of several cell types, although they may be characterised by the presence of certain specific cells (Fahn 1979, 1990). For example, xylem tissues are made up of various cells including xylem parenchyma and fibre cells, yet they are characterised the presence of the tracheary elements (Figure 2.4). The properties and functions of different tissues are therefore determined by the constituent cell types, which vary in detail from species to species

Tissue type	Function
Meristematic	Tissues where cell division occurs allowing growth
Parenchyma	Tissues with relatively little specialisation concerned with various physiological functions; their exact structure varies according to their function
Collenchyma	Supporting tissues in young organs and herbaceous plants
Sclerenchyma	Sclereids and fibre cells with thickened secondary walls provide supporting tissues in various organs and tissues
Xylem	Part of the vascular system which transports water and minerals from the soil
Phloem	Part of the vascular system which transports the products of photosynthesis throughout the plant body
Secretory tissues	Tissues involved in the accumulation and elimination of substances which are often *autotoxic*—toxic to the producing plant itself
Epidermis	The outermost layer of cells of all organs which have not undergone considerable secondary growth
Wood	Secondary xylem of dicotyledons and conifers
Cork	An impermeable barrier protecting internal tissues following lateral expansion associated with secondary growth

types, namely *collenchyma* and *sclerenchyma*. Collenchyma occurs largely in tissues which have been exposed to light, usually forming immediately below the epidermis, and is characterised by more or less elongated cells with unevenly thickened cell walls (Fahn 1990). The cells of sclerenchyma tissues also have thickened—usually lignified—walls and are generally elongated. Botanically, they are divided into *sclereids* and *fibre cells*, although in practice differentiating between the two can prove difficult. While both sclereids and fibre cells are distributed throughout many parts of the plant, the latter occur most commonly in the vascular tissues.

The vascular system, which is responsible for transporting water and

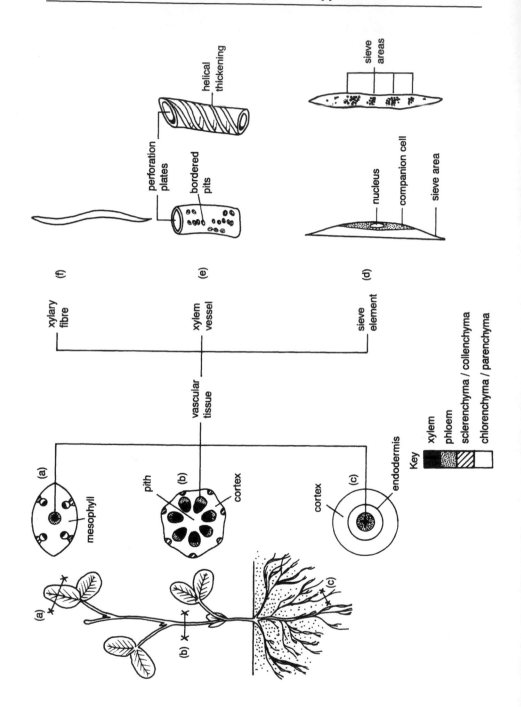

nutrients, is largely composed of *xylem* and *phloem*, both of which form discrete bundles of highly specialised cells which extend throughout the plant body. The xylem consists of a complex network of elongated cells called *tracheary elements*—non-living, lignified cells which lose their protoplast during the course of development and whose principal function is to transport water and solutes from the soil. The phloem also consists largely of elongated cells, the *sieve elements*, which are involved in the conduction of photosynthetic products. They are characterised both by their *sieve areas* which connect them to the protoplasts of neighbouring cells and by the absence of a nucleus which is lost during cellular differentiation.

In addition to these basic tissues, many species also exhibit specialised secretory structures which secrete a range of toxins and other biologically active phytochemicals. These structures include *secretory ducts* and *cavities, laticifers* and *glandular hairs* (Figure 2.5). Ducts, cavities and glandular hairs are common in many aromatic species such as *Citrus, Eucalyptus, Pinus* and *Artemisia*, where they isolate cytotoxic resins and essential oils from the living cytoplasm. In other species such as the opium poppy (*Papaver somniferium*) and rubber (*Hevea brasilien-*

Figure 2.4   Distribution of plant tissues. Various types of parenchymatous tissues make up the bulk of the primary plant body, including the *mesophyll* layers of the leaf whose main functions are photosynthesis and starch accumulation, the *pith* which occupies the central part of the stem, and the subepidermal *cortex* which lies external to vascular tissues in both roots and stems. The more specialised vascular tissues are comprised largely of phloem, xylem and fibre cells each of which fulfil specialised functions as outlined in the text. (d) The characteristic cells of phloem are the *sieve elements* which are arranged vertically in the plant to form *sieve tubes* which conduct photosynthetic products between different parts of the plant. The sieve elements are interconnected via the cells' characteristic *sieve areas*—regions where groups of pores allow continuity between the protoplasts of neighbouring cells, and are intimately associated with nucleate *companion cells*, which are necessary to the functioning of the anucleate sieve elements; (e) the tracheary elements or conducting cells of xylem may be divided in the *tracheids* and *vessel members*, both of which are elongated cells with cell walls which are thickened in characteristic patterns, and which are *pitted* in places allowing the passage of water. However, unlike tracheids which have unperforated end walls, vessel members exhibit *perforation plates*. Xylem *vessels* are composed of large numbers of vessel members which are connected via these perforation plates; (f) fibre cells are normally relatively long cells, which can reach up to 55 cm in length in certain species (*Boehmeria nivea*). Although they occur in various plant tissues, fibre cells occur commonly in the xylem, providing support and flexibility (Fahn 1990)

(a)

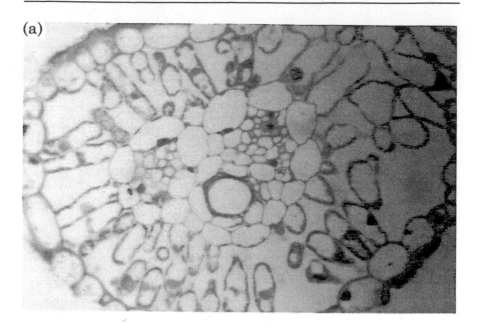

(b)

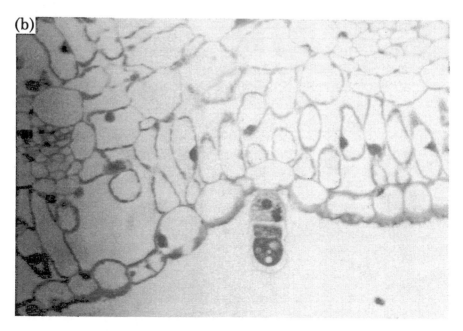

sis), which produce *latex*, the exudate is commonly accumulated in specialised elongated structures known as *laticifers*.

Protecting all of these internal tissues, the epidermis forms the outermost layer on all plant organs which are not woody. A complex tissue, it is made up of several types of cell including those of *trichomes*—which may be glandular or non-glandular—and the *guard cells* of *stomata* (Figure 2.6). The ordinary cells of the epidermis are always densely packed and covered by the *cuticle*—a water-repellent layer of *cutin* which effectively restricts water loss and helps to protect the plant against certain pests and pathogens.

Lateral Expansion

In herbaceous plants meristematic activity is largely restricted to the apical and intercalary meristems, producing the cells of the *primary plant body* which has been described above. However, in woody *gymnosperms* and dicotyledonous trees the *vascular cambium* develops with age initiating extensive *secondary growth* and allowing considerable lateral expansion (Figure 2.7). As the cambial cylinder continues to divide—often seasonally—throughout the life of the plant, it gives rise to the production of *wood*, which is composed largely of strong lignified xylem tissues capable both of supporting large trees and of transporting water and nutrients throughout the plant body. As extensive secondary growth invariably leads to the disruption of the existing epidermis, this protective function is eventually taken over as a new tissue known as the *cork* develops. Like epidermal cells, the cork layer is water impermeable due both to the presence of *suberin* in its cell walls, and to the tightly packed arrangement of its cells (Figure 2.8).

In monocotyledenous plants secondary growth as outlined above is

---

Figure 2.5   The secretory structures of French tarragon (*Artemisia dracunculus*). Light micrographs of (a) a leaf secretory duct, and (b) a leaf glandular hair in French tarragon (*Artemisia dracunculus*), where the plant's characteristic essential oils are accumulated and secreted. Accumulation of these *cytotoxic* components in non-living ducts and in short-lived epidermal hairs permits plants to produce these biologically active oils without causing damage due to autotoxicity (see Cotton *et al.* 1991a). Other essential oil-producing plants accumulate oils in a similar way, although in some cases secretory cavities may be more or less round as in species of *Citrus* and *Eucalyptus*, while secretory cells may occur internally as in the sweet bay, *Laurus nobilis* (see Fahn 1979)

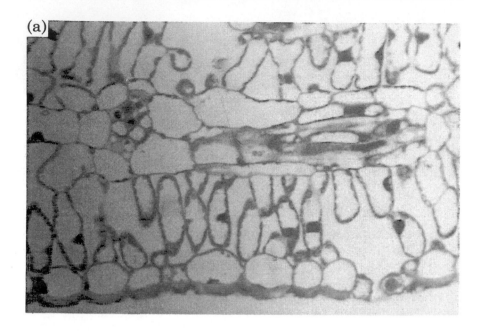

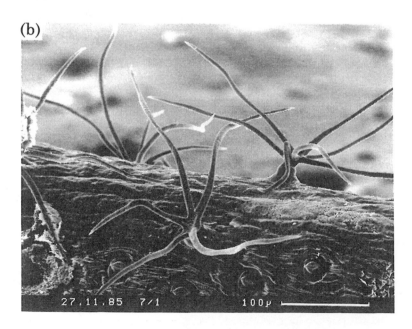

generally absent, yet several species exhibit thickened organs including the stems of palms such as coconut (*Cocos nucifera*) and the bulbs of onions (*Allium cepa*). This unusual type of thickening occurs due to intensive cell division immediately below the apical meristem *before* stem extension growth ceases, and is considered part of primary growth (Fahn 1990). However, secondary thickening from a lateral meristem—the *secondary thickening meristem*—does occur in certain mono-cotyledons, including species of *Yucca* and *Dioscorea* and *Aloë arborescens*.

Basic Biochemistry

The differentiation of plants into specialised organs and tissues is inti-mately associated with the range of specialised physiological pro-cesses which plants exhibit and are fundamental to their biological and ecological roles. For example, the differentiation of chlorenchyma is essential to photosynthesis, while the development of secretory ducts and glandular hairs facilitate the accumulation of specific toxins, many of which are fundamental to mutualistic relationships between plants and other organisms (Harborne 1988; Hay & Waterman 1993). These physiological functions are mediated by biochemical pathways, which can be broadly divided into two groups: those such as photosynthesis, respiration and protein synthesis, which are involved in *primary metabolism* and are considered essential to normal cellular function, and those specifically involved in *secondary metabo-lism* whose products have no known *direct* function in basic metabo-lism.

Figure 2.6 **The leaf surface of French tarragon (*Artemisia dracun-culus*).** (a) Light micrograph of the leaf structure of French tarragon (*Artemisia dracunculus*) showing the thick waxy cuticle and stomatal struc-ture of the epidermis; (b) electron micrograph illustrating the complex architecture of the leaf surface. While the stellate, non-glandular hairs or tri-chomes may protect against excess transpiration or small predators, the stomatal complex allows the gas exchange necessary to photosynthesis, respira-tion and transpiration. The stomatal apparatus comprises an epidermal pore bounded by specialised guard cells, which are able to alter their shape in response to changes in light and water availability. Hence they function to restrict water loss when closed, yet allow the entry of carbon dioxide for photo-synthesis while open

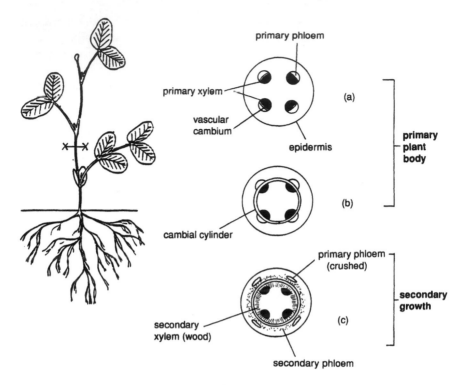

Figure 2.7   Primary and secondary growth of the plant body. (a) The vascular cambium of the primary plant body develops until (b) the cambium forms a continuous cylinder of meristematic tissue. (c) As this meristem divides laterally during secondary growth, the original phloem of the primary body becomes crushed while new *secondary phloem* is produced external to the cambium, and the *secondary xylem* or *wood* is produced internally

Nutrient Assimilation

By virtue of a range of specialised biochemical pathways, plants are capable of using very simple substances, such as atmospheric gases and dissolved minerals, to synthesise the enormous variety of complex organic molecules required for normal growth and reproduction. For example, in the photosynthetic pathway, carbon dioxide and water are used in the synthesis of carbohydrate, while dissolved nitrates—and in some cases gaseous nitrogen—may be absorbed and used in protein synthesis. In total there are 16 mineral nutrients which are essential to the growth of most plants and provide the raw materials for the synthesis of all the necessary

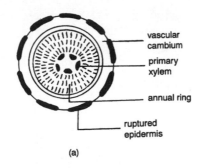

(a)

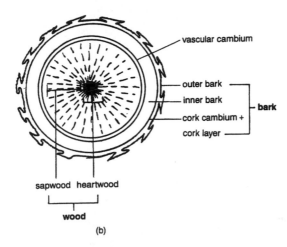

(b)

Figure 2.8 Wood and bark development in the secondary plant body. (a) As secondary growth continues the original epidermis ruptures due to expansion in stem girth; meanwhile *annual rings* occur as differential growth throughout a single growing season produces concentric rings of large and small xylem cells; (b) later still, a further lateral cambium, the *cork cambium* develops and divides to produce the *inner bark* which contains the functional phloem of the trunk, and the *outer bark*, which consists largely of dead phloem cells, and the protective *cork layer*, which takes over the function of the epidermis. The walls of the cork cells are impregnated with *suberin*, a waxy substance similar to cutin which protects the plant from water loss and pest attack. Hence, these densely packed cells again form a water-resistant barrier, while gas exchange occurs through less densely packed groups of cells known as *lenticels*. By this time, many of the original xylem vessels are no longer functional, forming the *heartwood*, while younger active xylem forms the functional *sapwood* (see Simpson & Conner-Ogorzaly 1986)

cellular components, including proteins, membrane and cell wall constituents, hormones, vitamins and many other complex substances (Table 2.5).

Respiration and Photosynthesis

All biological organisms require a continual source of energy to drive the metabolic processes necessary to maintaining living tissues, and to allowing growth and movement. Most obtain their energy supply from chemical sources such as carbohydrate which can be broken down during cellular respiration to release a usable form of energy according to the following generalised equation:

$$C_6H_{12}O_6 \quad + \quad O_2 \quad \rightarrow \quad 6CO_2 \quad + \quad 6H_2O \quad + \quad \text{useable energy}$$

$$\text{glucose} \qquad \text{oxygen} \quad \text{carbon dioxide} \qquad \text{water} \qquad \text{(via ATP—adenosine}$$

$$\text{(energy stored)} \qquad\qquad\qquad\qquad\qquad\qquad \text{triphosphate)}$$

The usable energy produced during respiration can then be used to drive energy-requiring processes such as biosynthesis, stomatal movement and active transport.

The initial formation of carbohydrate itself is carried out in green plants which, with the exception of a small number of photosynthetic bacteria, are the only living organisms capable of photosynthesis. The photosynthetic process represents a complex series of reactions during which photons of light provide the energy required to stimulate movement of electrons from one molecule (water) to another (carbon dioxide); a process which results in the production of carbohydrate according to the following equation:

$$\text{light energy} \quad + \quad 6H_2O \quad + \quad 6CO_2 \quad \rightarrow \quad C_6H_{12}O_6 \quad + \quad 6O_2$$

$$\text{(normally sunlight)} \qquad \text{water} \quad \text{carbon dioxide} \qquad \text{carbohydrate} \qquad \text{oxygen}$$

$$\qquad\qquad\qquad\qquad\qquad\qquad\qquad\qquad \text{(energy stored)}$$

This reaction is entirely dependent on the presence of the green pigment chlorophyll, which initiates electron movement in response to the absorption of light energy (e.g. Goodwin & Mercer 1983; Bryce & Hill 1993).

Secondary Metabolism

Unlike primary metabolites, the vast array of plant *secondary compounds* appear to have no direct role in normal cellular function. The *secondary metabolic pathways* often diverge from *primary pathways* with which they share common precursors (Figure 2.9) and as such it

Table 2.5 Nature and function of plant nutrients. There are 16 mineral nutrients which appear to fulfil the nutritional requirements of most plant species. Provided as simple elements or ions in soil water and the atmosphere, these nutrients are assimilated into complex organic molecules, through the activity of the specialist metabolic pathways which plants possess (Salisbury & Ross 1978). By virtue of these pathways, plants are able to synthesise not only carbohydrates, but also a range of amino acids, vitamins and other substances which are required by animals, yet which animal species themselves are unable to synthesise (Goodwin & Mercer 1983; Marschner 1986)

Essential elements	Major primary functions
Nitrogen (N)	Required for many essential compounds, most significantly in all *proteins* and *nucleic acids* as well as chlorophyll
Carbon (C) Hydrogen (H) Oxygen (O)	Form the basic backbone of all organic molecules
Sulphur (S)	Required for the synthesis of the essential *amino acids* methionine and cysteine which are present in most proteins, and forms an essential component of membranes and of the vitamins thiamine and biotin
Phosphorus (P)	Present in nucleic acids and membrane components
Calcium (Ca$^+$)	Calcium ions are involved in cementing cell walls at middle lamella, and in the control of many metabolic processes
Copper (Cu)	Essential to photosynthesis and the synthesis of certain plant hormones
Molybdenum (Mo)	Functions in pathways leading to the assimilation of nitrogen into essential organic molecules
Potassium (K$^+$) Magnesium (Mg^{2+}) Iron (Fe^{3+}, Fe^{2+}) Chlorine (Cl$^-$) Manganese (Mn^{2+})	These ions are involved in many activities including controlling of stomatal movements, facilitating essential enzymic reactions and stimulating photosynthesis
Boron (B)	Essential to higher plants, although its role is uncertain
Zinc (Zn)	Role uncertain, but is thought to activate certain enzymes

has often been argued that they represent no more than waste products, which function by preventing the accumulation of potentially toxic intermediates (Haslam 1986). Nevertheless, there is widespread evidence for the biological activity of many of these secondary products—ranging from the antibacterial properties of essential oils, to the

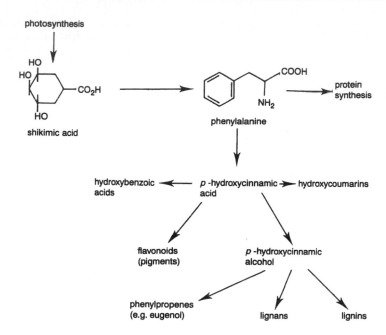

Figure 2.9   Divergence of primary and secondary metabolic pathways.
In many cases secondary biosynthetic pathways share key intermediates with
those involved in primary metabolism. In this example, the aromatic amino acid
phenylalanine can be seen to act as a precursor both in protein synthesis
(primary pathway) and in the synthesis of a range of plant phenolic compounds,
many of which perform a secondary function. For example, several hydroxybenzoic acids and hydroxycinnamic acids have been found to exhibit *allelopathic*
properties, while the flavonoid pigments are responsible for many of the orange,
red and blue petal colours which are commonly involved in attracting specific
animal pollinators (Harborne 1980, 1989)

abortifacient activity of oestrone-like *isoflavonoids*—and chemical
ecologists believe that they confer considerable benefit to the producing plant (Harborne 1988; Hay & Waterman 1993). So far, more than
100 000 biologically active secondary plant compounds have been isolated from higher plants (Howe & Westley 1988), with most of these
diverse structures falling into four main chemical classes, the *phenolics, terpenoids, sulphur compounds* and *nitrogen compounds* as illustrated in Table 2.6.

The plant phenolics constitute a diverse group of compounds all of
which possess an aromatic ring with one or more hydroxyl substituents

(Figure 2.10). Several thousand phenolic structures have been identified (Harborne 1984) showing a range of biological activity. For example, many *flavonoids*, which form the largest group of plant phenolics, can influence the appearance or taste of plants, while others exhibit insecticidal or antifungal activity; a number of hydroxycinnamic acids have been shown to stimulate growth both in whole plants and in tissue culture (George & Sherrington 1984); in contrast certain volatile phenolics such as allylanisole—the major component of tarragon essential oil (*Artemisia dracunculus*)—have inhibited growth in cultured leaf tissue (Cotton *et al.* 1991a).

Plant terpenoids constitute an enormous range of substances all of which are based on the isoprene molecule and contain one or more of these basic units (Figure 2.10). Some terpenoids, such as the chloroplast pigment ß-carotene and the plant hormones gibberellin and abscisin, are involved in growth and photosynthesis; others apparently perform a strictly secondary function. For example, the low molecular weight (C_{10} and C_{15}) compounds, which are generally volatile, are widespread in the aromatic oils of plants such as mint (*Mentha* spp), basil *Ocimum basilicum*, rose (*Rosa* spp) and geranium (*Pelargonium* spp); less volatile diterpenes (C_{20}) are common in the resins of *Pinus* and other gymnosperms; while the non-volatile triterpenes (C_{30}) are found in various parts of plants, including the waxy coatings of leaves and fruits, the cardiac glycosides of species such as *Digitalis lanata*, and as toxins which often accumulate in seeds (Harborne 1988). In addition, a number of important plant pigments such as the light harvesting carotenoids are tetraterpenoid substances (C_{40}), while the rubber polymer—polyisoprene—consists of a large number of isoprene molecules (C_n).

Although nitrogen is present in only about 2 per cent of the dry weight of plants—compared with 40 per cent of carbon (Harborne 1984)—this crucial element is present in an enormous number of primary and secondary metabolites throughout the plant kingdom. The nitrogen-containing secondary metabolites include the largest single class of secondary compounds—the alkaloids, of which around 20 000 have been identified so far (Howe & Westley 1988). This highly heterogeneous group includes chemicals ranging from the simple coniine of hemlock (*Conium maculatim*)—a traditional poison thought to have been used in the execution of Socrates—to the more complex strychnine of *Strychnos* spp (Figure 2.10). Many of these, including the familiar nicotine and caffeine are well-known for their diverse psychoactive and toxic properties (Ott 1993).

Table 2.6 Nature and activity of secondary products. Most plant secondary chemicals may be classified into one of these four main groups, although some other compounds such as certain organic acids and polyacetylenes may also exhibit considerable toxicity towards a range of organisms (Harborne 1988; Howe & Westley 1988). On account of their often autotoxic nature (Cotton *et al.* 1991b) many accumulate in specialised secretory tissues or else occur in the plant as non-toxic glycosides

Chemical class	No of structures (approx.)	Occurrence	Biological activity/characteristics
Phenolics			
Phenols	200	Universal in leaf and other tissues	Antimicrobial
Flavonoids	4000	Universal in vascular plants	Often pigments
Quinones	800	Widespread in higher plants	Pigments
Tannins	indefinite	Widespread in plants	Bind to proteins
Lignins	indefinite	Universal in vascular plants	Indigestible to animals
Terpenoids			
Monoterpenes	1000	Widespread in essential oils	Aromatic, antimicrobial
Sesquiterpene lactones	1500	Mainly in Compositae	Bitter, toxic, allergenic
Diterpenes	2000	Widespread in latex and resins	Toxic, allergenic
Saponins	600	In more than 70 families	Haemolytic, surfactants
Others	90 000+	Widespread	Various
Sulphur compounds			
Glucosinlates	80	Cruciferae (mustard family)	Acrid, bitter
Disulphides		*Allium* spp	Acrid, lachrymatory
Acetylenic thiophenes		Compositae (daisy family)	Toxic
Nitrogen compounds			
Alkaloids	20 000	Widespread in angiosperms	Toxic, bitter, psychoactive
Amines	100	Widespread in angiosperms	Repellent, hallucinogenic
Non-protein amino acids	400	Widespread, especially legumes	Toxic
Cyanogenic glycosides	30	Sporadic in fruits and leaves	Toxic (hydrogen cyanide)

(Data modified from Howe & Westley 1988; Harborne 1988); ns not specified

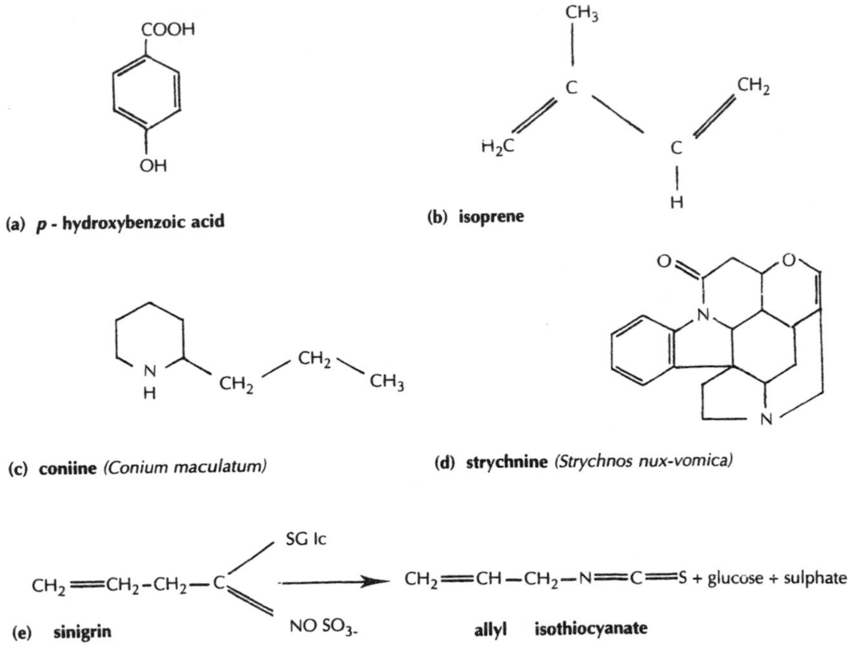

(a) *p* - hydroxybenzoic acid

(b) isoprene

(c) coniine *(Conium maculatum)*

(d) strychnine *(Strychnos nux-vomica)*

(e) sinigrin

allyl isothiocyanate

Figure 2.10 Basic structure of major groups of plant chemicals. (a) *p*-hydroxybenzoic acid—phenolic compounds are characterised by the presence of an aromatic ring plus one or more hydroxyl substituents; (b) isoprene—free isoprene is not accumulated *in vivo* but acts as a precursor to the synthesis of the plant terpenoids; (c) coniine—the major alkaloid of hemlock (*Conium maculatum*), which was used in the execution of Socrates; (d) strychnine—a heptacyclic alkaloid from the seeds of *Strychnos nux-vomica*, often used in the preparation of fish and arrow poisons; (e) enzymic release of the mustard oil, allyl isothicyanate from sinigrin, the major glucosinolate of cabbage (Simpson & Conner-Ogorzaly 1986; Harborne 1988)

Finally, the secondary sulphur-containing metabolites include the glucosinolates or mustard oils of the Cruciferae, the disulphides of *Allium* spp including onion, and acetylenic thiophenes which occur in the roots of some Compositae. Many of these compounds are volatile, exhibiting characteristic acrid flavours and an often obnoxious smell, and they are often associated with determining feeding choice in a number of organisms. For example, to most herbivorous insects, allyl isothiocyanate, the sharp-tasting principle of mustard is not only a repellent but can actually be toxic, and most therefore refuse to feed on diets containing its chemical precursor, sinigrin. However, to certain resistant species, such as the cabbage white butterfly *Pieris brassicae*, sinigrin acts not as a repellent but as a positive feeding stimulus (Harborne 1988).

The Genetic Basis of Biochemical Pathways

All *biochemical pathways*, whether primary or secondary in nature, consist essentially of a series of biochemical reactions mediated by special proteins called *enzymes*. For example, in the synthesis of *estragole*, a phenolic constituent of many aromatic oils, one of the first reactions in the pathway (Figure 2.11) involves the chemical conversion of one compound (the aromatic amino acid phenylalanine) into another—cinnamic acid. This reaction is specifically mediated by the activity of the enzyme phenylalanine ammonia lyase (PAL).

Like all enzymes, PAL owes its characteristic activity to its specific shape which allows it to bind to phenylalanine (Figure 2.12). Through genetic mutation, this shape may become altered, resulting in the modification or inhibition of the enzyme activity, in which case the mutant organism may no longer be able to synthesise estragole. In this case, the *phenotype* of the plant—its physical and chemical properties—is altered. If this *mutant* phenotype is adaptive or neutral, the organism will normally survive; if it is maladaptive survival is unlikely. The variation in both physical and biochemical phenotypes discussed below is therefore determined largely by differences in genotype which occur in all organisms as a result of random mutation. Adaptive or desirable phenotypes are then selected either naturally, giving rise to new ecotypes and species, or artificially, which results in the evolution of domesticated plants.

VARIATION IN PHYSICAL STRUCTURE

The physical structures of plants vary enormously from species to species while fulfilling the same essential functions: shoot systems may be creeping or upright; roots range from the taproots of carrot (*Daucus carota*) and radish (*Raphanus sativus*), to the fibrous systems of monocotyledonous species; leaves vary from entire lamina of several metres in length to highly dissected or tiny overlapping surfaces. This morphological variation largely reflects the ways in which plants have evolved physical adaptations in response to a range of environments, and has ultimately proved of immense value to humans, providing the raw materials for a wide range of commodities.

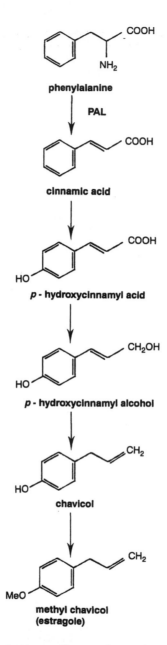

Figure 2.11 Biosynthetic pathway for estragole. The shikimic acid pathway produces the aromatic amino acid phenylalanine, which is then converted to *trans*-cinnamic acid by the action of the enzyme phenylalanine ammonia lyase (PAL). In the synthesis of the phenylpropene estragole (aka allylanisole or methyl chavicol) cinnamic acid is further metabolised via a number of specific steps, each of which is catalysed by the activity of a specific enzyme

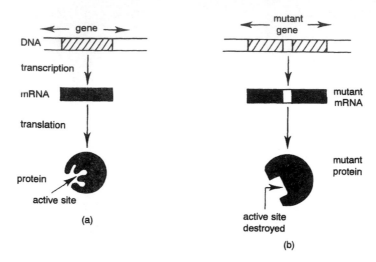

Figure 2.12   Genetic basis of biosynthetic pathways. (a) In a normal plant, the gene (DNA) which codes for the enzyme PAL is *transcribed* into a chemical message (mRNA); this is in turn *translated* into the active protein or enzyme. This final gene product has a specifically shaped *active site* which allows it to bind phenylalanine, thus allowing phenylalanine's chemical modification to cinnamic acid. (b) Where the gene coding for PAL is modified through mutation, if the processes of transcription and translation are able to occur at all, the final protein is likely to have an altered shape which prevents the binding of phenylalanine. Hence the production of cinnamic acid is inhibited and the synthetic route to estragole is no longer functional. (DNA—deoxyribonucleic acid; RNA—ribonucleic acid; mRNA—messenger RNA)

Leaves

In all but a few exceptions, leaves provide the major sites of photosynthesis, a process for which there are four basic requirements: the absorption of light energy from the sun, the uptake of carbon dioxide, and the maintenance of both a suitable moisture content and an appropriate temperature for enzymic function (Figure 2.13). In many habitats, balancing these conflicting needs may prove problematic—for example, in opening stomata to allow carbon dioxide uptake, plants invariably start to lose water, while excessive absorption of light energy may lead to leaf temperatures which are unsuitable for photosynthesis. In some plant species, such problems have been ameliorated through the evolution of specialised metabolic pathways—as in the case of the succulent plants of arid zones; yet frequently the specific physical structure and arrange-

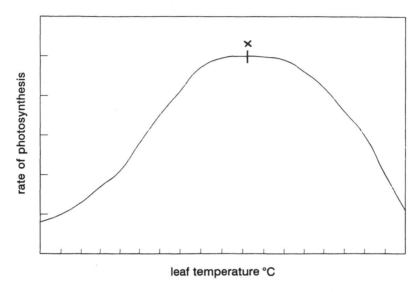

leaf temperature °C

Figure 2.13 Leaf temperature, water stress and photosynthesis. As leaf temperature increases up to point X, the rate of photosynthesis increases as the enzymic reactions involved are stimulated; however, at temperatures higher than X, the rate of photosynthesis begins to decrease. This reduction in photosynthetic rate may be due to a number of factors, such as the heat-induced *denaturation* of photosynthetic enzymes, or stomatal closure in response to excess transpiration. For example, the *photosynthetic optimum* for the C3 plant *Leucopoa kingii* is about 22°C, while that for the C4 plant *Spartina pectinata* is considerably higher at about 32°C (Salisbury & Ross 1978)

ment of leaves can also play a significant part in reaching a suitable compromise. For example, the leaves of those species normally exposed to high levels of light are often small or dissected and may be held at a steep angle, thus maximising self-shading and minimising excess transpiration. In contrast, understorey species—which are less likely to suffer from overheating or excessive moisture loss but need to maximise light capture—are more likely to exhibit a single large lamina positioned for maximum light interception. These profound differences in leaf morphology are also paralleled by variability in anatomical features: smaller leaves may require relatively little structural support, while large leaves—such as those of *Agave* spp and *Musa textilis*, or those which are held upright, are strengthened by groups of sclerenchyma cells, known collectively as *fibres*.

Not surprisingly, such variation in leaf structure has had a profound effect on the way in which plants have been exploited by humans, for while smaller leaves are generally of little structural value, the characteristics of larger leaves, have found a huge range of applications in *material culture*. For example, the palm-like *Carludovica palmata* (Panama hat palm) of Central America has huge, fan-shaped leaves of up to 4 metres long and 1 metre wide which are supported by petioles of up to 2.5 metres in length (Bennett *et al.* 1992). Supporting such enormous leaves, both the leaf laminae and petioles contain large quantities of strong fibres which have allowed this and similar plants to be used in a number of ways: entire leaves may be used in cooking or thatching and for making waterproof containers, the waxy cuticle providing excellent water-proofing qualities (Davis & Johnson 1987; Bennett *et al.* 1992; Fong 1992); petioles such as those of the sago palm (*Metroxylon warburgii*) have been used in weaving walls of houses (McClatchey & Cox 1992); strips of many fibrous leaves, such as those of the palm *Jessenia bataua* may be used for weaving items such as baskets, or as cord for lashing (Bennett *et al.* 1992, Milliken *et al.* 1992); isolated leaf fibres of sisal (*Agave sisalana*) and henequen (*A. fourcroydes*) are commonly used in ropes (Simpson & Conner-Ogorzaly 1986).

Stems

The structural characteristics of stems of both herbaceous and woody plants also vary considerably and their different features have been widely exploited by humans. For example, flexible stems which protect plants against damage during high winds have long been used in making rope—as in the case of jute (*Corchorus capsularis*) and hemp (*Cannabis sativa*)—and for twines used in weaving floor matting (Densmore 1974; Gilmore 1991); the hollow, lignified stems of rattan (*Calamus* spp) and bamboo (Bambusoideae), which provide support and limit herbivory, are of great value in both the local and commercial production of furniture, and in providing buoyancy for fishing floats and rafts (Flood 1983; Balick 1984; Simpson & Conner-Ogorzaly 1986; Avé 1988; Peluso 1992). Finally, as the trunks of large trees and the thickened stems of palms are capable of supporting heavy leaf canopies which are beneficial in the competition for light, they can provide strong timber for building and tool manufacture, as well as fibres suitable for making paper and twine (Dinwoodie 1981; Wilcox *et al.* 1991; Bennett *et al.* 1992; Milliken *et al.* 1992).

The characteristics of wood vary from species to species, the strength of wood, and therefore its value as timber, being directly related to its structure. For example, differences in both the thickness of the constituent cell walls and the species-specific distribution of certain cell types can affect the mechanical properties of a given tree species (Dinwoodie 1981), while even within a single tree, significant defects may result from changes in the deposition of lignin or cellulose in response to environmental stress (Wilcox *et al.* 1991). Within the timber trade, the wood of gymnosperms such as pine (*Pinus* spp) and spruce (*Picea* spp) is referred to as *softwood*, in comparison to the *hardwood* of dicotyledenous angiosperms. This distinction is based largely on anatomical differences, as in gymnosperms the xylem is composed largely of long *tracheids* and is relatively uniform, whereas the wood of dicotyledons contains large numbers of shorter *xylem vessels* and is much more *heterogeneous* (Simpson & Conner-Ogorzaly 1986). As a consequence of these structural differences, gymnospermous softwood often provides the most suitable material for paper-making due to the relative uniformity of its constituent cells (McLaughlin & Schuck 1991).

While the development of woody tissue is generally accompanied by the formation of the suberised cork layer, the exact nature of these cork cells and the way in which they are laid down varies considerably between plant species exerting a profound effect on both the appearance and properties of the outer bark. For example, in certain species of *Eucalyptus, Melaleuca, Bursera* and *Betula* the cork tends to peel like sheets of paper as the girth of the trunk increases, while in many species the outer layers of bark are sloughed off in much smaller pieces. Again, these variable characteristics can determine the ways in which different types of tree bark can be used. Commercially, the cork of a number of species—particularly *Quercus suber*—is stripped off in large sheets, providing material for bottle stoppers and other items such as tiles and floats (see Fahn 1990); in traditional societies, large sheets of pliable, waterproof bark have been used in making water containers (Gottesfeld 1992) and dishes (Densmore 1974), in roofing (Driver 1969), and in building walls (Milliken *et al.* 1992), canoes and rafts (Driver 1969, Flood 1983). Many barks also yield fibres which can be isolated for weaving and making ropes (Milliken *et al.* 1992).

Roots

The root system usually develops within the soil, although certain plant species may develop aerial roots which are often involved in aeration of waterlogged plants (Figure 2.14). Within the soil environment roots also

Figure 2.14 Aerial roots of tropical mangroves—Cape Tribulation, Australia. Tropical mangroves (*Rhizophora* spp), which are periodically inundated by seawater, are characterised by the production of both aerial *stilt roots* and of root projections known as *pneumatophores*. These specialised organs both produce a type of tissue known as *aerenchyma* which is recognised by the high proportion of air space between its constituent cells. These elongated air ducts facilitate gas exchange in waterlogged soils where oxygen is lacking, thus maintaining respiratory functions even in submerged roots

vary enormously according to the species and prevailing conditions, their morphological and anatomical differences maximising water and nutrient uptake, while minimising metabolic cost. The physical characteristics of these organs too have proved valuable to humans—for example their strong fibrous nature, which allows them to push through hard soils in search of nutrients and moisture, has been exploited in making ropes (Whistler 1988) and baskets (Turner 1988); the aerenchyma of the aerial roots of mangroves (*Rhizophora* spp) has found a use in providing buoyancy in rafts (Flood 1983); the supportive *buttress roots* characteristic of many tropical trees have been used in the manufacture of items such as boat paddles and shields (Milliken *et al.* 1992).

Reproductive Structures

The reproductive structures of plants differ enormously, varying from the tiny *capsules* of mosses to the colourful petals and often edible fruits of the flowering plants. Responsible for the perpetuation of the species, these reproductive organs may exhibit structural features, such as exposed anthers, or long inflorescences, which promote the dispersal of pollen and seed. In many cases, the uses of these structures are dependent on their chemical characteristics—the flowers of angiosperms produce a range of valuable oils and pigments, while the seeds of several plant species provide the staple diet of much of the global population. Yet again, there are a number of plant species whose reproductive structures play a direct role in human material culture. For example, the long straight inflorescences which facilitate seed dispersal in the giant grass *Gynerium sagittatum* are used by the Waimiri Atroari Indians of Brazil to make arrows (Milliken *et al.* 1992); large seeds—which can confer a competitive advantage during germination—have been used as buttons and beading (Gilmore 1991; Milliken *et al.* 1992) and for making craft items such as match boxes and perfume bottles (pers. obs.).

In several species, fruits and seeds yield useful hairs or fibres which function by trapping air and thus provide buoyancy during aquatic dispersal (Simpson & Conner-Ogorzaly 1986). In particular, the fibrous husks of coconut have been used in making cordage and items such as reef shoes, as well as for transporting fire (Whistler 1988); the water-resistant seed hairs of kapok (*Ceiba pentandra*) have been used in floats; most significantly, the seed hairs of cotton (*Gossypium* spp), have provided a vital source of plant fibre throughout the world, the longer hairs being used to produce thread for textiles, while the shorter hairs have been used in the manufacture of high-quality paper (Simpson & Conner-Ogorzaly 1986).

Microscopic Structures and Emergences

In addition to variation in their gross morphology, plants also exhibit a number of microscopic features and surface structures such as spines,

Table 2.7 Nature and functions of microscopic structures and emergences. In addition to the major plant organs, plants exhibit a number of characteristic microscopic or superficial structures which have a range of functions in reproduction and defence (Fahn 1979, 1990; Howe & Westley 1988; Pearsall 1989). Like other plant features, many of these also play some part in the material culture of both traditional and industrial societies (Simpson & Conner-Ogorzaly 1986)

Structure	Characteristics
Spore	Microscopic structure involved in the reproduction of lower plants—those which do not produce seed. Sporopollenin prevents desiccation during dispersal.
Pollen	Contains the male gametes in seed plants, and is transferred to the ovule (usually of another individual) of the same species during the process of *pollination* allowing *fertilisation* to take place. The pollen grain is protected by a coating of sporopollenin—the *exine*—which is sculpted into characteristic patterns which may assist in long-distance dispersal.
Phytolith	Opaline silica bodies of $SiO_2.nH_2O$, which occur largely in the epidermal cells and trichomes of stems, leaves and inflorescences of certain species—particularly grasses. They apparently function as a defence against herbivory, reducing both nutritional quality and palatability (Howe & Westley 1988)
Trichome	Often microscopic, epidermal hairs have various functions as described earlier. In some, such as the stinging hairs of the nettle *Urtica diocica*, the tips become silicified and may function as a kind of epidermal syringe, injecting the active irritant into the skin of potential predators. In many cases, leaf hairs provide important mechanical protection from small insects and mites (Juniper & Jeffree 1983).
Emergences	Unlike epidermal hairs, the spines and thorns of cacti and other plants include subepidermal as well as epidermal tissues. Often strengthened by the presence of silica bodies, these structures can function in, for example, the reflection of excess light and the deterrence of herbivores.

which also exhibit specific adaptive modifications. Microscopic features include spores, pollen, many types of trichome and phytoliths; emergences include the thorns of roses, all of which play a considerable part in plant survival (Table 2.7).

While the species-specific characteristics of many of these structures have proved of enormous value in plant identification in scientific studies (Chapter 4), some have found a more direct role in human culture. For example, the silica bodies of plants such as *Licania* spp have been used to lend strength to pottery by several Amazonian peoples (Milliken *et al*. 1992) while the commercial uses of silica *per se*, include applications in products ranging from toothpastes and detergents to heat insulators and absorbents for hazardous materials (Vilarem *et al*. 1992); even pollen has found an application, providing a yellow pigment for the distinctive sand paintings produced for ceremonial purposes by the Pueblo Indians of the American Southwest (Driver 1969). There are many examples too, of the use of spines in various aspects of material culture, such as the petiolar spines of the palm *Jessenia bataua*, which are used to make darts for blow guns, and the sharp leaf spines of *Agave sisalana* used traditionally as needles in Central America (Balick 1984; Simpson & Conner-Ogorzaly 1986).

Additional Uses of Plant Structural Characteristics

It is clear from the preceding discussion that adaptive variations in basic plant structures have led to the use of a wide range of plants in the material culture of both traditional and industrial societies. However, the diversity of plant leaf and root systems are also exploited on the basis of the environmental functions they perform, particularly in traditional agriculture where plants producing shade may protect light-sensitive crops such as coffee (Bunch 1989), or else can provide effective biological control of the voracious, shade-sensitive tropical weed *Imperata cylindrica* (Lightfoot *et al*. 1989), and where soil stability may be maximised on the basis of the binding properties of certain root morphologies (Richards 1985).

VARIATION IN CHEMICAL CHARACTERISTICS

Just as the morphological and anatomical attributes of different plant species vary enormously, so too do their chemical properties. To some extent, chemical composition will fluctuate according to nutrient availability, yet often chemical variability reflects the activity of distinct biochemical pathways which different species exhibit in coping with different types of environmental stress. For example, plant species growing in arid environments may use modified photosynthetic pathways which minimise water loss through open stomata; carnivorous plants in phosphorus-poor acid environments secrete enzymes capable of digesting their animal prey; gymnosperms accumulate resinous compounds which can assist in protecting against pathogens and pests (Table 2.8).

Table 2.8 Biochemical adaptations to stress. Different species of plant exhibit a number of biochemical adaptations which allow them to tolerate a considerable range of environmental stresses, including both abiotic and biotic factors (Salisbury & Ross 1978; Harborne 1988; Deans & Waterman 1993). Some of these adaptations involve modifications of primary pathways—as in the case of alternative photosynthetic pathways or the synthesis of specialised proteins; others are dependent on the synthesis of bioactive chemicals which are generally regarded as having a secondary function

Stress factor	Examples of biochemical adaptation
Drought	Specialised photosynthetic pathways increase water-use efficiency or carbon dioxide uptake in C4 plants such as sugarcane (*Saccharum officinarum*) and CAM plants such as the succulent *Kalanchoe* spp; volatile oils may restrict excessive transpiration.
Salt	Plants accumulate high levels of *osmotica* such as the amino acid proline, which facilitates continued water uptake as in the glasswort (*Salicornia europa*).
Waterlogging	Waterlogged plants are subject to oxygen stress, under which conditions the toxin ethanol is usually accumulated; modified biochemical pathways allow the synthesis of non-toxic alternatives—e.g. malate synthesis in *Juncus effusus*.
Freezing	Plants whose living tissues are tolerant of freezing temperatures can accumulate glycerol or other chemicals which act as an antifreeze by lowering the freezing point of the cell sap.
Extreme temperature	Adaptation to high or low temperature may be reflected in protein structure, specific modifications of which can ensure relative stability under temperature extremes.
Fire	Low-molecular weight monoterpenes (C_{10}) are widespread in plants found in areas prone to natural fire; it has been suggested that by burning at relatively cool temperatures these volatile oils may be important in the survival of plants, such as *Eucalyptus* spp.
Ultraviolet light	Anthocyanin accumulation can filter out harmful rays, protecting the nuclear material from potential damage.
Toxins	Metal-tolerant plants often produce specialised proteins which bind to heavy metals such as lead, rendering them non-toxic.
Nutrient deficiency	Carnivory depends on the secretion of digestive enzymes; nitrogen fixation depends on specialised biochemical pathways.
Biotic	Plants produce a wide array of chemicals which are toxic, repellent or otherwise detrimental to a range of organisms, including other plants, herbivorous animals, and microbial pathogens.

Information taken from Harborne 1988; Deans & Waterman 1993.

Chemical Adaptation to Abiotic Stress

Plants inhabit almost every type of environment, ranging from arid deserts, through tropical wetlands, to the permanent frost of the arctic tundra. Living under these often extreme conditions, plants must tolerate high levels of stress due to abiotic factors, including drought, extremes of temperature, nutrient deficiency and fire. We have already discussed how differences in the physical characteristics of plants may operate to alleviate some of these stress factors, and in many cases these are accompanied by modifications in plant biochemistry. For example, those plant species which remain functional in extreme temperatures often exhibit modified enzymes which are able to retain their structural integrity—and therefore their biochemical activity—under such conditions; while the volatile oils of plants in semi-arid regions may protect them from damage in these fire-prone environments (Deans & Waterman 1993).

Chemical Mediation of Biological Relationships

The one type of stress that all plants constantly face is that posed by other organisms, including herbivores, microbial pathogens and other plants which compete for limited resources. Faced with this perpetual threat, many species accumulate secondary chemicals which reduce their nutritional quality or render them toxic to potential parasites or competitors. For example, many volatile oils demonstrate biocidal activity against a wide range of bacteria, fungi, insects and other plants (Deans & Waterman 1993), while certain psychoactive alkaloids have induced aversive behaviour in experimental animals (Goldstein & Kalant 1990).

Although the role of phytochemicals in plant competition and pathogen control has received considerable attention in recent years (e.g. Bell 1980; Haslam 1986; Kleiman *et al.* 1988), perhaps the most widely studied defensive chemicals are those involved in protection against herbivory. There appear to be two basic chemical approaches to defence against grazers—the accumulation of high levels of substances such as cellulose and lignins which are difficult to digest, and the production of low concentrations of specific toxins which exhibit powerful biological activity (Table 2.9).

However, while all plants protect themselves against excessive herbivory, whether by physical or chemical characters, or behavioural modifications such as *ephemeracy*, there are also times when plants need to *attract* herbivores in order to facilitate processes such as cross-pollination and seed dispersal. Clearly therefore, many plants face a dilemma between restricting the loss of photosynthetic or other tissues to

Table 2.9 Phytochemical defence against herbivory. The chemical defences of plants apparently fall into two main categories: complex polymers or silica crystals reduce the digestibility to animals, while phytotoxins kill or repel herbivores and pathogens at very low concentrations (Howe & Westley 1988). It has been postulated that those plants which are easily found by herbivores—such as perennial species which are easily located by animals—tend to defend themselves using digestibility reducers which provide generalised protection against all herbivores; in contrast, short-lived or rare species are more likely to produce qualitative toxins that are effective against all but specialist herbivores and constitute a slower metabolic cost (Feeny 1976). There is some evidence to support this theory of *plant apparency* (Futuyma 1976), although it seems unlikely that any one theory could fully explain patterns of defence exhibited by the 250 000 angiosperms which are thought to exist

Chemical	Nature and defensive role
Digestibility reducers	
Cellulose, hemicellulose	Sugar polymers of cell walls, require gut flora for digestion
Lignins	Phenolic polymers, digested only by a small number of micro-organisms
Tannins	Phenolic polymers, bind with digestive enzymes of predator
Silica	Inorganic crystals; indigestible
Phytotoxins	
Alkaloids	Nitrogen-containing structures with a range of activity, including psychoactivity and inhibition of DNA and RNA production
Non-protein amino acids	Analogues, which compete with protein amino acids during protein synthesis producing inactive enzymes
Cyanogenic glycosides	Molecules releasing hydrogen cyanide, which inhibits cellular respiration
Glucosinolates	Mustard oils; cause disorders of the endocrine (glandular) system
Terpenoids	Dimers and polymers of C_5 units; range of antimicrobial and other activity, possibly by interference with respiration

Data modified from Howe & Westley 1988

herbivores, and attracting animals which fulfil a beneficial role, and in some cases the solution to this type of dilemma has been found in the evolution of complex *mutualistic* relationships between specific plant and animal species (Table 2.10). For example, in certain species of *Acacia* there exists a close relationship with ants of the genus

Table 2.10 Phytochemical mediation of plant–animal mutualisms.
While plants must defend themselves against excessive herbivory, the majority
of angiosperms are reliant on animals for pollination and seed dispersal, and in
many cases these conflicting needs have resulted in the evolution of very specific relationships between specific plants and animals. In some examples, both
organisms benefit from the relationship, the animal receiving a source of energy
or some other essential substance while the plant fulfils its dispersal requirements; however, in some cases where the plant resorts to mimicry the dispersal
agent may be attracted under false pretences

Chemicals	Relationship
Methyleugenol	The golden shower tree *Cassia fistulosa* attracts a specific pollinator (*Dacus dorsalis*) using the volatile phenylpropanoid methyleugenol—a sex pheromone involved in the insect life cycle; the tree is reliably pollinated, while the oriental fruit fly receives energy from the pollen.
Schottenol	The cactus *Lophocereus schotti* produces the steroid shottenol, which specifically attracts the fruit fly *Drosophila pachea*; the fly receives an essential moulting hormone which allows it to complete its life cycle, while it, in turn, removes unwanted dead tissues from the plant.
Sugars (nectar)	Certain *Acacia* spp attract *Pseudomyrmex* ants by virtue of their sugar secretions; the ants receive a reliable source of energy while the ants attack other potential pests.
Amines	Certain species of *Arum* produce amines similar to those produced in faecal matter, which they volatilise using free energy from respiration to generate heat. Dung beetles and flies are attracted to these odours and become trapped in the flower, as they try to escape they pollinate their temporary host. The insects themselves, however, receive little from this encounter.

Data modified from Harborne 1988

Pseudomyrmex: attracted by the nectar that the plants secrete, the ants
shelter in these leguminous trees, viciously attacking any other pests
which approach; the ant receives a reliable energy source, while the
plant is protected from foliage-eating herbivores which might considerably reduce its photosynthetic surface.

Exploiting Phytochemical Variation

Like their physical structures, the chemical characteristics of different
plants have also been exploited by humans over the millenia, and the
variations in their chemical makeup are clearly reflected in the ways in

which different plants and plant organs have been used. As they are able to synthesise several essential nutrients which humans are unable to synthesise themselves—including certain amino acids and vitamins (Goodwin & Mercer 1983; Marschner 1986), many have been adopted as sources of food; others produce lethal toxins used in hunting or fishing; others still contain psychoactive substances which have formed an integral part of traditional ceremonial life, or have provided the basis of certain allopathic drugs.

The nutritional value of different plant species varies according to differences in the type of carbohydrate they store, the kinds of proteins they produce, and the secondary compounds they accumulate. For example, while cereal grains store large quantities of starch as an energy source for the developing embryo, other plants such as oil palms (e.g. *Orbignya oleifera*) and olive (*Olea* spp) accumulate large quantities of oil which fulfils the same function (Khanna & Singh 1991); leguminous crops such as peas and beans accumulate only low levels of the essential amino acid methionine, while cereal grains are generally deficient in lysine and tryptophan (Shewry & Kreis 1991); many members of the Solanaceae family such as deadly nightshade (*Atropa belladonna*) and hebane (*Hyoscyamus niger*) are rendered inedible by the powerful toxic alkaloids they produce, whereas woody species and grasses are generally avoided due to their high content of lignin or cellulose. These variable nutritional characteristics have had a profound effect on which plants have been selected as foods during human evolution (Johns 1989), and although about 75 000 of the world's plant species are believed to be edible, of these a mere 30 or so species provide 90 per cent of the global nutritional requirements (Walters & Hamilton 1993).

Yet while relatively few plants are currently of widespread use as food, many other species are exploited for the biological activities exhibited by their varied secondary chemicals. Many of these have found roles in a wide range of applications, including the production of insecticides and fish poisons, medicines and hallucinogens, food preservatives and flavour compounds; they also provide raw materials for the fragrance and cosmetic industries, as well as for alternative sources of fuel, rubber, plastics and a large number of other valuable commodities for both local and commercial use (e.g. Simpson & Conner-Ogorzaly 1986).

Finally, the specialised biochemical pathways of plants are also of considerable direct value in agricultural and other activities. For example, the nitrogen-fixing properties of legumes are exploited to enrich soils with valuable nitrates, while the repellent properties of plants such as neem (*Azadirachta indica*) and *Chrysanthemum* spp, can afford biological protection against certain pests (Ley 1990). More recently, it has been suggested that the biochemical capacity of certain plants to detoxify heavy metals may prove useful in the reclamation of mining wastes where heavy metal concentrations are high (Howe & Westley 1988).

Genetic Consequences of Human Activity

Plants have played an integral part in the evolution of human cultures, their physical and chemical properties providing not only an invaluable source of food, but also a wealth of raw materials which fulfil many of our medicinal and material requirements. Yet as plant characteristics have influenced human development, so too have humans played an important part in plant evolution. For as humans have selected certain species and individuals with desirable phenotypes, this has resulted in the evolution of both *domesticated* plant species, and of *anthropogenic* vegetation types (Harris & Hillman 1989a, McNeely 1994).

The variation in the physical and chemical phenotypes discussed above have therefore evolved under the influences of both natural and artificial selection. Yet while natural selection tends to favour genetic variability, the plant breeding programmes of industrial agriculture generally lead to dependence on a very narrow genetic base, and in some instances this has precipitated serious problems such as devastating outbreaks of disease (Myers 1992). However, in traditional farming systems, both natural and artificial selection continue hand-in-hand, maintaining genetic diversity while enhancing desirable traits, and in recent decades, formal plant breeders have become increasingly aware of the need to conserve the germplasm of both wild plants and traditional crop varieties, if our long dependence on plants and their products is to continue.

SUMMARY

During the course of plant evolution, genetic variation together with both natural and artificial selection, have resulted in a range of physical, chemical and ecological characteristics all of which have a vital part to play in plant adaptation and survival. Physical characters such as leaf size may contribute to their ability to survive under conditions such as drought or shade; the accumulation of secondary compounds can determine survival against pests and disease; while ecological functions such as nutrient cycling are fundamental to biological sustainability. The recognition of these biological characteristics has frequently led to plants' exploitation by human societies, not only as raw materials and sources of new germplasm, but also as important tools in successful land management; in turn artificial selection and vegetation management have exerted a considerable influence on the evolution of plants and vegetation types.

Chapter 3

Traditional Botanical Knowledge

We are losing our ancestral knowledge because the technicians only believe in modern science and cannot read the sky.

Andean peasant expression (cited in Salas 1994)

INTRODUCTION

While plants have adapted to the diverse habitats of the world through their physical and biochemical modifications, human populations have adapted largely through the generation and application of knowledge— both ecological and technological, practical and theoretical. Today, traditional societies throughout the world possess a wealth of such knowledge which they have accumulated during prolonged interaction with the natural world, and which remains fundamental to their physical, spiritual and social well-being.

Throughout the course of this century, generations of biologists and anthropologists have attempted to study various aspects of ethnoscientific knowledge, and a range of terms describing these different aspects now exist (see Table 3.1). Ethnobotanists themselves have attempted to study that ethnoscientific knowledge which pertains specifically to plants, and so far they have presented a vast body of data concerning the ways in which plants are used, managed and perceived by different peoples. Although much of this research has focused exclusively on the knowledge held by *indigenous peoples*, more recent work has begun to study the ethnobotany of certain non-indigenous societies such as the *caboclo* farmers of Brazil, and the *mestizo* populations of Central America. For these *traditional* communities share a common ancestry

Table 3.1   Modern terminology describing ethnoscientific knowledge. Ethnoscientific knowledge has been described so far using a variety of terms, each of which can be interpreted in slightly different ways. As a number of these terms have been coined in response to recent changes in Western under-standing of the nature of traditional knowledge, some are more all-encompass-ing than others. For example, while the concept of ITK attempts to interpret traditional knowledge in isolation from its specific socio-cultural context, more recent concepts recognise the importance of examining knowledge within the framework of local ideology and social controls. Some of the more common of these terms—and the relationships between them—are outlined here (see [1]Rocheleau *et al.* 1989; [2]Scoones & Thompson 1994; [3]Johnson M 1992; [4]Millar 1993)

Nature and description of traditional scientific knowledge

Indigenous technical knowledge (ITK)[1]—refers largely to the technical knowledge held by traditional farmers, and includes knowledge about practices such as the use of beneficial crop combinations, the use of plant toxins in pest control, and the application of various processing methods in food preparation. In this sense, ITK is essentially regarded as a tangible stock of information which can be extracted and applied outside its original cultural context.

Indigenous agricultural knowledge (IAK)[2]—incorporates not only the relevant ITK, but all forms of indigenous knowledge which are pertinent to agriculture, including methods of crop management and livestock production, ethnoveterinary practices and methods used in crop or animal breeding. This term recognises the roles of indigenous experimentation and innovation in the continued development of IAK, and, unlike ITK, it attempts to consider the influence of local cosmology in shaping the development of local knowledge (see Box 3.1).

Traditional ecological knowledge (TEK)[3]—encompasses the total ecological knowledge of both indigenous and other traditional peoples. Like IAK, TEK attempts to consider the formative influences of local environmental perceptions on the nature and development of local knowledge.

Rural people's knowledge (RPK)[2,4]—again this is similar to IAK, and like IAK it is used largely in relation to the knowledge held by traditional farmers. However, this term has been expanded to include both indigenous *and* other traditional peoples, and to consider the influence of social structure and institutional organisation on the generation and distribution of knowledge within a given community (see Box 3.1).

Traditional botanical knowledge (TBK)—this is defined here as the total botanical knowledge held by any non-industrial community and incorporates all utilitarian, ecological and cognitive aspects of both plant use and vegetation management. This term therefore encompasses all types of knowledge outlined within this text, including that concerned with the identification, processing and management of plants used in subsistence, material culture and medicine, while considering this knowledge within its original spiritual and sociological context.

Table 3.1 (*continued*)

Nature and description of traditional scientific knowledge
Integrated knowledge system (IKS)—borrowing from Millar's concept of an agricultural knowledge and information system (AKIS),[4] which represents a synergistic integration of IAK and Western science, an IKS is defined here as any synergistic integration of traditional and Western knowledge. Within this text, this is concerned primarily with the integration of TBK with empirical Western knowledge from both the natural and social sciences.

with indigenous peoples and in many cases have coexisted with native cultures for hundreds of years.

However, the study of traditional knowledge has been fraught with difficulties, as cultural and linguistic differences between Western researchers and local participants have led to widespread misunderstandings—many of which have yet to be resolved. This chapter introduces the main approaches used in modern investigations of *traditional botanical knowledge* (TBK), and discusses how an increased understanding of traditional knowledge systems can not only facilitate more rigorous ethnobotanical investigation, but can also lead to a wider acceptance of the value of local knowledge.

THE DOCUMENTATION AND INTERPRETATION OF TRADITIONAL BOTANICAL KNOWLEDGE

One of the main problems associated with many ethnobotanical studies lies in their frequent failure to distinguish between ethnobotanical *evidence*, and traditional *knowledge*, a failure which has led to widespread misconceptions about the nature and validity of TBK. For where observed evidence such as the planting of a certain crop or the use of a particular tool appears to be sub-optimal from a Western scientific perspective, this has often been assumed to reflect limited knowledge of more suitable alternatives. In practice, however, traditional peoples have often reached a compromise based on a much broader range of influences than those considered by the scientists. This point has been illustrated clearly in a recent study in Zambia, where a series of formal agricultural trials demonstrated that beans planted in flat areas could produce higher yields than those established on artificial mounds. Yet while local farmers in the Copperbelt Province did plant their beans on the flat, those in the Central Province always planted them in specially constructed mounds of compost—a traditional practice which is extremely labour-intensive. Assuming that the farmers' behaviour is representative of their *knowledge*, this behavioural *evidence* might therefore suggest that the Central Province farmers are unaware of the potential benefits of flat land cultivation; in reality the mounds fulfil a number of distinct ecological functions which are not apparent from the formal trials. For example, their physical properties function to conserve sufficient heat and moisture to extend the local growing season by up to 2 months, while mound-grown crops are additionally protected against unpredictable climatic extremes (Drinkwater 1994). Indeed, even the farmers of the Copperbelt Province agreed that they would actually prefer to grow beans on mounds, but explained that in this region, seasonal patterns in labour demand unfortunately preclude this practice. Such examples, clearly demonstrate that in any ethnobotanical study, it is crucial that any evidence collected is interpreted strictly within its specific ecological and socio-cultural context.

Basic Approaches to the Study of Traditional Botanical Knowledge

Today there are three main approaches to the study of TBK: economic or utilitarian studies record how different plants are used, and in an increasing number of cases may seek to explain these uses on the basis of scientific analysis; cognitive and socio-cultural analyses attempt to determine how plants are perceived by different peoples and to explore how this perception is influenced by spiritual beliefs and other socio-cultural controls; ecological and cultural ecological studies investigate how the management and exploitation of plants can influence—or be influenced by—the characteristics and dynamics of the local environment. Used together, these distinct approaches can provide a much more meaningful understanding of TBK than can be achieved on the basis of utilitarian evidence alone (see Box 3.1).

A Utilitarian Approach: Identifying Useful Plants

Following Harshberger's strictly utilitarian definition of ethnobotany, a considerable proportion of early ethnobotanical research was motivated towards discovering how plants could be used as sources of food and medicine or how they were used in traditional material culture (see Berlin 1992). Like the informal treatments such as those of the sixteenth century friars Bernardino de Sagahún and Juan de Torquemada and of later New World settlers including Josselyn (Griggs 1981; Taube 1993), these studies described how different plant species were identified, processed and used by native peoples (e.g. Gilmore 1919; Densmore 1928; Zingg 1934; Zigmond 1941). Most of these early reports were organised into taxonomic lists of useful plant species (Gilmore 1991, or into sections which dealt with particular uses of plants—such as plants as foods, plants as medicines and plants used in decoration. In a few cases, comparisons between native and non-native uses were presented, as in Densmore's *Uses of Plants by the Chippewa Indians*, which details the principal active constituents of 69 Chippewan medicinal plants, 16 of which were officially recognised either in the US Pharmocopoeia or the National Formulary at that time (Densmore 1928).

Box 3.1 Integrated studies and the interpretation of traditional knowledge. As ethnobotanists have begun to realise the limitations of strictly utilitarian studies, increasing attention has focused on understanding how both cognitive and ecological factors have influenced the ways in which traditional peoples interact with plants. For example, a plant species with strong magico-religious associations is likely to be protected regardless of any obvious utilitarian value; meanwhile, the forces of natural selection may determine the long-term success of certain cultural practices. These different considerations have led to three very different approaches to modern ethnobotanical study: the *utilitarian*, the *cognitive* and the *ecological*. Yet only through the integration of all three approaches are meaningful ethnobotanical data likely to prove forthcoming.

Today, there are three main approaches to the study of traditional botanical knowledge: utilitarian, cognitive and ecological. Yet while each has its own aims, theories and methods, the integration of these distinct approaches is vital to conducting meaningful ethnobotanical investigation.

UTILITARIAN ETHNOBOTANY

objective plant characters	→	*empirical knowledge*	→	*sources of evidence*
physical structures		objective 'reality'		behaviour
chemical content	observation and			information
behavioural traits	experimentation			artefacts

The utilitarian approach to ethnobotanical study involves the collection of information about the uses and management of different plants, and includes the identification of useful species and the elucidation of methods used in the production and processing of these plants. The assumption here is that plant uses are based largely on objective plant characteristics and have developed—in many cases over generations—through observation and experimentation. However, while such assumptions are often valid, this approach generally fails to consider

_ continued _

---

continued

that different cultures invariably perceive the natural world in significantly different ways, and that this can modify considerably, the way in which certain plants
may be used.

COGNITIVE ETHNOBOTANY

subjective plant characters	→	_cultural modification_	→	_sources of evidence_
magico-religious	altered perception	subjective 'reality'		symbolic
social				linguistic
				sociological

In contrast to the strictly utilitarian approach, cognitive ethnobotany involves the
study of cultural symbolism and social structure to examine the ways in which
different plants or vegetation types are perceived by a particular individual or
community. Such studies are critical in interpreting some utilitarian data, for as
local ideology—both cosmological and social—can have a profound influence
on the ways in which people may view the natural world, this must equally influence how they might use or manage the plants around them.

ECOLOGY AND CULTURAL ECOLOGY

plant–human	→	_anthropogenic_	→	_adaptive behaviour_
interactions		_environment_		_maintained_
	artificial selection	plant domestication	natural selection	
		vegetation structure		

While plants are used and managed according to their perceived characteristics,
artificial plant selection and resource management can in turn affect the floral
characteristics of the local environment. As ecological scientists investigate the
nature and extent of these environmental effects, cultural ecologists seek to
demonstrate that the long-term success of the anthropogenic environment is ultimately determined by natural selective forces, such that sustained cultural practices are assumed to be ecologically adaptive.

More recently, the scientific evaluation of this type of ethnobotanical
information has become much more common, particularly as a number
of drug discovery programmes have begun the regular screening of
traditional herbal remedies. For example, in 1984, ethnobotanist Paul
Cox (Cox 1994) began an intensive investigation of Samoan ethnopharmacology in collaboration with the US National Cancer Institute (NCI).
Since then numerous traditional plants have been analysed not only for
their pharmacological properties, but also for other qualities, such as
insecticidal or repellent activity and nutritional quality (Table 3.2). The
effects of traditional processing methods have also received considerable attention, often revealing the empirical advantages of particular

Table 3.2   Chemical analyses supporting traditional plant use. In recent years, chemical analyses and biological assays have begun to play an important part in ethnobotanical studies, and there are now numerous examples where scientific analysis has provided objective evidence to validate traditional plant use. While the most common approach so far has involved the pharmacological analysis of traditional plant medicines, other studies have examined the chemical components of commodities ranging from traditional foods and beverages to insect repellents and hunting poisons. In several cases, such analyses have led to the identification of novel bioactive phytochemicals including those such as the insect antifeedent azadirachtin which is reported to control at least 125 species of insects, mites and nematodes (Hepburn 1989), and the antiviral compound prostratin which strongly inhibits the killing of human host cells by HIV, and is currently considered as a candidate for drug development by the US NCI (Cox 1994)

Scientific validation of traditional plant use

Pharmacological properties
Homalanthus nutans (Euphorbiaceae)—used by Samoan healers against the viral disease yellow fever; extracts have been found to exhibit potent antiviral activity, particularly against the human immunodeficiency virus HIV-1 (Cox 1994).

Melaleuca alternifolia (Myrtaceae)—used by Australian Aborigines for treating cuts and wounds and relieving nasal congestion; the oil has been shown to exhibit broad-spectrum antimicrobial activity and is widely used in cosmetics, antiseptics and in the treatments used by alternative therapists (Thursday Plantation nd).

Insect repellent activity
Azadirachta indica (Meliaceae)—used as an insecticide for centuries throughout India and other Asian countries (Isman 1994), extracts of the neem tree have recently yielded the most potent antifeedant yet discovered (Ley 1990).

Citrullus colocynthis (Cucurbitaceae)—used in post-harvest protection in granaries of Mut (El Dakhla), Egypt (Parrish 1994); contains triterpenoid cucurbitacins, which are among the most bitter and distasteful of the plant terpenoids and which have proved repellent to most insects (Harborne 1988).

Nutritional quality
Acacia spp (Leguminosae)—the seeds were collected from the wild by the Aborigines of Australia's desert regions; in many cases these have been shown to contain higher levels of energy, protein and fat than crops such as wheat and rice (Brand & Cherikoff 1985).

Lycianthes moziniana (Solanaceae)—the fruits have been collected, and more recently cultivated by traditional inhabitants of Mexican highland plateau and Oaxaca; recent analyses suggest high levels of vitamin C and low tannin concentrations—both of which are desirable nutritional characteristics (Williams 1993).

techniques. For example, while the complex processing of important food plants such as bitter manioc (*Manihot esculenta*) is known to remove potentially lethal toxins (Stahl 1989; Beck 1992), specific methods of drug preparation can have a profound influence on pharmacological activity (see Chapter 8).

Today, the compilation of *ethnobotanical inventories* still represents a common approach in the investigation of TBK and many have formed the basis of subsequent chemical, physical or behavioural analyses. Indeed, in 1994, more than 70 per cent of the papers published in the journal *Economic Botany* outlined the traditional or folk uses of various plant species, most of which included information on the chemical compounds or other characteristics related to their use (Figure 3.1). Yet despite the continued enthusiasm for this type of compilation-style of approach, some ethnobotanical studies have been criticised recently for their failure to provide crucial information (Farnsworth 1990, 1994). For example, in 1990 the *NAPRALERT database*—a computer facility collating the world literature on natural products—contained 1249 records of plants traditionally used as contraceptives (Farnsworth 1990), yet few of these reports stated the dosage used, the preparation required, or even whether the treatment should be used by men or women. Clearly, where researchers are interested in identifying plants with specific pharmacological activity, this type of information is essential (Croom 1983; Lipp 1989). In response to such criticisms, there has been a strong movement to develop more rigorous ethnobotanical methodologies, some of which use numerical analysis to predict the most likely drug candidates from a range of useful plant species (see Chapter 11).

Cognition, Context and Behaviour: Understanding Traditional Plant Use

Although utilitarian studies have provided considerable empirical evidence to support the traditional applications of many plants, there remain numerous examples where no rational basis for plant use has been identified using conventional scientific techniques. In such cases, only an understanding of how local people perceive particular plants can help to explain behaviour which may otherwise appear irrational. Since the mid-1950s a distinct branch of ethnobotanical study—that of *cognitive ethnobotany*—has begun to address such issues, by using symbolic, socio-cultural or ethnotaxonomic analyses, to explore not how plants are *used* so much as how they are *perceived* by traditional peoples.

Symbolic analysis basically involves the identification of recurring themes in art, myth and ritual, the interpretation of which can provide

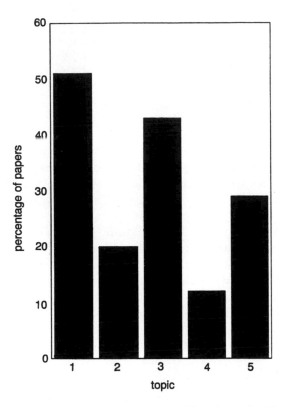

Figure 3.1 Empirical investigation into utilitarian ethnobotany. In 1994, a total of 41 papers were published in the journal *Economic Botany*—a journal which is devoted to the 'past, present and future uses of plants by people'. Of these articles, 29 (71 per cent) were devoted to the *traditional* uses of plants, 23 of which (79 per cent) included empirical data on the chemical or other characteristics of the plant species involved. These data indicate a strong interest in the orthodox scientific analysis of traditional plant use—an area of ethnobotanical enquiry which has developed enormously in recent years. (1 = papers concerning the traditional uses of particular plant species; 2 = papers concerning the traditional use of plants among a given culture; 3 = papers which include phytochemical data supporting traditional plant uses; 4 = papers which include physical or behavioural data supporting traditional plant uses; 5 = papers which are not concerned with traditional plant use)

insight into certain aspects of behaviour. For example, in British folklore a number of plants such as cow parsley (*Anthriscus sylvestris*) and privet (*Ligustrum ovafolium*) are considered unlucky if picked or taken indoors, and consequently are avoided even now in some areas of the UK (Vickery 1985); others such as lilies (*Lilium* spp) and palms still play an important symbolic part in Christian festivals (Goody 1993). In some of these cases, the origins of such floral symbolism have become obscured even where superstitions remain; for example, in many parts of Britain, it is considered unlucky to introduce broom (*Sarothamnus scoparius*) into the home, yet this widespread belief has so far defied explanation (Vickery 1985). In contrast, the origins of other examples are well documented—as in the case of the palm which has been used within the Christian church from the late twelfth century as a sign of the victory of Christ (Goody 1993). Similarly, plant species with religious or spiritual connotations among traditional communities are often associated with ritual or symbolic behaviour, which can be recognised or understood only when considered within its appropriate cultual context.

As local attitudes towards particular plant species or vegetation types are also influenced by a range of social factors, *socio-cultural analyses* can also play an important part in examining traditional plant knowledge. These studies involve the assessment of those rules, within a given community, which govern factors such as land ownership, gender restrictions and access to specialist or esoteric knowledge. For all of these influences can have a profound effect on an individual's perception of—and behaviour towards—the local environment. One example of this is seen among the Yolngu of northeastern Australia, where traditional principles of land allocation demand that when individuals first enter an area of land owned by another group, they must always ask permission before using any of the resources available. Should they fail to do so, they would fully expect their efforts to be unsuccessful, as it is believed that land will not yield its plants or animals to strangers (Williams 1982). Clearly, such rules of social behaviour therefore not only affect the ways in which a given individual may act within a given environment, but can also play an important part in exerting practical controls pertaining to effective resource management.

A further approach which has been adopted widely in investigating local perceptions of the natural world is that of *ethnotaxonomy*—the study of traditional systems of classification. In ethnobotanical terms, this involves the study of how different peoples categorise and name the plants around them, and can reveal where relationships are perceived between different plant species. For example, the Tzeltal Maya of Mexico recognise two plants known as *skelemal tul pimil* and *yach'ixal tul pimil*. Yet although recognised as distinct from each other, these two *taxa* are clearly regarded as related as both are considered to be *tul pimil*. In conventional Linnaean classification also, these plants are recognised as distinct, yet closely related species—namely *Mentha spicata* and *M. citrata*—two members of the mint family which, although morphologically very similar, can be distinguished on the basis of the presence or absence of leaf hairs (see Berlin 1992).

The origins of ethnotaxonomic study can be traced to Harold Conklin's influential doctoral dissertation *The relation of Hanunóo culture to the plant world*, which examines the structure and content of plant classification in the Philippino island of Mindoro (Conklin 1954b). The Hanunóo, who are indigenous to Mindoro Island, are essentially small-scale horticulturalists whose subsistence is based largely on upland swidden agriculture, producing mainly rice, maize, bananas and various root crops. Living as a number of distinct groups, the individual Hanunóo settlements are referred to using the name of the nearest prominent geographical feature. In Conklin's study, the focus of his enquiries were the Yāgaw Hanunóo, who are named after the nearby Mount Yāgaw. Unlike previous reports, which at best simply recorded native plant names, this monograph draws from earlier work on systems of colour classification (Bartlett 1929; Lenneberg 1953) to present a full description of the hierarchical nature of Hanunóo plant classification (Box 3.2). Moreover, Conklin's careful analysis also reveals a detailed vocabulary of specific botanical terms ranging from individual plant structures to entire plant communities, from which a considerable botanical knowledge can be inferred (Table 3.3).

Box 3.2   Plant classification among the Yāgaw Hanunóo. Conklin's pioneering work on the taxonomic systems of the Philippino Hanunóo provided the first real insight into both the hierarchical nature of ethnobotanical classification and the general features common to many plant names. Within the Hanunóo system of plant classification there exist several categories which are recognised yet which are *unlabelled*, including the plant kingdom itself, most intermediate groups and a number of distinct life-form categories such as 'mosses' and 'ferns'. In addition, certain folk species may include a range of phenotypes which are considered different yet are not distinguished linguistically. Of the 1625 terminal taxa reported by Conklin 571 are labelled by a basic plant name only, 961 have a basic plant name expanded by one attribute, while the remaining taxa are named using up to four attributive terms (Conklin 1954b). Following the presentation of these early ideas, similar work has since been carried out in many parts of the world. Much of this work has now been synthesised by Brent Berlin and his co-workers, to produce a general scheme of folk classification which is discussed in more detail in Chapter 9.

In his landmark doctoral thesis *The relation of Hanunóo culture to the plant world*, US anthropologist Harold C Conklin demonstrated clearly for the first time the hierarchical nature of ethnobiological classification and the main fea-

continued

---

continued

tures of traditional plant nomenclature. He also emphasised that individual plant taxa are often perceived as related to other, distinct taxa, and that these relationships may often be reflected in plant names (Conklin 1954b).

HIERARCHICAL ORGANISATION IN HANUNÓO TAXONOMY

At the highest level of organisation—the *kingdom*—plants are recognised as an *unlabelled* group of organisms which grow upwards yet which lack the power of self-locomotion; this general category is then divided into four major *life-form* groups which are distinguished as *trees, herbs, vines* or *others* on the basis of stem habit.

Plant category	Hanunóo label	Numbers
Kingdom (plants)	unlabelled	1
Life form (stem habit)		4
ligneous plant; tree growth; wood	*kāyu*	38%
herbaceous plant; weed	*ʾilamun*	35%
vine-like plant; vine	*wākat*	24%
plants with hollow centres (bamboo)		3%
Intermediate	unlabelled	NA
Conklin Type 1 (Folk generic)	basic plant name	571
Conklin Type 2 (Folk specific)	basic plant name plus attributive(s)	> 1000
Further distinctions (Folk varietal)	often unlabelled	NA

NA, data not available.

At the lowest level of organisation, the Hanunóo apparently recognise more than 1600 mutually exclusive *terminal taxa* or *specific plant types*, each member of which exhibits similar distinguishing characteristics, while differing in at least one component from the members of all other terminal taxa. At this level of categorisation, distinctions between members of different taxa are based largely on features such as shape, colour, size, taste, smell and habitat; these distinctions are often reflected within the plant name. Plant names themselves generally consist of one *basic plant name*, which is often accompanied by one or more *attributives*—whole word modifiers which are frequently descriptive. In some cases, a terminal taxon may be divided further but without linguistic recognition; in others a number of terminal taxa may be recognised as members of an *intermediate* category which is also often *unlabelled* (see Conklin 1954a, b, 1957; Berlin 1992).

Yet while ethnotaxonomic studies can be used to reveal that knowledge which is encoded in specific terminology, several workers have warned against the dangers of relying on lists of terms as sole evidence for TBK (Morris 1976; Alcorn 1981; Ellen 1994). Indeed, as Conklin's work and many subsequent studies have revealed, many folk classifica-

Table 3.3 Botanical terminology among the Philippino Hanunóo.
Conklin has demonstrated that, in addition to their extensive system of plant
classification, the Hanunóo also use at least 147 different terms to refer to specific plant structures, while local vegetation types are distinguished on the
basis of a range of factors such as location, dominant life forms and the predominance of particular species. Within the area inhabited by the Hanunóo, the
local flora is very complex—ranging from open grassland to climax rainforest,
and from cultivated swidden sites to largely uncontrolled, wild vegetation. This
complexity is reflected clearly in the detailed terminology used by local agriculturalists (see Conklin 1954a, b, 1957)

Examples of Hanunóo botanical terms

Plant structures
Stems—at least 64 terms are used to describe specific structures, including
aerial tubers, buttresses, the central lumen of rattan stems, outer and inner
bark, and sap

Roots/underground parts—17 terms have been recorded, referring to
features such as rhizomes, taproots, and the specific qualities of edible root
crops

Leaves—29 terms are used to describe different types and forms of leaf,
including terms for structures such as leaf sheaths and specific patterns of
venation

Fruits and flowers—37 different terms are used to distinguish structures
ranging from pollen and seed, to maize tassels and fruit juice

Vegetation types
Ecological zones—4 major vegetation types are recognised according to
location:
   baybay—vegetation along sandy coastal beaches
   dalaʔagan—tidal swamp vegetation (mostly mangrove)
   oīli—exposed mountain vegetation
   Paŋpaŋ sāpaʔ—streamside vegetation

Level of management—10 major vegetation types are distinguished on the
basis of how they are managed ranging from *parāyan*, which is
predominantly herbaceous and maximally managed, through to *kapurūʔan*
(primary forest) and *kakugnan* (uncontrolled grassland vegetation). An
additional 12 terms are used to describe vegetation at particular stages in
the swidden cycle, from a *selected swidden site* to a *long-term swidden fallow.*

Secondary forest—at least 8 subtypes of secondary forest are commonly
recognised as specific plant associations and are named on the basis of their
dominant species. These distinctions are particularly important in
agricultural terms, as the time required for forest regeneration varies from
one subtype to another.

tion systems include a number of *covert taxa*, which although recognised are given no specific label. In addition, Ellen (1994) argues that the knowledge encoded in formal vocabularies may reflect only that knowledge which is *commonly shared* among various members of a given community, while specific knowledge gained through personal experience is less likely to receive lexical recognition. The notion of the existence of knowledge which is not easily articulated is also highlighted in Alcorn's discussion of an agricultural *script*—essentially a series of routine steps which have evolved locally and are carried out largely unconsciously (Alcorn 1989). For example, in Mexico and Central America, many traditional farmers maintain certain leguminous trees in their fields—trees which are known in conventional science to fix nitrogen, and are therefore likely to contribute to the long-term maintenance of soil fertility. The result of generations of observation and experimentation, this practice has become a part of local custom and is supported and sustained within local ceremonial life. As a result, farmers are now uncertain about *why* the trees are retained or why they are maintained at certain densities and the knowledge encoded within this traditional practice has become difficult to articulate (Wilken 1987). Clearly then, while some traditional knowledge is recognised in lexical terms and can be revealed relatively easily, other knowledge—such as that which is encoded in *behavioural scripts*—has frequently been overlooked in ethnobotanical data collection.

The Ecology of Culture and Cultural Ecology

In this, and the previous chapter, we have seen how the traditional use and management of plants can be influenced not only by the physical and chemical characteristics of the plants themselves, but also by the ways in which plants are perceived within a particular socio-cultural context. Hence, it is argued that certain practices—such as harvesting taboos or food preparation rituals—may be understood only when relevant cultural influences are known. Yet despite their culture-specific nature, many practices which at first glance may appear irrational are now thought to have important functional consequences. For example, the Ka'apor Indians of Amazonia habitually bury their dead in certain parts of old swidden sites known as *kağwer-rena*. However, believing that the *earthbound souls* of the dead are inherently dangerous to the living, the Ka'apor deliberately obstruct any trails to these sites by planting deterrents such as spiny palms, vines and brambles, in order to prevent the return of earthbound souls to the settlement. Subsequently, the *kağwer-rena* are deliberately avoided during hunting or gathering trips despite the high density of game associated with old swidden fallow—a practice which is thought to permit the stabilisation of the regional density of game animals (Balée & Gély 1989). During the late 1940s, similar observations led American anthropologist Julian Steward to propose his *cultural ecological* theory, in which he suggested that human behaviour itself—and the development of the knowledge which underlies it—is ultimately the product of natural selection. This concept has been further extended in a number of related theories which are outlined in Table 3.4.

However, despite many examples where culturally determined practices have been interpreted in functional ecological terms (Table 3.5), this *determinist* stance has frequently been criticised. Some scholars argue that a *synchronic*, or short-term account of a given practice is not sufficient to distinguish between behaviour which is genuinely adaptive and that which simply exists at the time of the study; others observe that cultural ecological models often neglect the fact that through deliberate environmental management, human beings are able to alter significantly the selective forces in play (Balée 1989). Many workers therefore maintain that the key to the development of human knowledge lies not so much in the passive selection of sustainable behavioural traits, as in the human ability to assess and solve specific problems (Keesing 1981; Alcorn 1989; Balée 1989).

Collecting Ethnobotanical Evidence: The Dynamics and Distribution of Traditional Botanical Knowledge

It is clear from the discussion above, that meaningful study of TBK requires the interpretation of symbolic, linguistic and behavioural evidence within the context of local spiritual and political ideology. However, the collection of useful ethnobotanical evidence requires some preliminary understanding of the structure of a knowledge system itself, including the ways in which information is passed on to successive generations, and how knowledge may vary in both spatial and temporal terms. For while it has often been assumed that the knowledge of a given traditional community is an essentially static, uniform body of shared information, recent research demonstrates that traditional knowledge is not only intrinsically dynamic, but that much of it is distributed according to specific social and personal factors. Indeed, traditional knowledge is now recognised as changing in response to local innovation and experimentation as well as to changes in environmental constraints, while knowledge distribution can be influenced by numerous factors, including age, gender and social class.

Table 3.4   Developments in the theory of cultural ecology. Since the middle of this century, cultural ecologists and ecological anthropologists have studied the relationships between human societies and their environment, while attempting to provide a materialist explanation for socio-cultural behaviour, which they regard as a product of *adaptation* in response to the pressures of *natural selection*. Central to these ideas is the concept of carrying capacity as the major determinant of population size in a given environment, and the notion that culturally defined practices which maintain population size at an appropriate level will therefore be selected for (see: [1]Balée 1989; [2]Burch & Ellanna 1994; [3]Edwards *et al.* 1994; [4]Ludvico *et al.* 1991; [5]Keesing 1981; [6]Seymour-Smith 1986; [7]Orlove 1980). More recently, however, it has become increasingly clear, that human populations can significantly modify the carrying capacity of a given environment, and many argue against the validity of determinist explanations and the concept that human populations passively adjust their religious and social institutions in adapting to natural environmental constraints (Keesing 1981)

Major stages in the development of cultural ecological theory

1949	Julian Steward, founder of the school of *cultural ecology*, first suggests that soil characteristics and environmentally determined protein levels may ultimately constrain socio-cultural expression.[1]
1954–80	Cultural ecologists such as Meggars and Harris formalise the *limiting factor theory* (LFT) which suggests that population density is ecologically determined by the *carrying capacity* of the local environment. LFT cites *soil limitation* and/or *protein limitation* as the ultimate determinants of socio-cultural behaviours which keep populations low (e.g. endemic warfare, food taboos, and social structures which result in small, dispersed settlements).[1]
1973–81	Ecologists interested in the economics of animal grazing activities develop the *optimal foraging theory* (OFT), which predicts that organisms should maximise their mean rate of food intake during a given food quest (Emlen 1973*). In addition, the *marginal value theorem* (MVT), predicts that a foraged patch should be left at the point when returns drop to the *mean return rate* for a given environment (Charnov 1976*). More recently, these theories have been applied in the investigation of hunter-gatherer subsistence strategies (Hames 1980*; Durham 1981*).[1,2,3,4]
1974–79	Harris presents his theory of *cultural materialism* which stresses that natural selective forces acting on population density represent the principal determinant of cultural change. Within this model of *environmental determinism*, he and other proponents of the theory attempt to demonstrate a hidden ecological rationality behind many of the bizarre and/or irrational practices which are associated largely with symbolic or religious expression (see Table 3.5).[5,6]
1969–95	Some critics of these *determinist* explanations of human behaviour point out that not only is carrying capacity itself extremely difficult to estimate, but that it can be modified as a result of changes in technology and the deliberate manipulation of the environment. Others demonstrate that many cultural practices may actually be *maladaptive*. More recent models in cultural ecological theory

Table 3.4   (*continued*)

Major stages in the development of cultural ecological theory

> eliminate the assumption of maintaining environmental equilibrium (see Orlove 1980) and recognise that different individuals within a single group may adopt alternative strategies according to their own knowledge and desires.[1,5,6,7]

* Original publications are cited in the texts indicated.

Table 3.5   Cultural ecological interpretations of behaviour. Since early developments in the field of cultural ecology, many culturally determined practices have been interpreted from a functional, ecological perspective, frequently emphasising the role of protein supply as the overriding factor in population control (see: [1]Keesing 1981 and references therein; [2]Gould 1982; [3]Balick 1984; [4]Concar 1994; [5]Sidik 1994). However, while it seems likely that practices such as the protection of sacred groves can fulfil a practical function, many workers argue that to demonstrate the consequences of a given social institution is not to explain its existence (e.g. Keesing 1981; Ellen 1994)

Functional interpretations of culturally determined behaviour

Human sacrifice—Harner (1977*) suggests that the Aztec practice of human sacrifice was a consequence of population pressure in the valley of Mexico. He proposes that large-scale cannibalism (forbidden to commoners) disguised as sacrifice would ensure a protein source for the elite.[1]

Male supremacy—several workers have identified correlations between population density, protein availability, settlement patterns and intergroup conflict (see Siskind 1973*; Harris 1974*; Gross 1975*). They postulate warfare as integral to the *male supremacy concept* and suggest that population increase is limited by cultural selection against female children through infanticide or neglect.[1]

Pollution taboos—fear of pollution by, for example, a mother whose child is not yet weaned, has been described as 'a form of ideological birth control' (see Lindenbaum 1972).[1]

In-law avoidance—in the Western Desert of Australia, in-law avoidance behaviour is strictly observed, yet exchange of goods and food is constant between in-laws on meeting. Gould suggests that this may guarantee foodsharing regardless of sentiment, in times of stress.[2]

Evil spirit homes—Mike Balick reports that certain species of palm, including the corozo palm (*Scheelea zonensis*) are avoided by some groups of people, believing that they are inhabited by evil spirits. In recent years this palm has been found to harbour the insect vectors of Chagas' disease (Whitlaw & Chaniotis 1979*); hence cultural avoidance may protect populations from this debilitating disease.[3]

Sacred groves—in Australia, fragments of rainforest are protected from burning as myths relate that careless land owners will become blind. Consequently valuable fire-sensitive food species such as yam (*Dioscorea* spp) are maintained.[4]

(*continued overleaf*)

Table 3.5 (*continued*)

Functional interpretations of culturally determined behaviour

Drug preparation—within traditional medicinal remedies, certain ingredients which are included on strictly symbolic grounds may actually possess a pharmacological function. For example, in the *jamu* tradition of Indonesia, charcoaled mouse nests and powdered egg shells are used to ensure an easy birth (like that of the mouse) and a healthy baby (like a newly hatched chicken). In practice, however, these ingredients provide essential nutrients such as calcium and carbon which may be beneficial to both mother and fetus.[5]

* Original publications are cited in the texts indicated.

Sources of Knowledge: The Dissemination of Traditional Botanical Knowledge

Until recently, most traditional societies have had little or no dependence on written culture, such that the myths, legends and stories of these *oral traditions* represent a major source of ethnobotanical evidence. However, symbolic representations in art and ceremony can also prove informative—for example, carved gourds produced by peasants of the Andean highlands portray important facets of Andean agricultural knowledge which children are taught to 'read' (Salas 1994). Meanwhile, in Central America much of the ecological knowledge associated with *milpa* farming—the local system of shifting cultivation—has become tightly encoded within ceremonial activities which mark each stage of the *milpa* cycle (Alcorn 1989). A great deal of environmental knowledge as well as many practical skills are also learned *in situ*, as children work alongside parents or elder siblings (Ulluwishewa 1993), and common tasks associated with subsistence—such as hunting techniques or making baskets for gathering foods and medicinal plants—are generally learned from an early age. For example, among the Maasai pastoralists, adolescent boys are taught about the grasses on the rangelands, while young girls more frequently learn about the nature and uses of traditional herbal medicines from their mothers, grandmothers and older siblings as they work in and around the home (Sindiga 1994).

Sources of Knowledge: The Differential Distribution of Traditional Botanical Knowledge

Within any community, there are a number of social factors which can affect how knowledge is distributed among individuals (Table 3.6). One of the most common of these is gender, and while conventional ethno-

Table 3.6 Socio-cultural factors affecting the intracultural distribution of knowledge. In early ethnobotanical studies, the question of intracultural differences in knowledge distribution has been largely overlooked. Yet while it is obvious that knowledge will vary according to gross cultural factors such as the nature of a group's subsistence activities, even within a single community different individuals hold different types and levels of knowledge. Many of these differences in knowledge distribution are determined by specific socio-cultural factors such as age and gender, and significantly, this may have a profound influence on the successful elicitation of specific types of information during the ethnobotanical investigations

Socio-cultural influences on knowledge distribution

Intercultural influences
Mode of production	Ethnicity
Biological environment	Religion
Level of external contact (acculturation)	

Intracultural influences
Gender	Occupation
Age	Migration for work or marriage
Class	Age at marriage
Place of birth	Kinship and marriage relations
Class	Number of children
Education	Number of generations in household
Literacy	Language ability

Information modified from Martin (1995).

graphies have tended to concentrate on the knowledge and behaviours of 'man the hunter', there has been increasing interest in women's roles since the 1970s. For example, among the Aguaruna of Peru, detailed knowledge of manioc cultivars is clearly held by, and communicated among, women (Boster 1985), while ethno-ornithological knowledge appears to be greater among Aguaruna males (see Berlin 1992). In this case, the differential distribution of knowledge is clearly related to distinct gender roles among the Aguaruna—for it is invariably women who cultivate the gardens while the men are predominantly hunters. The distribution of knowledge can also be determined on the basis of factors such as age, class and social position, and while this can clearly influence the nature of information provided *by* individual informants, it can also have a profound effect on the type of information revealed *to* particular researchers. Indeed, where certain information is held specifically by, say, women, they are likely to talk only in rather general terms to male researchers on the assumption that men would neither understand nor require more detailed information (e.g. Boster 1984). This notion has been illustrated clearly during a study of the ecological knowledge of Maasai pastoralists. For while talking with local people it soon became evident to researchers that the pastoralists began to reveal the true nature of their sophisticated environmental knowledge, only when they realised that the researchers themselves were sufficiently knowledgeable to discuss relevant ecological concepts in considerable detail (Western & Dunne 1979).

The Dynamics of Knowledge: Observation, Experimentation and Adaptation

In recent years there has been an increasing number of reports detailing how informal experiments and information exchange contribute to the continual development of TBK. For example, in the forest region of Tupac Amaru in Peru, Rhoades (1989) reports the emergence of a new system of forest farming based entirely on local innovation, while Boster (1984) discusses how the Aguaruna Jívaro of the Peruvian Amazon make constant efforts to maintain and expand the genetic diversity of the important dietary staple manioc (*Manioc esculenta*). In his examination of the dynamics of Aguaruna manioc inventories, Boster (1984) points out that losses of cultivars through rejection or accidental loss may be compensated by the influx of new germplasm both through deliberate selection of rare *volunteers*, and through plant exchange with other communities—even over distances of up to several hundreds of miles. In many cases, these new crop cultivars are evaluated experimentally—for example in a recent survey of the crop repertoires of the Hopi farmers of Arizona, at least two informants were experimenting with new varieties of corn (*Zea mays*), while others had planted gourds (*Lagenaria sicer-aria*) obtained from neighbouring New Mexico Pueblos with whom new seed varieties are often exchanged (Soleri & Cleveland 1993).

In many cases, new practices or technologies are also evaluated and often modified prior to their incorporation into a given society. For example, during the early 1970s scientists implemented the 'introduction' of *diffused light storage* of potatoes to small-scale farmers in a number of countries—a practice first observed among traditional farmers in Kenya (see Rhoades 1989). However, during follow-up studies in several areas, it was found that at least 98 per cent of farmers had not adopted the technology as it had been presented in extension efforts. Instead, they had compared the effects of different storage methods on characteristics such as shrinkage, sprout elongation and overall seed quality, and adapted the method to suit their own needs. Similarly, in a recent study of irrigation devices used by the Jua Kali farmers of rural

Kenya, widespread experimentation is evident in the variety of sprinkler models observed (Bedini & Masera 1994). Here, following an earlier introduction of commercial sprinklers into the area the Jua Kali artesans began to repair the imported sprinklers until, with the rapid development of the new skills required, they were able to design and produce their own, cheaper versions based on local recycled materials.

TRADITIONAL BOTANICAL KNOWLEDGE AND WESTERN SCIENTIFIC KNOWLEDGE: CONFLICT AND CONCILIATION

As new approaches to the study of ethnobotanical knowledge have developed, and Western understanding of traditional knowledge has improved, TBK has become increasingly recognised as a valuable source of information on the use and ecology of many plant species. Hence, during recent years there have been a number of attempts to integrate this valuable ethnobiological knowledge with Western scientific knowledge (WSK) both on the basis of its scientific value (Richards 1985; Johnson 1992) and as an acknowledgement of the basic rights of aboriginal peoples. Indeed, since the 1980s an increasing number of projects have begun to adopt a participatory approach to research, where local people are actively involved in research projects, and in some examples formal collaborations between traditional and non-traditional researchers have been initiated (Baines & Hvinding 1992; Martin 1995). However, despite some success with this approach, a number of important philosophical barriers continue to impede more widespread acceptance of the concept of these *integrated knowledge systems*.

Learning from Traditional Knowledge

The idea that Western science can learn from traditional methods is hardly new. Indeed, by 1943 a report from the West Africa Commission not only observed that traditional farmers were often excellent judges of soils, but also argued that local practices contained elements which researchers might usefully investigate further (see Richards 1985). Later, in 1967, the monograph *Ethnopharmacological Search for Psychoactive Drugs* emphasised the potential role of traditional medical knowledge in the search for novel therapeutic agents (Efron *et al.* 1967). Since these early observations there have been many ethnobotanical reports describing traditional practices which have been 'discovered' only recently by Western science. These 'discoveries' include both the uses of plants—as in the identification of therapeutic and other bioactive agents (see Table 3.2)—and the methods used in managing plant resources. For example, relatively recent Western developments such as the concept of biological pest control have long been used by traditional societies (e.g. Frechione *et al.* 1989; Parrish 1994), while 'modern' ecological ideas concerning plant–animal interactions appear to have formed the basis of ethno-ecological knowledge (Box 3.3). It is examples such as these which have led to increased interest in *participatory ethnobotanical research* (PER).

Box 3.3 Traditional ecological knowledge among the Aleut of Alaska. While Western scientists have begun to consider the complexity of the interactions between different species occurring within a given habitat, ecological data remain relatively limited so far, and Western views remain essentially reductionist in nature. In this example, the scientist's job was to collect field data on moose, while the native groups provided information that went beyond the scope of the scientist's field assignment. In this case the native voice was never heard, yet recently, examples such as this have caught the attention of some ecologists who now believe that there is much for Western scientists to learn from traditional environmental knowledge (e.g. see Johnson 1992).

Illarion (Larry) Merculieff is a member of the Aleut, one of three distinct aboriginal races in Alaska. Born and raised on the Pribilof Islands, a group of islands in the Bering Sea, he has been certified by the state of Alaska as an expert on Aleut history and culture, and he estimates that his people have occupied the same region for about 10 000 years. Keenly aware of the depth of knowledge and experience about the environment which is inherent in his culture, he has struggled to convince the intellectual powers behind the influential *World Conservation Strategy* (WCS), of the practical utility of such knowledge; a struggle which, after 5 years of effort, resulted in the UN agreeing to consider amending the WCS folio in recognition of indigenous people's potential contribution to the goals of the strategy. In a recent paper (Merculieff 1994), Larry recounts an event which underlines the value of traditional environmental knowledge, and illustrates how aboriginal concepts of the natural world have pre-empted modern ecological ideas of Western science.

MOOSE POPULATION RESEARCH IN ALASKA

In 1993, a remote village in Alaska formed the site of a meeting between scientists, land and resource managers and tribal chiefs. The aim of the meeting was to discuss local subsistence, the seven tribal chiefs acting as representatives for local villagers who were highly dependent on hunting and trapping for their survival. During the course of the meeting, one of the state representatives made a 45 minute presentation, describing how they were about to conduct field research into the health of local moose populations which were thought to be at a critical threshold of sustainability. In response, the lead spokesperson of the tra-

continued

— continued —
ditional governing group gave a 45 minute dissertation during which he described how local villagers had noticed a distinct drop in marshland water levels, which had in turn affected adversely, the food sources for the moose. He added that he had noted at least 20 small tributaries to the Yukon River had been dammed by beavers, an observation which perhaps could account for the changes in the marshland water levels. He finally pointed out that while the Alaska Department of Fish and Game might propose to cut villagers' subsistence take of moose as their solution, it might be more appropriate to plan studies on beaver—a much stronger contender as the cause of the problem.

Traditional Botanical Knowledge in Rural Development: The Origins of Participatory Research

The concept of participatory research evolved initially as *agricultural extension workers* collaborating with traditional farmers in several countries identified three important barriers to successful rural development. First, in many cases, the problems accorded high priority by development agencies do not necessarily coincide with those prioritised by the farmers themselves (see Chambers *et al.* 1989); secondly, practices imposed by external agencies are much less likely to be adopted than those based on local systems (e.g. Dialla 1994); thirdly, practices which are not based on local systems frequently prove to be ecologically unsustainable in the long-term (e.g. Porter *et al.* 1991). Participatory research therefore developed originally as a kind of 'self-help' approach in rural development, where the prioritisation of problems is based on local knowledge, and where much of the research into possible solutions is carried out by local participants (Richards 1985).

Although a fairly recent innovation, the scientific literature advocating this *populist* approach is already vast (e.g. Richards 1985; Chambers *et al.* 1989; de Boef *et al.* 1993; Scoones & Thompson 1994), much of which emphasises that local people's science is basically good science and can play a significant part in developing sustainable management strategies, particularly in fragile habitats worldwide. Since its initial conception in a largely agricultural context, the idea of participatory research—or *native collaboration*—has now been embraced by ethnobotanists and local peoples worldwide, in regions ranging from sub-Arctic Canada to the tropical rainforests of Amazonian Brazil.

Partnerships in Practice

Collaborative research between scientists and local people has already been proved productive in applications ranging from *cultural impact assessment* to the identification of *non-timber plant products* (NTPPs) as candidates for commercialisation (Berlin 1984; Johannes 1989; Stoffle *et al.* 1990; Baines & Hviding 1992; Johnson 1992; Martin 1995). Common to all such projects is the belief that traditional scientific knowledge (TSK) has generated a vast reservoir of environmental and utilitarian knowledge which can make an important contribution to Western science and *vice versa*. For example, while many traditional medicines have proved efficacious in modern pharmacological analyses, problems can arise in practice where variable dosage or the presence of toxins can lead to serious poisoning incidents and even fatalities. This problem has recently been addressed in Madagascar, where a pioneering project has been initiated as part of the People and Plants Programme (outlined in Box 3.4). Here, an *appropriate development programme* is aimed not at replacing traditional medical practices, but at linking traditional medicine with modern technology to develop an *integrated health care system*. For example, a number of local people have been trained to carry out phytochemical analysis, and are working towards providing more quantitative information on factors such as seasonal variations in the bioactive principles accumulated by certain species (Haman 1991; Quansah 1994).

Box 3.4 People and Plants—ethnobotany and the sustainable use of plant resources. Ethnobotanists involved with the People and Plants initiative work with local people on a range of projects relating to the conservation of both plant resources and traditional ecological knowledge. They organise participatory workshops and discussion groups pertaining to the development of sustainable plant use, and generate literature on ethnobotany, traditional ecological knowledge and sustainable management of plant resources. Specific projects are currently underway in various parts of the world, including Bolivia, Brazil, the Caribbean, Central Africa, Malaysia and Mexico (WWF *et al.* 1993).

The People and Plants initiative was initiated in July 1992 as a joint project between the World Wide Fund for Nature (WWF), the United Nations Educational, Scientific and Cultural Organisation (UNESCO) and the Royal Botanic Gardens, Kew, to promote the sustainable and equitable use of plant resources. The prime aim of the initiative is to provide support to ethnobotanists from developing countries, in order to develop methods for the use and conservation of local plant resources.

continued

continued

LOCAL KNOWLEDGE AND COLLABORATIVE RESEARCH

The People and Plants initiative stems from the recognition that people in traditional communities often have extensive, detailed knowledge of their natural environment, including the physical, chemical and ecological properties of the local flora. The organisations involved in this international, multidisciplinary project are concerned that much of this knowledge is being lost as a result of habitat destruction and acculturation, and support a range of projects aimed at the long-term conservation of both biological and cultural diversity.

Projects supported within the People and Plants initiative

Beni Biosphere Reserve (Bolivia)—an inventory of useful plants in the Beni Biosphere Reserve is being compiled by local communities and Bolivian students.

Projeto Nordeste: local plants for local people (Brazil)—field and laboratory research is being carried out on plant biodiversity and economic botany in northeast Brazil. Plants used locally are identified and evaluated while the sustainable use of these resources is being encouraged.

Conservation status of useful plants (Caribbean)—this project was launched to identify plant species used by people in the Caribbean that are vulnerable to over-harvesting, and to identify action which should be taken to ensure their long-term survival.

Harvesting of Prunus africana (Cameroon)—since 1972, commercial harvesting of the bark of wild _Prunus africana_ (extracts of which are now used worldwide in the treatment of prostate gland hypertrophy) has resulted in serious conservation problems in Afromontane forest. The possibility of cultivating trees for bark production is being examined.

Development of health care and forest conservation (Madagascar)—this project aims to combine conservation with local health care. Useful plants and traditional remedies are inventoried, tested for toxicity and efficacy, and an integrated health care system combining traditional and orthodox medicine is being developed.

Many of the projects supported by the People and Plants initiative combine the exploitation of traditionally useful plants with the development of strategies for plant conservation. In addition, these projects are committed to the training of local people at various levels of expertise, and ethnobotanical data are made available to local people. For example, the Beni Biosphere Reserve Project combines both training and support for local ethnobotanists, while providing data which facilitate the estimation of the potential contribution of non-timber forest products to local development. Similarly, as part of Projeto Nordeste, a network of independent databases is being used to link details of plant use with inventories of plant resources in order to facilitate local access to this valuable information. In addition, the Madagascan health care project has allowed students of medicine, pharmacology, biochemistry, geography and botany from the University of Antananarivo to be trained in the field.

There are now many examples of this type of collaborative research (e.g. see Chambers *et al.* 1989; Johnson 1992; Warren *et al.* 1995), and while earlier collaborations were largely initiated by Westerners, an increasing number are now being instigated by their traditional counterparts. For example, in the Marovo region of the Solomon Islands, 'The Marovo Project'—an integrated project aimed at the systematic examination of local resource potential—has developed from an initiative of the Marovo community itself. Similarly, in Australia, collaboration between native and non-native researchers has become the norm, and in several cases indigenous groups have simply hired Western scientists to assist in local research projects (Posey 1994).

Working Towards Greater Integration

Although there now exists considerable evidence illustrating the potential benefits of integrated knowledge systems, as yet this concept has not been widely accepted in conventional scientific circles. The commitment to integration is therefore concentrated currently in the hands of a small number of ethnobiologists, development workers and of non-governmental organisations (NGOs) such as the US-based *Cultural Survival*, and the international conservation organisation the World Wide Fund for Nature (WWF). However, as much of the hostility of Western academics towards traditional knowledge is rooted in basic differences in traditional and scientific ideologies, it is anticipated that an increased understanding of the nature and dynamics of different knowledge systems will facilitate greater, and more effective, cross-cultural communication in the future (Merculieff 1994).

Reconciling Disparate World Views

Despite the awe inspired in many Western ethnographers who encountered the 'phenomenal' TSK held by traditional peoples (Berlin 1970), the Western view of indigenous and other traditional knowledge has largely been one of scepticism. In the middle of this century, most anthropologists regarded traditional peoples as 'primitive' (see Maybury-Lewiss (1992), and even today, many regard TSK as not only 'primitive' and 'unscientific' but as simply 'wrong' (see Scoones & Thompson 1994). Yet even as recent studies have increasingly demonstrated the empirical validity of many aspects of TSK, mutual misunderstandings across both cultural and disciplinary boundaries persist.

The continued reluctance of mainstream Western scientists to recognise the value of TSK is based on two main factors. First, Western scientists have historically regarded their 'objective' and largely quantitative approach to natural science as inherently superior to the 'spiritually based' knowledge systems of traditional peoples. Indeed some workers have suggested that development specialists and ethnobotanists are simply 'romanticising' the virtues of traditional knowledge (Borlaug 1992). However, as research into both traditional and Western knowledge has continued, a number of important parallels have been recognised. For example, both traditional knowledge, and WSK, are inherently context-determined and both are continually reinforced through sustained interaction between theory and practice (Table 3.7). The second major barrier relates to the science of ethnobotany itself, which has often been criticised as vague, imprecise and unscientific. However, such criticisms fail to take into account either the intrinsic value of careful qualitative analysis, or the increasingly quantitative comparative techniques which have been developed in recent years (see Chapter 4).

Finally, although perhaps more willing to accept TSK than other scientists, ethnobotanists too are beginning to realise that the influence of their own cultural background can lead to inaccurate interpretations of ethnobotanical data (Scoones & Thompson 1994). For example, Alcorn (1989) points out that Westerners tend to describe what is observed in the present—such as the layout of crops or the location of fields, while traditional agriculturalists are more likely to conceptualise a fluid complex of managed fields which form an ecological continuum between field and forest. Similarly, in a discussion of the complex intercropping systems used by traditional farmers of West Africa, Paul Richards illustrates how a synchronic assessment of a farmer's crop layout might assume it to be the product of a carefully planned system designed to limit problems of competition and pests; in practice, however, the final crop mix arises as a result of sequential adjustment to the unpredictable conditions which occur throughout a given growing season (Richards 1989). Such observations have now led to the expansion of the original concept of participatory research (Table 3.8), and through developing a greater understanding of traditional knowledge itself, it is hoped that more meaningful interpretation of ethnobotanical evidence may be achieved in the future.

Protecting Traditional Botanical Knowledge

Proponents of integrated knowledge as a key to sustainable development, envisage this new collaborative approach both as a new opportunity for western science, and as a means of reaffirming traditional cultural iden-

Table 3.7 Comparing traditional and orthodox knowledge systems. This table outlines some of the more common ideas regarding the differences between traditional systems of knowledge and their Western scientific counterparts. Traditional knowledge is often characterised as highly specific and context-bound, with knowledge emerging simply from localised practical experience. This is contrasted with Western scientific knowledge which is regarded as theoretically based, providing objective, generalisable knowledge. With the long-running philosophical bias in favour of theoretical knowledge, science has generally been regarded as 'superior' to traditional knowledge (Scoones & Thompson 1994). However, these ethnocentric views fail to recognise several important points regarding both types of knowledge system. First, in common with traditional world views, modern ecologists are now beginning to support the view of people as part of nature; secondly like conventional scientific knowledge, traditional knowledge is often generated through the formation and replicable testing of general hypotheses; thirdly, traditional knowledge is not a commonly held body of knowledge, but is generated, held and controlled by certain individuals—just as in Western science (see Johnson 1992; Drinkwater 1994; McNeely 1994; Scoones & Thompson 1994)

Traditional knowledge systems	Western scientific systems
All parts of the natural world are regarded as animate, and all life forms as interdependent	Human life is generally regarded as superior, with a moral right to control other life forms
Knowledge is transmitted largely through oral media	Knowledge is transmitted largely through the written word
Knowledge is developed and acquired through observation and practical experience	Knowledge is generally learned in a situation which is remote from its applied context
Knowledge is holistic, intuitive, qualitative and practical	Knowledge is essentially reductionist, quantitative, analytical and theoretical
Knowledge is generated by resource users on a diachronic (long-term) time scale	Knowledge is generated largely by specialist researchers on a synchronic (short-term) time scale
The nature and status of particular knowledge is influenced by socio-cultural factors such as spiritual beliefs, and is communally held	The nature and status of particular knowledge is influenced by peer review, and is held by individual specialists
Explanations behind perceived phenomena are often spiritually based and subjective	Explanations behind perceived phenomena are essentially rational and objective
Knowledge is used to make suitable decisions under variable conditions	Knowledge is used to put forward hypotheses and to verify underlying laws or constants

Date modified from Johnson (1992).

Table 3.8 Recent developments in the theory of participatory ethnobotanical research. The initial populist approach to participatory research is well reflected in Paul Richard's concept of an *indigenous agricultural revolution*, which advocates the application of indigenous technical knowledge in rural development. This paradigm has been developed on the basis that local people's scientific knowledge is good science. However, while this essentially positivist approach regards local knowledge as a stock of scientific information which can be extracted and applied within different cultural contexts, this view has been modified in recent years with the development of the post-positivist movement. Perhaps the most significant aspect of these more recent ideas, lies in the fact that local knowledge is recognised not as a unified body of information, but as a dynamic, and fragmentary system of knowledge, whose generation and mainte-nance is strongly influenced by the specific socio-cultural context in which it exists. These changing views by no means reject the major ideas of the original populist approach, both carrying a similar agenda aimed at active participation, empowerment and the alleviation of poverty. However, the post-positivist ideas emphasise that the effective integration of local knowledge and Western ideas can occur only where the socio-cultural context of a given knowledge system is fully considered (see Scoones & Thompson 1994)

Participatory research: a populist approach	Beyond the populist approach
Assumes common goals, interests and power among individuals of traditional communities	Recognises different interests, goals and access to resources between individuals
Envisages a uniform body of knowledge which can be readily extracted and incorporated into Western scientific systems	Recognises diffuse knowledge which does not necessarily fit directly into Western scientific systems
Searches for community consensus solutions to a given identified problem	Attempts to bridge negotiation and mediate conflict between different interest groups
Involves local peoples in planning and implementation of Western-designed solutions with planned outcomes	Envisages local empowerment through planning based on collabora-tive work which allows the dynamic im-plementation of negotiated outcomes
External researchers are viewed as invisible information collectors and managers, or more recently as facilitators and catalysts	External researchers viewed essentially as facilitators and catalysts
Local participants are viewed as essentially passive, reactive respondents	Local participants are viewed as essentially active investigators and analysts
Investigations involve positivist research often centred on the empirical validation of traditional knowledge	Investigations involve post-positivist research which consider knowledge within its specific socio-cultural context

Data modified from Scoones and Thompson (1994).

tity. However, a major concern associated with participatory research is that of exploitation, either through the acquisition of local skills 'on the cheap' (see Richards 1985) or by using traditional knowledge for commercial gain without drawing up agreements to ensure the equitable sharing of any resulting benefits. In attempting to address such matters, a number of ethnobiologists (e.g. Posey 1990) have been involved in developing a series of ethical guidelines for researchers, while investigating possible legal mechanisms aimed at protecting the *scientific and cultural property* of traditional communities (see Posey 1994b). Meanwhile since the beginning of the 1980s, indigenous organisations have increasingly taken matters such as these into their own hands.

There are now over 1000 indigenous organisations worldwide, established not only at local levels, but also at national and regional levels, and as numbers have grown, these organisations have become increasingly effective. Many have campaigned internationally to gain support for the Declaration of Indigenous Rights (Burger 1990), and in 1992 the world's indigenous peoples and their political organisations received formal recognition at the UN Convention on Environment and Development—the so-called *'Earth Summit'* (UN 1993). Over the last 20 years, indigenous peoples in Australia, Canada, Colombia, Mexico, India, Aotearoa (New Zealand), Argentina, Venezuela, USA and USSR have revived their languages, established their own schools and are re-educating the young in traditional practices, history, religion and social customs. Nevertheless, erosion of cultural diversity persists due to continued loss of traditional lands and restrictions on access to essential natural and spiritual resources. For example during the last three decades, the explosion of more than 650 nuclear weapons in the Nevada Desert (USA) has seriously damaged the territories of the western Shoshone communities (Burger 1990), while in Brazil, the *Carajás Development Programme* in eastern Amazonia has encouraged a series of violations of indian lands within the Greater Carajás region (Hall AL 1991). Hence, while political and social policies continue to marginalise native peoples throughout the world, ethnobiologists hope that more widespread recognition of the value of TSK may not only contribute to the sustainable economic development of traditional lands, but might also help to limit further loss of cultural diversity itself.

SUMMARY

Throughout the first hundred years of ethnobotanical study, Western understanding of TSK has changed considerably as various approaches to its study have been developed. Much of the early research was of a predominantly utilitarian nature, and discussed only the ways in which particular plants were used within traditional communities; later, more cognitive studies began to explore how plants were perceived by different peoples, providing greater insight into traditional rationale behind the uses of certain plants. Meanwhile, cultural ecologists began to suggest that culturally determined practices often concealed functional ecological concerns. Using these various approaches to ethnobotanical study, it has become evident that TBK is neither static nor uniform as is often assumed, but is generated, maintained and modified according to local ideology, external social or practical influences and changing resource availability. It is also clear, that in order to elicit and understand fully the nature and extent of the TBK of a given culture, the local knowledge system itself—its socio-cultural constructs, its distribution and its modes of dissemination—must also be considered.

As early as 1939 a new ecological approach to environmental management was proposed in a report produced by the West Africa Commission. Here it was suggested that farmers and scientists should pool their skills in order to improve rather than to replace traditional farming systems (see Richards 1985). However, overshadowed by the onset of the Second World War, this innovative option was never adopted. More than 50 years later, evidence to support the concept of integrating these disparate knowledge systems is now considerable, yet even after the formal recognition of TSK at the Rio 'Earth Summit' (UN 1993), the future of TBK still remains uncertain as negative attitudes to TSK remain deeply entrenched within the Western scientific community.

Chapter 4

Methods in Ethnobotanical Study

... the modern social anthropologist can, and sometimes does, benefit from collaboration with other 'outside' specialists in the field; such as agriculturalists, economists and medical research workers. The difficulty is that generally there are no such specialists available in the field ...

John Beattie 1966 in *Other Cultures*

INTRODUCTION

The study of ethnobotany has clearly changed enormously since the compilation of the first 'laundry lists' of traditionally useful plants, and as interest in ethnobotany has expanded, so too has the methodology used in its study. There now exists a wide range of sophisticated techniques which may be used to investigate the relationships between plants and humans, and as the study of ethnobotany is inherently multidisciplinary, these appropriate methodologies are extremely diverse. They draw from both the biological and social sciences, and include techniques from disciplines as diverse as economics, linguistics, ecology, anthropology and pharmacognosy, and together they allow the various lines of investigation associated with ethnobotanical study. Anthropological or archaeobotanical field methods, together with plant taxonomic and linguistic expertise are fundamental to any ethnobotanical investigation; phytochemical or molecular biological analysis can play a vital part in applied studies; while ecological methods are essential in determining the environmental impact—and level of sustainability—of plant–human relationships both in the past and present. Although detailed discussion of each of these methods is beyond the scope of this text (but see Jain 1987; Bellamy 1993; Martin 1995) this chapter will outline those methods which are commonly used in ethnobotanical field studies, discussing their potential applications, their problems and highlighting any recent progress in their development. More specific techniques, such as those used in ethnopharmacological work or economic evaluations are only briefly outlined here and are discussed in more detail in later chapters.

GENERAL ETHNOBOTANICAL TECHNIQUES

Any ethnobotanical enquiry into the *traditional botanical knowledge* (TBK) of extant peoples is dependent on the effective application of a number of key anthropological and botanical methodologies. For example, anthropological field techniques including *participant observation* and *structured surveys* permit the collection of both qualitative and quantitative data related to plant use and subsistence practices, while orthodox plant *taxonomic* methods are crucial to any accurate ethnobotanical investigation. In addition, techniques such as linguistic and other *symbolic analyses* can prove invaluable in investigating the ways in which different peoples cognise their natural world, while *archaeobotanical* methodology is crucial to the study of historic and prehistoric peoples.

Anthropological Field Methods

In most cases, the successful collection of anthropological data requires a close and sustained observation of a people, which can be achieved only by long-term participation in local customs and daily life. Whether studying the social institutions, subsistence strategies or environmental perceptions of a given culture, studies carried out in a few weeks or months can be very misleading (Beattie 1966), and external researchers may need to spend a considerable length of time with a given group. In addition, there has been an increasing trend towards formal employment of, or collaboration with, local researchers who are better qualified to collect certain types of ethnobotanical field data (Berlin 1984; Johnson 1992).

However, although a range of data collection techniques are now available, anthropological studies remain inherently difficult. For the study of people involves a number of unique practical, cultural and ethical considerations, which prevent social scientists from designing the type of controlled, replicable experiment so favoured by natural scientists. This is partly because of the influence of uncontrollable variables such as an individual's personality and decision-making powers; more importantly, however, it is essential that the research methods used do not compromise either the physical or spiritual well-being of any member of the host community. A particularly important aspect in respect to this last point lies in the way in which information is obtained. Research teams must be careful to enlist the co-operation of the entire community and at the outset must provide the community with a complete description of the research group's aims (see Srivastava 1992). Equally, it is imperative that any informant's desire for confidentiality is respected, whether on the basis of personal, professional or political considerations. Finally, it is also important not only that local contributors receive a 'fair return' for their time or information, but that this compensation is provided in a way which is in keeping with local concepts of general reciprocity (Baines & Hviding 1992).

Quantitative and Qualitative Approaches in the Field

Anthropological methods used in ethnobotanical field studies are based largely on long-term participant observation, which forms a fundamental element of most modern anthropological field work (Seymour-Smith 1986). Yet during the course of long-term immersion into the daily life of a given people, the anthropologist may collect information using a range of qualitative and quantitative methods, which will vary according to both the aims of the project and the nature of the relationships established between external researchers and local peoples. As these different approaches involve distinct methodologies and generally have rather specific applications, many researchers choose a combination of quantitative and qualitative methods to ensure the collection of data which are both accurate and complete (Table 4.1). This section considers how different *interview techniques* may be used either quantitatively or qualitatively and outlines the types of method used in eliciting information; it also describes a range of analytical tools which can be used in the quantification and verification of data collected.

Interview Techniques and Elicitation Methods

There are four basic interview techniques which are used commonly by field ethnobotanists: *open-ended* and *semi-structured* interviews, which are used in qualitative data collection, and *structured* interviews and *questionnaires*, which may be used for quantitative analyses (Martin 1995). Open-ended interviews are essentially casual conversations which can reveal detailed life histories. They may also include a more practical element whereby external researchers are encouraged to participate in subsistence and other practices in order to learn about traditional techniques through direct experience. For example, an active 'hands-on' approach has been

Table 4.1 Qualitative and quantitative approaches to anthropological study. During the course of *participatory ethnobotanical studies*, information is gathered from selected participants, primarily through observation, casual conversation and the use of various types of analytical tool. Informal or qualitative methods such as open-ended interviews, generally yield responses which can be used in compiling general ethnographic accounts of a community and its culture. More systematic or structured methods (that is formal or quantitative methods) yield data which may be used to calculate a range of numerical indices such as the relative usefulness of a given plant species. Ethnobotanists are finding increasingly, that a combination of qualitative and quantitative methods is proving most useful in the collection of data which are both accurate and complete (see Martin 1995)

Qualitative approach	Quantitative approach
Methods	
Open-ended and semi-structured interviews	Structured interviews and questionnaires
'Hands on' learning of traditional techniques	Free-listing
	Pile-sorting and preference ranking—including triadic and paired comparisons
	Systematic surveys—e.g. of transects or hectare plots
Applications	
Reveal a range and depth of information which is difficult to elicit using more formal methods	Facilitates the cross-verification of data both within and between informants
Facilitate the development of informal relationships between local and external participants	Facilitates the numerical evaluation of factors such as the use-value or relative economic importance of a given species
Provide practical experience of using traditional methods	Facilitates the selection of participants who are particularly knowledgeable in certain areas

adopted in the Marovo Project (Solomon Islands), where visiting investigators learn to carry out traditional techniques under the supervision of local experts (Baines & Hviding 1992). Semi-structured interviews, on the other hand, while remaining much more flexible than formal, structured interviews, are based around a checklist of topics or questions which the researcher wishes to cover. Where quantitative data are required for analytical purposes, structured interviews are conducted using a series of predetermined questions which form the basis of the *interview schedule*, or in some cases where local *participants* are literate, the formal verbal interview may be replaced by a written questionnaire.

Whatever the interview method employed, ethnobotanical information must be *elicited* using one of several devices. Most commonly,
various visual or other stimuli can be used to elicit information about
particular organisms. These stimuli may be provided by fresh plant
material—either *in situ* or recently harvested; otherwise prepared
voucher specimens or photographs may be used. Alternatively,
ethnobotanical artefacts—items made locally from plant materials—may
be presented to informants who are encouraged to discuss all the species
and methods associated with the manufacture and use of each item
(Banack 1991). Appropriate stimuli are presented to local participants
under a range of conditions. They may be presented formally during
organised discussion sessions or systematic *utilisation surveys*, or they
may be discussed during informal forest walks and field excursions. In
other cases stimuli are not presented at all, but are collected by participants during hunting trips and other routine activities. In recent years,
the more systematic approach has proved increasingly popular among
researchers, and surveys of various types have been carried out. For
example, plant use has been assessed on the basis of discussing the
plants present in domestic gardens, open fields or local markets (Begossi
et al. 1993; Nicholson & Arzeni 1993). Forest plots and experimental
gardens have also been used in eliciting information about local plant or
vegetation types (Boster 1985; Phillips & Gentry 1993a, b).

The specific method of elicitation employed can have a significant effect
on the data collected (e.g. Ellen 1993), and it is essential that techniques
are standardised, particularly where information is collected for comparative purposes. Therefore, when carrying out structured surveys the
phrasing and sequence of questions on the interview schedule or questionnaire must be carefully considered, particularly as certain types of question can bias the data or else may prove offensive or simply alien to
participants. For example, questions such as 'What is the use of this
plant?' which implicitly demand a specific answer is more likely to lead to
improvised responses than questions such as 'Does this plant have a use?';
on the other hand, posing questions such as 'How many cattle do you have
in your herd?' might be tantamount to enquiring how much someone has in
their bank account (Barker & Cross 1992; Johnson & Ruttan 1992). Unlike
the open-ended questions of less formal interview methods, surveys generally pose questions requiring short answers only. These may be either
dichotomous—requiring a yes/no response, *multiple choice*—where a
limited range of answers are possible, or of a *fill-in-the-blank* nature, each
type possessing its own particular attributes and limitations (Table 4.2).

The information gathered during these discussions may be documented using photographic and recording techniques—except where

Table 4.2   Asking questions in anthropological research. Within anthropological methodology, different methods of asking questions may be used for a range of very different purposes. For example, while open-ended interviews can be useful in investigating detailed life histories they do not lend themselves readily to any quantitative analysis; in contrast, while structured surveys based on multiple choice or dichotomous questions are readily analysed statistically, the range of material collected is often very limited (modified from Martin 1995)

Question type	Level of detail	Subject range	Statistical analysis	Number of informants
Open-ended	Very high	Very broad	Difficult	Low
Fill-in-the-blank	High	Broad	Moderate	Moderate
Multiple choice	Moderate	Narrow	Easy	High
Dichotomous	Low	Very narrow	Easy	Very high

participants prefer otherwise (Balick 1990; McDonald Fleming 1992). In addition, field notebooks are commony used, although in some studies note-taking during open-ended interviews has led to loss of valuable eye contact (Barker & Cross 1992). A number of research groups involved in agricultural extension work have recently found that maps and diagrams constructed by traditional farmers provide both an effective means for the communication of detailed knowledge, and a useful means of recording information (Conway 1989; Gupta & IDS Workshop 1989; Lightfoot *et al.* 1989).

Quantification and Verification

In addition to the use of structured interviews and questionnaires a number of other *analytical tools* have been developed in order to facilitate not only the quantification and *cross-verification* of ethnobotanical data but also the choosing of local participants for particular projects. Most commonly, the quantification of ethnobotanical data allows the local significance of a given plant species or management practice to be estimated on the basis of various *numerical indices*, some of which are based on expressing simply the proportion of informants who use a certain plant or carry out a particular practice (Begossi *et al.* 1993; Soleri & Cleveland 1993). Other values are calculated on the basis of more specific data. For example, the *index of saliency* gives an idea of those plants which are in common usage, *preference ranking*, assigns a mean numerical value to plant species according to their perceived significance, while the *use-value* of a species, estimates the overall usefulness of a given plant.

Table 4.3   Methods used in the quantification and verification of ethnobotanical data. In addition to formal interviewing techniques which can lend themselves to numerical analysis, a number of analytical tools have been developed, allowing both the quantification and cross-verification of ethnobotanical data. Some of these methods also facilitate the selection of informants possessing accurate knowledge in particular areas (see [1]Martin 1995; [2]Phillips & Gentry 1993a; [3]Anderson 1991; [4]Phillips & Gentry 1993b; [5]Boster 1985; [6]Balick 1994)

Analytical tools used in quantification and data verification

Quantification of a species' local significance

Free-listing—this involves asking community members to list, for example, any organisms which might be used for a particular purpose, and works on the principle that those organisms which are more significant are more likely to be mentioned by several informants, and are likely to be mentioned earlier in each list. By assigning a numerical value determined by the order in which a given organism appears on each list, each organism's *index of saliency* can be calculated.[1]

Preference ranking—participants are asked to order a number of items according to specific criteria (such as personal preference, local economic importance or species scarcity), and a numerical value is assigned accordingly. By collating information from a number of informants a total value is calculated and an overall ranking is determined.[1]

Direct matrix ranking—a more complex version of preference ranking, in this method participants are asked to order a group of objects by considering several attributes together. Again an overall ranking is calculated based on total species usefulness.[1]

Utilisation surveys—here a given area or transect of vegetation is defined and participants are asked to identify each plant and its uses. Depending on the exact method used, the data generated may be used to calculate the *use-value* or *consumptive use-value* of each species present.[2,3]

Data verification and informant choice

Relative use-value—is calculated on the basis of data collected in utilisation surveys, and considers the consistency of an individual respondent's replies while comparing the answers given by different individuals.[4]

Paired and triad testing—here objects are presented to informants in a range of combinations in order to assess consistency of an individual informant's responses.[1]

Indices of agreement—developed by James Boster, these indices compare the level of agreement between different informants, providing an overall ranking based on informant accuracy.[5]

Multiple use curve—based on the principle of the *species–area curve*, this has been suggested as a suitable method for estimating the number of informants required to collect complete or representative data within a given community.[6]

These indices are calculated from data which are collected using a range of specific techniques (Table 4.3). For example, in methods such as preference ranking, participants are asked to arrange selected items according to personal preference, local importance or perceived similarity, and the appropriate indices are calculated (Box 4.1). In a utilisation survey, a given area of vegetation (normally about 1 ha) is defined and the plants to be studied (generally those greater than a certain diameter at breast height) are tagged. Informants are then asked to identify the tagged plants and their uses (Milliken *et al.* 1992; Phillips & Gentry 1993a), and depending on the exact method used, the data generated may be used to calculate the use-value or *consumptive use-value* of each species. These numerical values, which are used to estimate relative usefulness, vary significantly in nature and must be applied according to a researcher's specific requirements: the use-value described by Phillips and Gentry (1993a) is based on the *mean number* of uses ascribed to a given species by several informants, and takes into account any inconsistencies in data supplied by individual informants (Box 4.2); the consumptive use-value is determined on the basis of the *total number* of uses attributed to an individual species or taxon (McNeely *et al.* 1990; Anderson 1991). This concept of a *total use-value*, has recently been used by economic botanist Mike Balick to address one of the key issues in ethnobotanical field work—that of estimating the sample size (number of informants) required to provide complete information on a given species (Figure 4.1).

Box 4.1   Quantitative data collection—preference ranking. Using quantitative ranking methods, ethnobotanists are able to calculate a numerical index which gives an estimate of, for example, the local significance or economic value of a given plant. In preference ranking and direct matrix ranking, data may be cross-verified using paired or triad testing, and in both cases, an overall ranking may be determined on the basis of the combined (or mean) data from a range of respondents (see Martin 1995).

Preference ranking represents one of the simplest analytical tools used by ethnobotanists. This method allows the calculation of a numerical index which gives an estimate of factors such as the cultural significance or scarcity of a given plant species. In preference ranking each informant is asked to arrange a group of items according to a given criterion, such as personal preference or perceived importance. Each item is then assigned a value, with the most important ranking highest, while the least important is assigned a value of '1'. Data from a range of informants can then be collated to produce an overall ranking value. In the example shown here, data represent the relative value of five different forest products as perceived by five respondents from Motisingloti village in Gujarat.

continued

_____ *continued* _____

Plant	Ranking value for each respondent					Total score	Ranking
	A	B	C	D	E		
Sag seeds	1	2	1	1	1	6	5th
Bhindi seeds	3	1	2	2	4	12	4th
Timru leaves	5	4	5	5	5	24	1st
Mahua seeds	2	3	4	3	2	14	3rd
Mahua flowers	4	5	3	4	3	19	2nd

Data taken from Martin (1995).

A more complex version of preference ranking known as *direct matrix ranking* involves asking informants to order items according to several different criteria, rather than just one. For example, in a *participatory rural appraisal* (PRA) carried out in the Middle East and North Africa, participants were asked to define the good and bad characteristics of four important species, *Eucalyptus*, palm, *Acacia* and pine. The table below details the information provided by one informant, where again, the best species for a given use receives the highest ranking value, while the least useful is assigned a ranking value of '1'.

Plant	Ranking value							Total score	Ranking
	Fuel	Construction	Fruit	Medicine	Fodder	Shade	Charcoal		
Eucalyptus	4	4	1	4	3	4	2	22	1st
Palm	1	1	4	1	—	3	—	10	4th
Acacia	2	2	2	3	4	1	3	17	3rd
Pine	3	3	3	2	2	2	4	19	2nd

Data taken from Martin (1995).

As in preference ranking, in direct matrix ranking, data from a number of respondents can be collated to produce an overall ranking that is representative for the whole community.

Of course another key element in ethnobotanical research lies in the verification of the data provided, and this remains one of the most difficult aspects of accurate data gathering. We have already seen how internal inconsistences can be detected where an individual is presented with the same species on a number of occasions (see Box 4.2), and this basic concept has been used in the development of a number of methods aimed at quantifying the level of consistency in the data collected. Such methods may be used to verify either the data provided by a single respondent on different occasions, or the information obtained from a range of different informants. For example, *triadic* and *paired comparisons* can be used to cross-check the consistency of data provided by a given informant, as outlined in Box 4.3.

Box 4.2 Quantitative ethnobotany—the concept of use-values. A quantitative method recently developed by Oliver Phillips and Alwyn Gentry (1993a, b) allows the calculation of the use-value for a given species s (UV_s), which can be compared statistically with use-values of other species. In this way it should be possible to identify any species which are perceived as being particularly useful. These data can also be used to calculate further indices, such as the *family use-value* (*FUV*)—the overall use-value of a given plant family (FUV = Σ (UV_s)/number of species in that family found locally), and the *relative use-value* (*RUV*) which provides an estimate of the relative knowledge of a given informant (see Box 4.4). The authors have demonstrated the potential applications of this method in a range of areas, including the identification of plant families which are particularly useful for specific purposes, the investigation of patterns of knowledge distribution and the assessment of the extent to which various taxonomic, *physiognomic* and ecological traits can influence a species' usefulness.

In a recent paper in the journal *Economic Botany*, Oliver Phillips of Washington University and the late Alwyn Gentry (formerly of the Missouri Botanical Garden) presented a new quantitative technique for estimating the relative usefulness of different plant species within a particular community. This new technique, developed using information obtained from *mestizo* populations in the Tambopata region of Peru, is based on systematic surveys involving a number of informants.

CALCULATING THE USE-VALUE (UV) FOR A GIVEN PLANT SPECIES

Informants were asked to identify the nature and uses of each plant occurring within a series of 1 ha forest plots (see Phillips & Gentry 1993a, b). While more than 116 different uses of plants were defined by local informants, Phillips and Gentry divided these into five broad categories: 'edible', 'construction', 'commerce' (not discussed here), 'medicinal' and 'technology and crafts'. Each informant was then asked about the uses of certain plants in order to determine the number and range of uses for each species. In each case, a single *'event'* is defined as the process of asking a single informant on 1 day, about the uses of a given plant species. Using this method, the information from each informant was used to produce, for each plant species, a data set similar to that shown below:

	Construct	Food	Medicinal	Technology	Total (U_{is})
Event 1	0	0	1	0	1
Event 2	1	1	2	1	5
Event 3	0	0	1	0	1
Total	1	1	5	1	7 ($\Sigma\ U_{is}$)
Mean	0.333	0.333	1.333	0.333	2.333 (UV_{is})

Data from each informant were then used to calculate the mean number of uses of a given plant species. In this way inconsistencies in the information given are

_____ *continued* _____

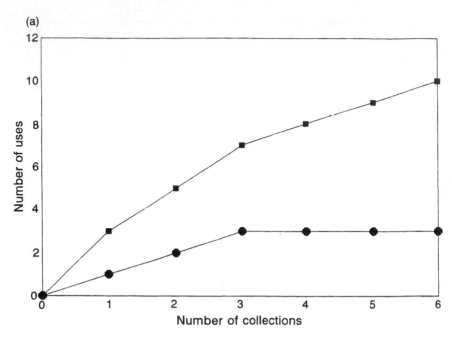

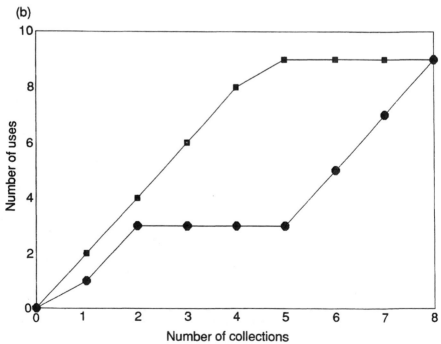

continued

taken into consideration, and the overall mean value UV_{is} represents the mean number of all uses of a given plant species (s), as recognised by a single informant (i). This information is then used to calculate the overall use-value for this species (UV_s) based on the information from the total number of informants using the following equation:

$$UV_s = \frac{\sum UV_{is}}{i_s}$$

where:

UV_s = the overall use-value of species s
UV_{is} = the use-value of species s as determined by informant i
i_s = total number of informants interviewed for species s

Choosing Participants

Clearly, a fundamental aspect of ethnobotanical study lies in choosing suitable *informants* or participants—local people who are involved in the provision and/or collection of ethnobotanical data. As with any aspect of field methodology, participants must be chosen according to the specific needs of the project in hand. For example, where the overall range and distribution of information is unknown, it is advisable to start with a random sample which provides a representative cross-section of the information held by a community as a whole. In random

Figure 4.1 Multiple use curves and sample size. In many early ethnobotanical field studies, the adequacy of the data collected is often questionable. However, in more recent studies, the construction of *multiple use curves* has been suggested as a suitable means of estimating the sample size required to ensure the collection of complete, or representative information for a given species. This method is based on the concept of the species–area curve which is used to estimate the size of survey area required to include all of the species present in a given sample site (Campbell *et al.* 1986). Where the curve reaches asymptote, it can be concluded that all of the available information has been recorded: (a) multiple use curves for *Vitex gaumeri* (●) and *Neurolaena lobata* (■) demonstrate that for *V. gaumeri* complete data are collected from interviews with three informants, while for *N. lobata* even the information from six informants appears incomplete; (b) multiple use curves for *Piper jaquemontianum* presented in 'chronological' format (●) and 'best case' format (■), illustrate how information in a multiple use curve may be misleading. Balick (1994) suggests that in many ethnobotanical studies the chronological approach may prove adequate. All data presented here are taken from Balick (1994)

Box 4.3   Cross-verification in ranking exercises—triadic and paired comparisons.
Paired and triadic tests can be used to verify information provided by a given informant. Where perfect *transivity* occurs between different tests, the data obtained can be viewed with considerable confidence. However, where particular taxa give rise to inconsistent data, these discrepancies are readily identified and, if necessary, can form the focus of further investigation. Paired and triad tests can be used in this way to check information provided both in preference ranking and pile sorting, as a means of cross-verification.

While preference ranking allows informants to sort a range of items into order based on criteria such as relative usefulness or economic value, triadic and paired comparisons allow researchers to check that this order remains constant, regardless of how items are presented (see Martin 1995).

CALCULATING RANKING INDICES

Triadic and paired comparisons involve presenting participants with a number of items, which are either in sets of three or in pairs. All possible combinations of the items are organised and presented to individual informants in a randomised order, and in each case, informants are asked to order the items on the basis of criteria such as personal preference, or the relative value of different plants. At the outset of a session, each participant is asked initially, to sort the whole group of items in order; later, the items are presented in the randomised groups shown below. In paired comparisons, each item is compared and assigned to either the 'more useful' (= 1) or 'less useful' (= 0) category; in the triadic comparison each item is assigned to 'most useful' (= 2), 'less useful' (= 1) or 'least useful' (= 0) categories.

Pairs		Triads	
Sequential	Randomised	Sequential	Randomised
ban, lem	lem, **or**	ban, lem, or	**ban**, *lem*, pear
ban, or	**or**, pear	ban, lem, pear	*ban*, **or**, pear
ban, pear	ban, **or**	ban, or, pear	*lem*, **or**, pear
lem, or	lem, **pear**	lem, or, pear	ban, *lem*, **or**
lem, pear	**ban**, lem		
or, pear	ban, **pear**		

oranges (or); lemons (lem); bananas (ban); pears (pear)—the informant's preference is indicated in **bold**; the least favoured option is in *italics*

In each case the data generated can be analysed to produce a numerical ranking of usefulness. For example, let us assume that in this case, the informant originally assigned a preference ranking of: oranges (3), bananas (2), pears (1), lemons (0), where a value of 3 was assigned to the most useful and of 0 to the least useful. However, when presented with these items in groups of two or three, the ranking

continued

___ *continued* ___

indices obtained from each test indicated certain inconsistencies in the information provided.

Fruit	Preference ranking index	Pairwise ranking index	Triad ranking index
Orange	3	1 + 1 + 1 = 3	2 + 2 + 2 = 6
Banana	2	0 + 1 + 0 = 1	2 + 0 + 1 = 3
Pear	1	0 + 1 + 1 = 2	1 + 1 + 1 = 3
Lemon	0	0 + 0 + 0 = 0	0 + 0 + 0 = 0

In both the paired test, and the triad test, the ranking index is calculated by summing the total values assigned to each item to produce the ranking indices as illustrated here. In this example, a comparison of the three indices obtained indicates that oranges are clearly the most useful of the fruits tested, while lemons are consistently demonstrated to be the least useful. However, data provided on bananas and pears are inconsistent throughout the three tests and their relative usefulness remains ambiguous.

sampling, researchers must ensure that their method takes account of sociological variables such as age, gender, occupation, education and class, to ensure that the data collected are not biased in favour of any one social group (e.g. Begossi *et al.* 1993; Maikhuri & Gangwar 1993; Soleri & Cleveland 1993). Where more specific information is required however, specialist informants may be chosen in consultation with community leaders and other community members (Barker & Cross 1992; von Geusan *et al.* 1992). In some cases where chosen candidates have been suggested by, and contacted through local leaders, this has facilitated greater openness during data collection, as informants feel that their participation has been locally endorsed (Barker & Cross 1992). Finally, a number of analytical tools have been developed to assist in identifying informants with expertise in particular areas of interest. For example, Phillips and Gentry (1993b) have developed the concept of a relative use-value for a given informant (RUV_i), which compares the information given by one informant with that given by all other informants (Box 4.4); alternatively, Boster (1985) has developed two *indices of agreement* which fulfil a similar function (Box 4.5).

Increasingly, participant choice is now no longer limited to choosing 'informants', but often involves the selection of local researchers (e.g. Barker & Cross 1992). As the dynamics of open-ended or semi-structured interviews depend as much on the knowledge of the researcher as on that of the informant, it is important that the process used for choosing

Box 4.4   Comparative analysis of informant knowledge—calculating relative use-value. In order to compare the knowledge of different informants, Phillips and Gentry (1993b) developed the *relative use-value* which provides a more accurate estimate of an individual's relative level of knowledge than simply comparing the mean UV_{is} for each informant. Using this method to compare the RUV_i of each of 11 informants these workers were able to demonstrate that certain informants were significantly more knowledgeable than others about the uses of 136 folk species of tree and vine. Some of these highly significant differences in RUV_i demonstrated that knowledge of forest species can vary enormously even between informants of similar age, suggesting that ethnobotanists and anthropologists must work with a large and representative sample of informants if their conclusions about traditional knowledge and its distribution are to prove valid.

The concept of calculating a *relative use-value* for a given informant (RUV_i), has been developed recently by Phillips and Gentry (1993b) using data collected from the *mestizo* people of Tambopata in Amazonian Peru. This index can be used to compare the use-value data supplied by one informant with that provided by the total number of informants.

CALCULATING THE RELATIVE USE-VALUE (RUV)

For any plant taxon the use-value data provided by one informant can be compared with the overall use-value data using the equation below. This numerical index thus facilitates the identification of informants who regularly provide anomalous information.

$$RUV_i = \frac{\sum \left(\dfrac{UV_{is}}{UV_s} \right)}{s_n}$$

Where:
RUV_i = the relative use-value for informant i
UV_{is} = the use-value of each folk species s as described by informant i
UV_s = the overall use-value estimated for each folk species s described in the study
s_n = the number of folk species described by informant i, for which data have also been provided by two or more other informants

Box 4.5 Comparative analysis of informant knowledge—calculating indices of agreement. Like Phillips' and Gentry's relative use-value, Boster's indices of agreement are based on the level of agreement between a given informant and the other informants involved. Boster has concluded from his research that a high level of agreement generally correlates positively with a high degree of accuracy, and proposes that indices such as those outlined here may therefore be employed in identifying suitable local participants for ethnobotanical research projects (see Boster 1985).

A sophisticated method for assessing the relative knowledge of a range of informants has been developed by American anthropologist James Boster using data collected from the Aguaruna Indians—a group of swidden cultivators in Amazonian Peru (Boster 1985). Boster used his data to calculate two indices of agreement: the *proportion of agreement* (PA), which indicates the amount of agreement between pairs of informants, and the *overall agreement* (OA), which indicates one informant's mean proportion of agreement with the rest of the population.

CALCULATING THE PROPORTION OF AGREEMENT

Data were collected by guiding informants around an experimental garden which had been planted with a range of manioc varieties, and at every specimen the informant would be asked 'What kind of manioc is this?' Using these data, Boster was able to identify pairs of informants who agreed with each other to a high degree.

$$PA_1 = \frac{a_1}{n_s}$$

Where:
PA_1 is the proportion of agreement for informant pair 1
a_1 is the number of times informant pair 1 agree on a given stimulus
n_s is the total number of stimuli presented to both members of informant pair 1

CALCULATING THE OVERALL AGREEMENT

Once the proportion of agreement has been calculated for each possible informant pair, the mean overall agreement between a given informant and all other informants can be calculated.

$$OA_i = \frac{\Sigma PA_n}{n_p}$$

Where:
OA_i is the overall agreement for informant i
PA_n is the proportion of agreement for informant i with each of the n members of the population
n_p is the total number of informant pairs in the population

continued

> *continued*
>
> Having calculated the indices of agreement for each informant, Boster used these values to construct a two-dimensional matrix which highlighted clusters of informants showing a high degree of agreement, both with particular individuals, and with the rest of the population as a whole. On further investigation of these clusters, Boster has concluded that a high degree of agreement might reasonably be used as an indicator of a high degree of knowledge.

researchers is equally rigorous. For example, while interviewers must be conversant with the concepts under discussion (Johnson & Ruttan 1992), their linguistic fluency, personality and social standing are crucial to establishing rapport between the participants involved (Martin 1995). Equally, factors such as cultural restraints on single women must also be considered, as members of different sociological groups may respond differently to researchers of a certain gender or social status (Barker & Cross 1992). Finally, the Western scientists who make up the external research team must also be carefully assessed; both Mike Balick of the New York Botanic Garden and Paul Cox of Brigham Young University, have experience of medical doctors who lack either the ability or the willingness to relate well in the field (see Farnsworth 1990). In an increasing number of cases, members of traditional communities have begun to adopt a much more active role both in directing the nature of ethnobotanical research and in deciding on who should be involved not only locally but also externally, rather than *vice versa* (Baines & Hviding 1992; Posey 1994). The methods which are used in participant choice are summarised in Table 4.4 along with the other field methods which have been discussed here.

Linguistic and Other Symbolic Analyses

A major key to the understanding of any ethnobotanical information gathered, lies in understanding the ways in which this information is communicated, and as we have seen in the previous chapter, much of this is through symbolic representation in art, ritual and myth. For example, in recent years researchers have attempted to explore the significance of symbols as diverse as the geometric designs found in palaeolithic rock art (Lewis-Williams & Dowson 1988) and of plants which are commonly portrayed in traditional mythology (Karim 1981). In addition, a variety of linguistic and experimental techniques have been employed in elucidating the botanical identities of artefacts such as the floral sculptures of thirteenth century English abbeys (Nelson & Stalley 1993), and the medicinal and other plants described in ancient texts (Biggam 1993).

Table 4.4 Summary of anthropological field methods discussed in the text. During extended periods of participant observation, a variety of methods may be used to collect ethnobotanical field data. The exact methodology used will vary considerably according to both the specific requirements of a given project and the relationships which have been established between local and external participants. For example, in some cases the range of knowledge held by a given community may be of interest in which case a random approach to participant selection may be useful; in other studies, where very specific types of information are sought, consultation with local leaders may provide the most practical approach to identifying informants with specialist knowledge (see: [1]Martin 1995; [2]Johnson & Ruttan 1992; [3]Baines & Hviding 1992; [4]Posey 1994, [5]Phillips & Gentry 1993a; [6]Boster 1985)

Anthropological methods used in ethnobotanical field studies

Interviews and surveys—interviews may be *open-ended, semi-structured* or *structured*, the latter of which may be carried out using *written questionnaires*. Unlike open-ended or semi-structured interviews, which are essentially casual conversations, structured interviews and questionnaires use a series of predetermined questions which allow for quantitative *comparative analyses*.[1,2]

Choosing participants—this may range from *random sampling*—which provides a representative cross-section of the information held by a community—to the identification of *local specialists* such as traditional healers. In addition, as members of different sociological groups may respond differently to researchers of different gender or social status, these factors must be considered when selecting researchers themselves—whether local or external.[3,4]

Elicitation and recording of data—a range of visual or other stimuli can be used in eliciting information regarding particular organisms including *voucher specimens*, photographs or *ethnobotanical artefacts*; more structured means of data collection may involve the use of defined areas of forest such as hectare plots, transect lines or experimental gardens. Data may be collected in field notebooks or as taped interviews, which must usually be *transcribed* and *translated* by linguistic experts.[2,5,6]

Analytical tools—these encompass a range of quantitative methods which may be used in calculating various *numerical indices*, such as the relative usefulness of a given organism or the reliability of a particular informant. These tools include methods such as *free-listing, pile sorting* and *preference ranking* which facilitate, for example, the identification of plant species of particular cultural significance, or the elucidation of the relationships which are perceived between distinct plant taxa.[1]

Clearly, a thorough understanding of the languages and dialects used locally is fundamental to the interpretation of most ethnobotanical data, and a number of authors have recently emphasised the dangers of inadequate communication, which can lead to a distortion in the translation and misunderstanding of particular terms and concepts (Posey 1984; Cox 1990; Farnsworth 1990; Barker & Cross 1992; Johnson & Ruttan 1992; Lentz 1993). However, the analysis of language itself can also yield considerable insight into TBK. For example, recent ethnotaxonomic studies suggest that detectable correlations may exist between the *linguistic structure* of a *plant name* or *label* and the cultural significance of that plant, while language reconstruction may prove useful in the identification of plants named in ancient texts (See Chapter 10).

Symbolic and Empirical Analysis of Myths and Folklore

While there is no set methodology for carrying out a symbolic analysis of a given text (Martin 1995), the comparative approach pioneered by French anthropologist Claude Lévi-Strauss has been adopted by researchers interested in subjects ranging from the origins of sacred plants in Polynesia (Lebot 1991) to the psychoanalytical study of modern myths (Carroll 1992). Using this method, researchers collect as many versions of a given myth as possible and identify their common *motifs*—elements which recur in all versions, as well as their *allomotifs*, where substitutions created during oral transmission are *symbolically equivalent* to the original (Carroll 1992).

A clear example of this approach can be seen in the analysis of a common Polynesian myth which recounts the origins of the *kava* plant—a psychoactive plant which is cultivated in many regions of the Pacific. The study (Lebot 1991 and see Martin 1995) involves the comparison of four variants of the myth and reveals the existence of a number of recurring themes (Box 4.6). For example, the repeated association of human burial with the appearance of the *kava* plant, emphasises the role of human activity in the plant's survival (it is now completely domesticated), while the intoxication motif relates to the perceived magical properties of the plant. In some cases, this type of analysis is also supported by empirical evidence—for example, an Australian Aboriginal legend which tells of a time when the earth blew up seems, almost certainly, to describe the volcanic eruption of Mount Wilson in the Blue Mountains near Sydney (Flood 1983), while many myths relating to castastrophic flooding may be traced to the rise in sea level at the end of the last ice age about 10 000 years ago. Similarly, the empirical analysis of the *kava* plant has revealed a chemical explanation behind its magical properties (Box 4.7). Such examples demonstrate clearly, the potential application of symbolic analysis in for example, the search for biologically active plants.

Box 4.6 Symbolic analysis of traditional folklore—Polynesian origin myths of _Piper methysticum._ Using the comparative method pioneered by Lévi-Strauss, some researchers have attempted to explore the ethnobotanical information embedded within the symbolic motifs of traditional myths. Many such myths have revealed information about the cultural significance or special properties of certain plant or animal species; others describe the consequences of failing to care for the natural environment—for example by cutting a tree from a sacred grove or by breaking some other taboo. Although there is no set methodology for interpreting the symbolism found in myths and legends, researchers generally collect details of numerous versions of a given myth from many informants, and will often confer with local people about the interpretations they make (Martin 1995).

Piper methysticum (Piperaceae) is a psychoactive plant which has been cultivated for centuries in many regions of the Pacific, and which is known by the Polynesian name _kava_, along with numerous local names. This species has long been used in the concoction of a drink which is consumed by many Pacific peoples at both public ceremonies and private gatherings (Martin 1995). A recent study by French botanist Vincent Lebot (1991) records a number of variants of the myth surrounding the origins of this culturally significant plant, all of which have several _motifs_ in common.

RECURRING MOTIFS IN POLYNESIAN _KAVA_ MYTHS

In each of the variants described by Lebot, the _kava_ myth recounts the tale of an unusual plant which appears spontaneously on the grave of a central character who has died recently; and in each case, the plant is found to have a significant effect on the behaviour of a rat which feeds on the mystery plant. In comparing the main points of these different myths, it soon becomes clear that certain features recur in all variants, while other details are more variable.

Place of origin	Identity of deceased	Burial	Growth of of _kava_	Consumption of _kava_	Physiological effect
Vanuatu	Sister	Yes	After 1 week; Solitary	Roots eaten by rat; roots eaten by brother	Death Intoxication
W Samoa	Brother	Yes	After 3 days; 2 plants	Plant eaten by rat	Intoxication
Tonga	Human flesh	Yes	n/s	Plant eaten by rat	Paralysis
Pohnpei 1	Human	Yes	n/s	Juice drunk by human	Intoxication
Pohnpei 2	n/s	n/s	n/s	Roots eaten by rat	Intoxication

n/s, not specified.

continued

____ *continued* ____

While these stories differ slightly in certain details, there are several elements common to each. For example, in almost every case it is a human grave which leads to the occurrence of the plant; in almost every case also, the bioactive properties of the plant are discovered through the observation of animal behaviour; finally, there is normally the presence of a rat—a wild animal which often occurs near human communities. Some of these motifs may be interpreted as symbolic of the important interactions which exist between humans and the natural world. For example, the magical *kava* plant is symbolically dependent on humans for its growth and survival, and grows only at the interface where culture meets nature (see Martin 1995).

Box 4.7   Empirical analysis of the *kava* myths—understanding traditional mythology as a source of ethnobotanical information. Recent evidence regarding the nature and properties of the *kava* plant and its extracts, may help to explain several features of the *kava* myths outlined in Box 4.6. For example, the plant's symbolic dependence on human intervention might be explained by the fact that *kava* plant is completely domesticated, while the physiological effects caused by ingesting plant tissues can be explained by the presence of a range of psychoactive chemicals. Equally, the presence of strongly active pipermethystin in the leaves of certain strains only, might explain the fact that some of the myth variants specify that intoxication stemmed from eating the roots, while others do not specify that the roots were eaten.

The *kava* plant, *Piper methysticum,* has been used for centuries by traditional peoples of the Pacific Islands, its use having been documented in 1768 when Captain James Cook visited Hawaii (Ott 1993). The inebriating beverage prepared from this plant—which is also referred to as *kava*—was used largely as both an intoxicant and as a ritual offering to ancestors or gods, although a number of medicinal uses have also been reported. The beverage itself was prepared traditionally by children who would chew the roots and lower stems of the plant, before expectorating the macerated tissues into a communal bowl; this mash was subsequently diluted and filtered prior to consumption. More recently, however, a less potent beverage has been prepared in some areas, by grinding, rather than chewing the plant tissues, illustrating clearly that salivary enzymes play an important part in releasing the plant's psychoactive constituents (Mann 1994). In relation to the myths surrounding the origins of *kava*, a number of features can be explained in the light of recent scientific studies.

THE DEPENDENCE OF *KAVA* ON HUMAN INTERVENTION

Although the origins of *kava* can apparently be linked with *P. wichmanni*—a wild species of *Piper* which is endemic to certain Pacific islands, *kava* itself is now

____ *continued* ____

— continued —

completely domesticated and its reproduction is entirely dependent on vegetative propagation by local cultivators (Martin 1995).

LEARNING ABOUT PLANT PROPERTIES FROM ANIMAL BEHAVIOUR

Recent observations of animal behaviour have led to the development of the controversial new science of *zoopharmacognosy*, or the study of self-medication by animals. There now exist several examples which suggest that early *Homo* might have learned to use plants for medicines or other purposes on the basis of observed animal behaviour (McRae 1994, and see Chapter 10).

PSYCHOACTIVE CONSTITUENTS OF *KAVA*

Phtyochemical and pharmacological analyses of *Piper methysticum* have revealed a number of non-nitrogen compounds which exhibit psychoactive properties. These chemicals—the *kava-lactones* or *kava-pyrones*—have been extracted primarily from the roots of the plant, and at least five of these compounds have been shown to possess sedative or relaxant properties. In addition, a *pyridone alkaloid*, pipermethystin, has been detected in the leaves of certain strains, and in some cases, the leaves of these strains have proved as potent as the roots of normal plants (see Ott 1993).

Constituent	Occurrence	Physiological effect
Methysticin	Roots	Muscle relaxant
Dihydromethysticin	Roots	Human anticonvulsant
Kawain	Roots	Muscle relaxant
Dihydrokawain	Roots	Human sedative
Yangonin	Roots	Muscle relaxant
Desmethoxyyangonin	Roots	n/s
Pipermethystin	Leaves	n/s

n/s—not specified; data presented in Ott (1993).

Of these active *kava* compounds, the first to be recognised was methysticin which was isolated initially in 1860. In addition to the psychoactive properties outlined here, several of these bioactive constituents also exhibit significant fungistatic activity—a finding which supports the plant's ethnomedical use as topical antiseptic.

Plant Labels and Cultural Significance

Another aspect of symbolic analysis that can be useful to ethnobotanists, lies in the field of *ethnotaxonomy*, which involves the study of traditional systems of classification. During ethnotaxonomic studies, researchers primarily collect two types of data: numerical evidence from techniques such as pile sorting which demonstrates the perceived relationships between different plants, and linguistic evidence in the form of plant names. While the physical grouping of organisms can provide considerable insight into local perceptions of the natural world, patterns revealed through the linguistic analysis of plant names or other taxonomic labels can provide clues to the uses and other characteristics of local plants.

Following the collection of linguistic data in the form of recorded speech, the spoken words must first be transcribed phonetically using the standard linguistic symbols of the International Phonetic Association (IPA). Data analysis may then begin by translating the transcribed terms. Translation may involve the use of a *gloss* or *free translation*, in which case the closest equivalent word available in English (or any other language) is used; alternatively terms may be translated literally, in which case the word is translated word-for-word. For example, *tsapxo'j* the name used by the Mexican Mixe for *Cydonia oblonga*, can be glossed as *quince* or translated literally as *heaven-oak* (Martin 1995). Further analysis of these names may then provide information on the characteristics of a given plant as it is perceived locally.

In the analysis of plant names, the identification of common structural features—recurring *suffixes, prefixes* or *root words*—can provide important clues about the meaning of certain words. Indeed in the case of *tsapxo'j* mentioned above, although the literal translation of *tsap* is heaven, an examination of other Mixe plant names reveals that the term is commonly used as modifier for objects which were originally introduced by the Spanish (see Martin 1995). Similarly, the suffix *dum* in many Mixe dialects appears to refer to objects which are round, such as the fruits of *aaydum* (*Annona cherimola*) and the inflorescences of *maydum* (*Mimosa albida*). Another common linguistic tool involves the detection of *cognates*—variations of the same word which refer to the same object. Through the comparative analysis of these variants, linguistics are often able to reconstruct a probable *ancestral term* from which each of the current terms has been derived (see Martin 1995), and this can have a number of implications in palaeoethnobotanical studies. For example, through recognising patterns in the spatial distribution of cognate plant names, it is anticipated that events such as the prehistoric dispersal of human populations may be investigated (Chapter 10); equally, the reconstruction of ancestral terms may prove useful in the retrospective identification of the plant species described in ancient texts (Biggam 1993).

Plant Collection and Taxonomy

The collection of plant specimens forms an integral part of any ethnobotanical study, both for the preparation of *voucher specimens* (permanent records of plants for which ethnobotanical data have been collected), and in studies which require the chemical or molecular analysis of plant materials. In either case, it is essential that permits allowing the collection and export of plant materials are obtained (Womersley 1981; Bridson & Forman 1992) and that all protected, or otherwise endangered species are treated with particular caution. Indeed for certain species it may be advisable to take representative photographs rather than to collect any living material at all. In any event, plant collection must always be carefully planned in order to produce the maximum information while minimising detrimental aspects such as habitat disturbance or overharvesting of particular species.

The Nature and Uses of Voucher Specimens

As in any botanical field study, accurate botanical identification plays a fundamental part in ethnobotanical research, and provides a vital link between two bodies of information—that of the Western scientist and that of the local people. For example, a plant known as *kiswe* or *kyswi* by the Waimiri Atroari Indians of Brazil is known to produce a pigment which is sometimes used in body painting; yet it is not until the species is identified formally as *Bixa orellana*, that we are able to see that the same plant is used for exactly the same purpose by the Brazilian Kayapó to whom it is known under several synonymous labels including *pỳ kumrenx*, and *pỳ krã re* (see Posey 1984; Milliken *et al.* 1992). Hence, if we are to make sensible comparisons between ethnobotanical studies, accurate botanical identification is of paramount importance.

In many studies, however, formal identification in the field may not always be possible, and certain species can require the attention of specialist taxonomists. In such cases, good quality voucher specimens may be sent to specialists for verification. These voucher specimens are essentially representative samples of a given plant species, which exhibit both the main features required for its identification and the range of variation in these features (such as leaf shape or size). While a good collection should ideally include samples of all the plant organs, and at all stages of development, it is particularly important that intact examples of reproductive structures are represented wherever possible, as identification can otherwise prove difficult and time consuming.

Samples must be selected and harvested with care—particularly where samples of roots or bark are required (see Martin 1995)—before they are preserved and mounted as outlined in Figure 4.2. Voucher specimens are ultimately stored in appropriate *herbaria*, and where duplicates are available, both the host country and that of visiting researchers should receive a sample.

While good voucher specimens are clearly crucial where identification in the field is not possible, they are also important even where an *in situ* identification has been made. First, in the event of a misidentification occurring in the field, the voucher specimen can be consulted and re-identified at a later date, such that valuable ethnobotanical information is not lost altogether. Secondly, representative vouchers can prove useful as stimuli in interviews or as props for sorting experiments such as those outlined above.

Taxonomic Studies: Plant Identification

Accurate plant identification requires the skills of an experienced plant taxonomist, who is involved in the identification, nomenclature and classification of plant species. In early studies, plant identification was largely determined on the basis of gross morphology, yet this area of research has been revolutionised in recent decades with the development of powerful technologies such as electron microscopy, molecular biology and computer-assisted *morphometric analysis*. Such methods now allow the investigation of plant identity at all levels of organisation—from the morphology of the whole plant to its microscopic detail and its molecular characteristics.

At the whole plant level, all plant parts may reveal characteristic features which assist in plant identification, including growth habit and form, plant size, leaf shape and venation, leaf arrangement and the branching of the root system. However, identification on the basis of vegetative growth alone often proves difficult, and where possible reproductive structures, including flowers, cones, seeds, fruits or spore-producing structures are examined. Morphometric analysis may be carried out either by eye, or, more recently using video and computer technology, where a computer image of the plant specimen is analysed automatically, allowing both precise measurement and the calculation of parameters such as length/width ratios, which are used in numerical taxonomic studies (Decker & Wilson 1986; Sánchez *et al.* 1993).

A range of microscopic plant features can also be used for diagnostic purposes, including surface characteristics such as stomatal anatomy, leaf hairs and the shape of epidermal cells, and internal characteristics

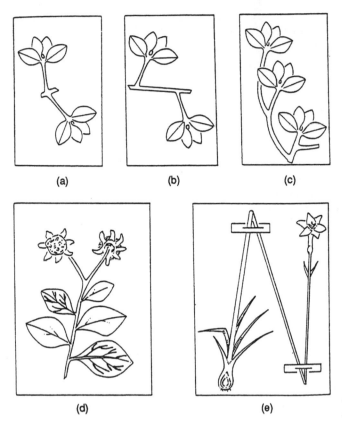

(a)　　　　(b)　　　　(c)

(d)　　　　　　(e)

Figure 4.2 Preparation of voucher specimens. The collection of high-quality voucher specimens forms an important part of any ethnobotanical study, allowing the permanent preservation of representative plant samples. These are prepared by drying and pressing suitable plant specimens, and arranging them on *herbarium sheets*, which are then deposited in suitable herbaria. Where large plants are sampled, the plant should be cut to conserve as much of the branching as possible as illustrated here, and where possible the stem apex should be retained. Where small plants are collected, sufficient individuals to fill a complete sheet should be obtained. In either case, plants should be arranged to display the maximum information possible. For example, leaves should be spread to avoid overlap, and at least one should be turned to expose the undersurface; likewise both aspects of the flower should be displayed. In addition, where available, additional loose flowers and fruits should be included, and placed in packets attached to the main sheet if appropriate. Here, (a), (b) and (c) illustrate examples of large plants which have been cut to conserve details of branching patterns: (a) opposite branching; (b) alternate branching; (c) sympodial branching; (d) shows how both aspects of leaf and flower are displayed; (e) illustrates how long, flexible specimens can be displayed effectively (from Bridson & Forman 1992). © Copyright The Board of Trustees of the Royal Botanic Gardens, Kew. *Artist: Sally E. Dawson*

such as wood anatomy, the nature and distribution of vascular systems, and other tissue types. Pollen grains and phytoliths can also prove useful as often they can be identified on the basis of features such as size, shape, and, in the case of pollen grains, the characteristic sculpting of the exine (Figure 4.3). In this respect, advances in electron microscopy—particularly in conjunction with computer-assisted morphometry—have provided powerful tools for microscopic analysis, scanning electron microscopy (SEM) allowing detailed examination of surface characteristics, while transmission electron microscopy (TEM) facilitates the study of internal features, such as starch grains, whose structure, shape and size can vary markedly between species (Cortella & Pochettion 1994).

However, despite these technological advances, detailed characterisation on the basis of morphological and anatomical features alone can prove difficult, and an increasing number of species or subspecies have now been identified on the basis of biochemical characteristics. For example, the white or yellow mustard *Sinapis alba* has been distinguished from the closely related genus *Brassica* on the basis of enzyme identification (Vaughn 1991), while a chemical key has been constructed to distinguish six species of *Khaya* (African mahogany), which were poorly defined on the basis of morphological characteristics alone (Adesida *et al.* 1971).

Once an organism has been characterised on the basis of its specific morphological, anatomical or chemical attributes, it must be assigned an unambiguous name according to a common classification system. In most cases this is simply a question of designating the existing scientific name to a given specimen, yet where a previously unrecognised species is identified, a new name must be assigned according to the require-

---

Figure 4.3 (opposite) **Pollen characteristics and plant identification.** In the living pollen grain of an angiosperm, the wall is made up of two layers; the outer layer or *exine* is composed of *sporopollenin*, which is strongly resistant to degradation, while the inner layer, or *intine* is made up of cellulose. The morphological detail of the resistant, sporopollenin-containing exine is strongly species-specific, allowing the recognition of plant species on the basis of their pollen alone. This is particularly pertinent in the examination of fossil plants, where often the pollen exine is the only tissue which is preserved. The identification of any pollen grain is based on the examination of a number of characteristic features—including the shape and size of the pollen, the shape and distribution of any *apertures* in the exine, and the *sculpturing* or fine structure of the exine. (a) Illustrates some of the distinctive shapes of angiosperm pollen in *polar* view; (b) illustrates characteristic shapes of pollen in *equatorial* view; (c) illustrates some common aperture patterns; (d) demonstrates the sculpturing types visible on the surface of a pollen grain (see Moore & Webb 1978. Reproduced by permission of Edward Arnold, London)

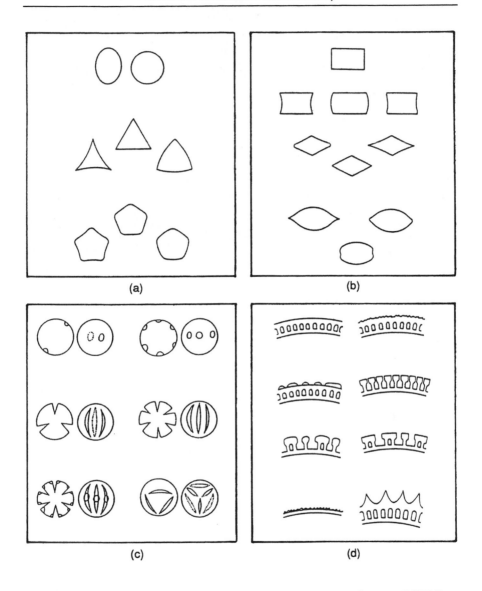

ments of the International Code of Botanical Nomenclature (ICBN) as outlined in Figure 4.4.

Taxonomic Studies: Classification

Plant taxonomic studies involve not only the identification and naming of a given botanical species, but also its *systematic classification*, which

Alysicarpus	**glumaceus**	**subsp. glumaceus**	**var. intermedius**
(generic name)	(specific epithet)	(subspecies)	(varietal name)

Figure 4.4   Plant nomenclature—rules of scientific name construction.
All scientific names are written in Latin or in Latinized form, in accordance
with the rules of the International Code of Botanical Nomenclature (ICBN).
The name of a *species* always consists of two words and is therefore known as its
binomial, for example, *Rosa canina*. In each *species name*, the first word is the
generic name or the name of the genus to which the species belongs, and this
always begins with an upper case letter; the second name is the *specific epithet*,
which begins with a lower case letter. The specific epithet typically has a Latin
ending and usually acts as an adjective, as in the case of *Caesalpinia pulcherrima* which translates as 'very beautiful Caesalpinia'. Occasionally, taxa below
the level of species are recognised—including subspecies (denoted *subsp.*) and
variety (*var.*), and where more than one of these is used, the correct sequence
must be followed as illustrated above. In addition to the name of the species
itself, the plant name when written in full, is followed by one or more personal
names, which are frequently in abbreviated form. This name (or sequence of
names) is known as the *author citation* for the plant, and is the name of the
person (or persons) who first validly published the plant's officially recognised
name. For example, *Commelina* L. (or Linnaeus); *Acacia schweinfurthii* Brenan
& Exell (see Bridson & Forman 1992)

allows the arrangement of plants into convenient groups of related
species. Like the methods used in plant identification, those used in classification have also expanded enormously in recent decades, and currently
include experimental investigations as well as phenotypic and genotypic
analysis. For example, artificial pollination experiments between individuals of the Mesoamerican *Jaltomata procumbens* have recently led to
reorganisation of this subgroup into six distinct species (Mione 1994).

The use of biochemical analysis represents a relatively new tool in
plant taxonomic studies (Harborne 1984), its value stemming from the
fact that biochemical characteristics can be identified and quantified
accurately, and that chemical evidence—particularly at the molecular
level—can be more readily located within a proposed evolutionary
sequence than a morphological feature such as leaf shape (e.g. Hormaza
et al. 1994; Joshi & Nguyen 1993; Vaughn 1991). As with any taxonomic
method, there are problems associated with a chemosystematic
approach (Harborne & Turner 1984); yet there is little doubt that chemical analysis can significantly enhance the information gleaned by
morphological characters alone.

Archaeobotanical Methods

Palaeoethnobotany, or the study of past relationships between plants and people, can involve the pursuit of many different lines of evidence including the analysis of historical texts and the interpretation of plants in ancient art. However, fundamental to any palaeoethnobotanical study is the study of the botanical evidence found in archaeological sites, which requires the application of *archaeobotanical* techniques. Archaeobotanical evidence includes both *plant fossils—* ancient plant materials which have survived normal biological degradation (see Pearsall 1989), and more indirect evidence such as grain impressions left in baked clay (Renfrew 1973) or microwear polishes found on teeth and stone tools (Lewin 1993). Careful collection and analysis of these sources of data can provide considerable insight into both the ways in which humans have exploited plants in the past, and how human activity has influenced changes in the local vegetation.

The major steps in any palaeoethnobotanical study involve careful sampling and collecting at archaeological sites; recovery of the plant remains from soil samples; identification and dating of the plant materials; and the analysis and interpretation of the data obtained. While the exact methods used will vary depending on whether remains are macroscopic, microscopic, relatively intact or extensively damaged, one problem common to all cases is that of possible *cross-contamination,* where exotic seeds, pollen or tissue fragments are inadvertently introduced into a given sample, either from a different archaeological context, or from modern plant remains. Possible precautions to minimise this problem have been discussed elsewhere (Pearsall 1989; Smith 1989) and must be considered in all archaeobotanical investigations.

Collecting Archaeobotanical Data

Pearsall (1989) recommends a 'blanket sampling' strategy for collecting from archaeological sites, in which an appropriate sample is collected from each *excavation context* identified. Excavation contexts include general sample sites which are determined by dividing the whole excavation area into a suitably sized grid, and sampling from a given point in each square; in addition *specific contexts* such as preserved artefacts or buildings are sampled separately. Once suitable excavation contexts have been identified, samples are collected using *composite, point* and/or *column sampling* techniques (see Pearsall 1989). Where soil samples are collected they are carefully bagged and labelled, and macroscopic remains such as seeds, fruits, and fragments of stems, leaves and wood, are recovered either by eye, by flotation or by sieving (Pearsall 1989). Microfossils such as pollen grains and phytoliths, are recovered from samples which have been processed to remove larger fragments and the desired residues are concentrated using standard techniques (Moore & Webb 1978).

The recovery of macrofossils by eye is carried out *in situ* at the site itself, and produces *naked-eye assemblages* which will obviously be biased towards larger plant remains; flotation and sieving produce less biased data, and are much more efficient in terms of overall recovery from a given sample. Flotation techniques exploit the differences in density between the organic and inorganic materials present within a soil sample, the lighter organic residues generally floating to the surface, while the inorganic soil particles sink. This method is particularly useful for the recovery of carbonised samples, and a range of manual and machine-assisted devices have now been developed to deal with particular problems relating to soil-type or water availability, which may be encountered at a given excavation site. However, where plant remains are either desiccated or water-logged, flotation may be inappropriate as dried tissues may become damaged during rehydration, while waterlogged materials show a tendency to sink. In such cases other techniques such as sieving, or separation by air pressure may be used as discussed by Pearsall (1989).

Where pollen grains are to be extracted from soil samples, the basic method involves the chemical degradation of other materials present, based on the fact that the sporopollenin of the pollen exine is highly resistant to degradation—a feature which accounts for the comparative durability of pollen grains in the archaeological record. Soil samples are therefore treated with a range of chemicals such as potassium hydroxide and hydrochloric acid, which degrade other organic and mineral residues found in soils, leaving resistant microfossils intact (Moore & Webb 1978). Alternatively, where phytoliths are to be isolated, samples are sieved to remove large residues, and the subsequent addition of a heavy liquid such as cadmium iodide solution causes the phytoliths to float to the surface (Pearsall 1989). In either case, the resulting suspension is centrifuged to precipitate a pellet of pollen or silicaceous residues which can then be stained and resuspended in a small volume of molten glycerol jelly. The resulting suspension is then spread on to microscope slides for examination.

Identification of Archaeobotanical Remains

Many of the methods which are used to identify living plant specimens are also applicable to the identification of archaeobotanical remains, and methods such as morphometric analysis and molecular characterisation are frequently employed in archaebotanical investigation. However, unlike living specimens, most plant fossils have been preserved only due to processes such as carbonization, desiccation or water-logging—all of which limit the activity of biodegrading microorganisms, and all of which can result in profound changes in morphology. For example, changes such as a decrease in size due to shrinkage, change in colour and distortion of shape are all common during the fossilisation process. Hence, comparison of samples with existing herbarium specimens may not be appropriate, and instead a *comparative collection* of plant parts which have been artificially dried, carbonised or otherwise processed provides a more useful alternative (Pearsall 1989).

A number of biochemical analyses have also proved valuable in archaeobotanical identification, particularly where charred or damaged residues reveal few, if any visible characteristics, and both mineral analysis and solvent extraction of organic residues have been used with varying degrees of sensitivity. For example, Hastorf and DeNiro (1985) have examined the ratios of carbon and nitrogen isotopes to discriminate between C4, C3, and CAM plants, as well as between leguminous and non-leguminous plants, while Hill and Evans (1989) have used infrared (IR) spectroscopy to produce diagnostic 'fingerprints' for several plant species (see Chapter 10). In addition, while the usefulness of protein and DNA analysis in archaeobotanical remains is often limited by their susceptibility to degradation, recent examinations of the DNA from mummified maize seeds have revealed that the relative proportions of characteristic genomic components of single kernels may offer an important new means of investigating ancient maize populations (Rollo *et al.* 1991).

Evidence from Specialised Archaeological Contexts

In many palaeoethnobotanical investigations, samples obtained from specialised archaeological contexts can provide more direct evidence of the ways in which plants were used, harvested and processed. For example, the analysis of specific contexts such as grinding tools and *coprolites* (preserved faecal matter) can reveal much about prehistoric diet. In many cases, analysis of these specialised contexts involves the identification of plant fossils such as phytoliths embedded in human teeth or grinding equipment (see Smith 1989) or carbonised residues on pottery shards (Hill & Evans 1989). However, in recent years, the examination of *microwear polishes* through *use-wear analysis* has provided an additional source of useful evidence in palaeoethnobotanical research. Here, experiments are carried out using replica tools to process plants in various ways, and the microscopic wear patterns produced can be examined. This type of approach used in the examination of stone tools from Kenya has suggested that certain tools dating from 1.5 million years ago were used for sawing wood (Keeley & Toth 1981); even more ambitious experiments have compared the striations produced through harvesting various herbacious plants on tilled *versus* untilled soils (see Harris 1989). If the efficacy of this kind of technique is confirmed, such evidence could ultimately provide an invaluable means of tracing when particular land management practices were first used in prehistoric plant cultivation.

Dating Methods and Data Presentation

Once archaeobotanical evidence has been collected and identified, the fossilised remains or artefacts must be dated, and the data analysed and interpreted. Before the development of radiometric techniques, fossil age was generally estimated on the basis of evidence such as the presence of known indicator species; today, however, the age of plant remains, other fossils and preserved artefacts may be determined either directly or indirectly using a variety of techniques, such as radiometric dating and thermoluminescence (Table 4.5). Accurately dated archaeobotanical data can then be used for both qualitative and quantitative analysis. Qualitatively, the data can provide information on factors such as the past uses of a given plant, its domestication and its pattern of spread; equally, data may be used to assess details such as the seasonality of site occupation, and the nature of human diet. However, palaeoethnobotanical studies are now becoming increasingly quantitative, with the development of analytical methods which allow, for example, the estimation of changes in *species diversity* as a consequence of human impact (see Chapter 10).

Specialist Ethnobotanical Methods

While the methods outlined above are common to ethnobotanical or palaeoethnobotanical data collection, they by no means include all the methodologies applicable to specific types of ethnobotanical research.

Table 4.5 Dating methods used in plant fossil analysis. Today, the age of both biological remains and preserved artefacts can be estimated using a range of techniques. In some circumstances, pollen analysis remains an important technique for dating archaeological sites, while new techniques such as phytolith thermoluminescence and accelerator mass spectrometry are providing methods for the direct dating of even very small samples (see Odin 1982; Pearsall 1989)

Dating methods used in palaeoethnobotanical studies

Pollen profiles and indicator species—the presence of specific *biological indicators* provided an early means of estimating the age of a given stratigraphic level. Comparative pollen analysis is still widely used to estimate the age of archaeological sites.

Isotope analysis—the demonstration that radioactive isotopes of certain elements such as potassium, rubidium and uranium, decay at a constant rate, irrespective of changes in environmental factors, has provided a method for estimating the absolute age of certain types of rock. However, while such dating methods may be used to date rocks of considerable age (greater than 100 000 years old), radiocarbon dating must be used for more recent fossils. Here the decay of the unstable carbon isotope ^{14}C can be used to measure the age of fossils up to 40 000 to 50 000 years of age, with a degree of accuracy of about 2–3 per cent. Most recently, radiocarbon dating by accelerator mass spectrometry has allowed extremely small samples, such as individual seeds, to be dated directly.

Thermoluminescence—in addition to radiocarbon dating it now seems that measuring the thermoluminescence of preserved phytoliths or pottery shards may provide an alternative means of direct dating.

For example, studies aimed at the discovery of new biologically active plant chemicals will require the additional analytical techniques of *phytochemists* and *pharmacognosists*, while the economic evaluation of tropical forests can involve ethnographic studies in *time allocation* and the *cost/benefit analyses* of environmental economists. In addition, where projects require information regarding the sustainability of certain management techniques or levels of plant exploitation, quantitative ecological methods can be used in conjunction with TBK. Each of these specific methodologies (Table 4.6) are discussed in the relevant chapters.

Information Systems

Whatever the aim of an ethnobotanical investigation, and whatever methods are employed, a wide range of information systems are currently available to assist in all aspects of such research. A large number of general databases such as Biological Abstracts and Current Contents (both available on CD-ROM) can help in general literature searches,

Table 4.6 Specialist methods used in ethnobotanical studies. There are many areas of study—both academic and practical—which are relevant to ethnobotanical study. Outlined here are the nature and applications of some of the more common of these, many of which are discussed further in later chapters

Nature and applications of more specialist methods
Languages and linguistics—in addition to languages which clearly are essential to communication, linguistic techniques may be useful in various areas. For example, in the identification of plant species in historical texts; in tracing patterns of human dispersal; in the analysis of plant names.
Art history—analytical techniques may help to elucidate the identity and significance of plants represented in contemporary, historic and prehistoric art.
Agricultural science—understanding of crop science and soil science may be useful in the analysis and evaluation of native crop plants and traditional farming practices.
Ecology—methods used in ecological impact assessments may be used in assessing the sustainability of traditional methods of biological resource management.
Phytochemistry—methods of plant analysis are useful in examining the phytochemical nature of traditionally useful plants; equally, an understanding of plant chemistry is useful in evaluating traditional methods of plant processing.
Pharmacognosy—methods used in assessing the bioactivity of plant extracts is invaluable in investigating traditional medical systems, particularly where data are to be applied in practice.
Molecular biology—molecular techniques are useful in a range of applications. For example, they can be used in the identification of both modern plants and fossilised materials; they can be used in investigations into plant domestication; they also provide a basis for examining the genetic consequences of traditional plant management practices.
Applied anthropology—including *economic anthropology* and *development anthropology*. While methods used in economic anthropology may be useful in constructing realistic cost/benefit analyses for development options, methods used in development anthropology might be useful in applying ethnobotanical data to contemporary development problems.
Environmental economics—like economic anthropology, methods used by environmental economists are needed for the accurate evaluation of ethnobotany-based development strategies such as the establishment of *extractive reserves*.
Ethical analysis and law—methods used in these areas are essential to developing mechanisms for protecting the owners of TBK and to ensuring equitable distribution of benefits from ethnobotany-based developments.
Communication and education—the future success of ethnobotanical research and its application in promoting both biological and cultural diversity is largely dependent on the effective communication about the local and global significance of TBK.

while more specialist databases are now available for researching specific topics. For example, the Royal Botanic Gardens, Kew (RBGK) are currently developing the SEPASAL (Survey of Economic Plants for Arid and Semi-arid Lands) database (Wickens *et al.* 1985), while the prototype PLANIMAL (plant–animal interactions) database aims to provide a facility for predicting plant–animal mutualisms based on the interactive manipulation of ethnobotanical and ecological literature (Cotton & Hodgson 1994). A relatively large number of organisations have now established databases of natural products and medicinal plants, including *The Chapman and Hall Chemical Database*, which has been produced recently by the London-based publishing house Chapman & Hall, NAPRALERT (Natural Products Alert) produced by a team of ethnopharmacologists at the University of Illinois (Farnsworth *et al.* 1981; Loub *et al.* 1985) and MEDFLOR, a database dedicated to ethnobiological data (Beecher & Gullenhaal 1993).

In addition to these facilities which allow rapid searching of the scientific literature, a number of further information systems are also of great value to ethnobiological research. Some of these are designed to facilitate communication between researchers in different parts of the world—for example, a number of organisations have set up worldwide directories documenting the people and institutions working in particular areas, including the *World Directory of Ethnobotanists* (Jain *et al.* 1986) and the *Directorio Latinoamericano de Etnobotánicos* (GELA 1994). In addition organisations such as the *Forests, Trees and People Programme/Network* (administered jointly by the Italian Forestry Department and the International Rural Development Centre in Sweden) and the *Indigenous Knowledge Resource Centres* throughout the world provide useful networks for individuals working in the field of indigenous knowledge. Other specialist databases provide powerful research tools—such as libraries of known gene or protein sequences; information on spectral characteristics of chemical compounds; or details required for the automated interpretation of remote sensing analyses—all of which allow rapid comparison of new data with existing datasets. Finally, Geographical Information Systems (GIS) allow the storage and manipulation of large and complex datasets, such as those produced in environmental assessments (e.g. Tabor & Hutchinson 1994).

SUMMARY

The methods outlined in this chapter include those which are most commonly employed in the collection of ethnobotanical field data, including those techniques—developed in both the social and biological sciences—which have facilitated the development of a much more rigorous ethnobotanical methodology. While qualitative data collection during participant observation allows in-depth exploration of the TBK of local experts, recent advances in quantitative methodologies can be useful in the verification or statistical comparison of the data supplied, and in the comparison of the efficacy of different data collection methodologies. Similar advances in botanical and archaeobotanical techniques have also occurred, and a range of innovative morphometric and biochemical methods now look likely to prove invaluable in the identification and classification of many plant species. The crucial methods determining the real success of any project are those of the ethnographer and the plant taxonomist, yet the contribution of many additional techniques—for example, those of phytochemists, ecologists and environmental economists—can complement each other to produce a broad range of data concerning the nature and value of traditionally useful plants or the sustainability of local land management techniques. Whatever the aim of a given project, a multidisciplinary approach can benefit not only from the wide range of expertise available from within different disciplines, but also from the broader view of the world which can arise as often rather disparate individuals work towards a common goal.

Chapter 5

Traditional Botanical Knowledge and Subsistence: Wild Plant Resources

For thousands of years, aboriginal peoples around the world have used knowledge of their local environment to sustain themselves and to maintain their cultural identity. Only in the past decade, however, has this knowledge been recognised by the scientific community as a valuable source of ecological information.

Martha Johnson 1992 in *Lore*

INTRODUCTION

Throughout the world, more than 250 million indigenous peoples plus many more peasant communities remain substantially dependent on traditional modes of production to fulfil their basic requirements for food and fuel. Most rely on a combination of subsistence strategies, and rely to varying degrees on the five major activities of hunting, gathering, fishing, herding and the practice of small-scale cultivation (Table 5.1). However, although few societies today depend on wild resources for more than 56 per cent of their total subsistence requirements (Figure 5.1), most exhibit a considerable knowledge of wild plants which are used as food, fodder or fuel. For example, despite obtaining their staple foods from cultivated plants such as manioc (*Manihot esculenta*), banana (*Musa* spp) and sugar cane (*Saccharum officinarum*), the Waimiri Atroari of Brazil recognise more than 90 wild species as important components of their diet, plus at least 189 species which are known to attract particular species of animal (Milliken *et al.* 1992). This

Table 5.1   Subsistence activities of traditional peoples today. Throughout the world, more than 250 million indigenous people rely on traditional modes of production for some proportion of their subsistence requirements. This table illustrates the location of the world's remaining indigenous peoples, and in each case provides an estimate of the number of individuals and their proportion of the total population. Examples of extant indigenous groups are also given for each country, and their major subsistence activities are outlined. Although few of these societies rely on wild resources for more than half of their total subsistence requirements, most have an extensive knowledge of wild plant resources which are used as food, fodder and fuel

---

Major subsistence activities of the world's indigenous communities

Arctic (100 000; majority) **Inuit**
<10 per cent hunting and fishing, the rest are largely urbanised

Europe (60 000; 0.1 per cent) **Saami**
<10 per cent reindeer hunting, the rest are largely urbanised

Canada (326 000; 4 per cent) eg **Blackfoot, Cree, Mohawk, Dene**
Mainly hunters with Mohawk and Iroquois now urbanised

USA (1.5 million; 0.5 per cent) eg **Apache, Navajo, Hopi, Sioux**
Mainly urbanised or peasant farmers

Central America (13 million; 1–50 per cent) eg **Maya, Mixe, Nahua, Miskito**
Mainly shifting cultivators and peasant farmers

South America (>15 million; 0.1–66 per cent) eg **Quechua, Kayapó, Yanomami, Guarani**
Mainly shifting cultivators, peasant farmers and hunter-gatherers

Africa (>25 million; majority) eg **Djibouti, Fulani, Maasai, Tuareg**
Mainly nomadic pastoralists and hunter-gatherers, plus some peasant farmers

South Asia (>100 million; majority) eg **Bhil, Vedda, Vasara**
Mainly peasant farmers, shifting cultivators and hunter-gatherers

Oceania (6.5 million; 2 per cent to majority) eg **Aborigine, Maori, Pacific islanders**
Many urbanised; others mainly shifting cultivators and hunter-gatherers

East Asia (>100 million; 1 per cent to majority) eg **Negriton, Karen, Hanunóo**
Mainly shifting cultivators and hunter-gathers, with some pastoralists and traditional farmers; others urbanised

USSR (1.4 million; 0.5 per cent) eg **Turkic, Kazakh**
Mainly either pastoralists or else urbanised

Data from Burger (1990).

chapter considers the nature of traditional knowledge pertaining to the use and management of wild plant resources, and, where possible includes conventional scientific evidence which supports the traditional practices observed.

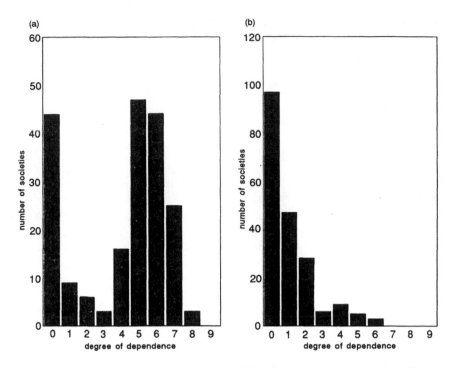

Figure 5.1 World dependence on wild plant resources. According to Murdock's *Ethnographic Atlas*, five categories of subsistence activity may be defined: gathering, hunting, fishing, herding and agriculture (Hunn & Williams 1982). From a representative sample of 200 distinct societies, Murdock estimates that (a) more than half are dependent on agriculture for at least 46 per cent (degree of dependence =5) of their subsistence needs; in contrast, (b) only about 10 groups depend on gathering (wild plants) for more than 46 per cent of their needs. (Degree of dependence: 0=0–5 per cent dependence; 1=6–15 per cent dependence; 2=16–25 per cent dependence; 3=26–35 per cent dependence; 4=36–45 per cent dependence; 5=46–55 per cent dependence; 6=56–65 per cent dependence; 7=66–75 per cent dependence; 8=76–85 per cent dependence; 9=86–100 per cent dependence)

NATURE, DISTRIBUTION AND PHENOLOGY OF WILD RESOURCES

Ethnobotanical research into modern modes of subsistence often reveals that large inventories of edible plants are recognised by traditional peoples. In many cases these inventories are supplemented by details of food storage or processing techniques, or else are expanded through empirical nutritional or economic analyses (Altman 1987; Kuhnlein & Turner 1991); in others, the local ethnoecological knowledge of factors such as spatial distribution or seasonal availability are considered. Whatever the specific approach adopted, most studies suggest that traditional botanical knowledge (TBK) relating to the use and management of wild plant resources is extensive, regardless of the dominant subsistence strategy employed.

Wild Plants as Food, Forage and Fuel

As devastating famine has affected many parts of the world in recent decades, interest in the wild plant resources used by traditional populations has escalated. For through the expansion of existing knowledge of food plants, it may be possible to develop the means to increase the availability of high-quality foods or forage—particularly in areas which are not well suited to cultivation. Equally, the frequent claim that the continued collection of firewood for local consumption constitutes a significant factor in environmental degradation, has motivated a number of studies into the exploitation of traditional sources of fuel.

Range and Quality of Wild Plant Foods

Within the last decade, there has been considerable debate regarding the range of wild food plants recognised by traditional horticulturalists compared with hunter-gatherers. Some workers believe that horticultural societies recognise more species than comparable hunter-gatherer groups, and suggest that while the wild resource base of the hunter-gatherers is likely to be well adapted to unpredictable climatic extremes, the relatively high risk of crop failure demands that farmers are familiar with a wide range of *emergency foods* (Brown 1985). However, as individual ethnobotanical studies vary considerably in their aims, methods and detail, it is virtually impossible to carry out quantitative comparisons of plant use on the basis of existing literature. The data presented here are not, therefore, intended as a detailed cross-cultural analysis, but merely illustrate the kind of data which are currently available.

It is revealed in Table 5.2 that the number of wild plant species and families which are represented in traditional diets can, in fact, be of about the same order of magnitude, whatever the local vegetation or dominant subsistence strategy. For example, the Seri hunter-gatherers of the Baja Californian Desert apparently recognise as many wild foods as the Amazonian Waimiri Atroari, who practise shifting cultivation; both groups collect foods from about 30 distinct plant families. Similarly, the Aborigines of the arid Western Desert of central

Table 5.2   Range of wild plant foods. The data presented here give some indication of the range of wild plant foods used by various traditional groups. It is clear that the numbers of wild plant species used as foods can be similar in both agricultural and hunter-gatherer groups. For example, the Seri hunter-gatherers rely on approximately the same number of species as the horticultural Waimiri Atroari. This indicates that traditional knowledge concerning wild food plants may be found across a whole range of subsistence groups (see [1]Felger & Moser 1985; [2]Smith & Kalotas 1985; [3]Milliken *et al.* 1992; [4]Schmeda-Hirschmann 1994; [5]Maikhuri & Gangwar 1993; [6]Densmore 1974; [7]Cane 1989)

Habitat and major food sources	No. of species	No. of families
Seri (Baja California, USA) *arid desert and coast* Rich and varied diet of marine and terrestrial animals; plant staples from several members of the Cactaceae plus seeds from a number of desert ephemerals[1]	84	35
Bardi (N. Australia) *tropical forest and coast* Approximately 50:50 animals and plants; maritime economy plus wild tubers such as yams (*Dioscorea* spp)[2]	63	35
Waimiri Atroari (Amazonia) *tropical rainforest* Hunting and swidden horticulture; main staple is cultivated manioc (*Manihot esculenta*)[3]	90	30
Ayoreo (Paraguay) *xeromorphic forest* Hunting, gathering and more recently the cultivation of about a dozen plants including maize (*Zea mays*) and manioc[4]	33	19
Khasi/Garo (NE India) *subtropical forest* Settled cultivators—dooryard gardens and swidden horticulture; cereals and tuber crops plus domesticated animals and fish[5]	45	30
Chippewa (Minnesota, USA) *mixed woodland* Subsisted largely on plants and fish with some hunting; staple plant—wild rice (*Zizania palustris*) plus cultivated maize[6]	39	ns
Western Desert Aborigines (Australia) *arid desert* Predominantly vegetarian, collecting a range of seeds and fruits, plus small mammals and more rarely, larger game[7]	54	ns

ns=not specified

Australia collect almost as many species as the Bardi Aborigines in the coastal forests of the tropical north despite profound differences in local vegetation. While it remains difficult to draw meaningful generalisations simply on the basis of the range of plants used, the examination of additional data can be revealing. Indeed, if the proportions of plants exploited for specific organs are compared (Table 5.3), it seems that while hunter-gatherers use a high proportion of storage organs from wild species, agriculturalists show a proportionally greater dependence on organs such as fleshy fruits and leaves. It is quite possible that these patterns of wild plant use are related to the nutritional characteristics of different plant parts—for example, while storage organs such as seeds and tubers generally provide high levels of energy, fruits and leaves are more likely to provide supplementary requirements such as vitamins and minerals (Table 5.4). However, the failure of many studies to discriminate between foods which are normally eaten and those which are consumed only infrequently, can severely limit the validity of such comparisons.

In many societies, a large number of edible plants, are regarded as *emergency* or *famine foods*, and although normally contributing little to the overall diet, these species can play a critical part in the survival of both individuals and entire communities during periods of food scarcity. According to Turner and Davis (1993) famine foods may be broadly divided into four main categories: those which are normally eaten, but which become more important during times of stress; less preferred foods, which are used only very occasionally under normal circumstances; starvation foods, which are eaten only during periods of stress; and hunger suppressants and thirst quenchers, which are used to alleviate short periods of food and water deprivation. A recent study of the famine foods used by indigenous peoples of northwestern North America has demonstrated the use of more than 100 species of higher plant, of which about 5 per cent are strictly starvation foods (Table 5.5). Plants recognised as sources of famine food have also been reported recently among both horticulturalists such as the Brazilian Waimiri Atroari (Milliken *et al.* 1992) and pastoralists such as those of Saharan and sub-Saharan Africa (Harlan 1989).

Processing and Storage of Wild Foods

Whether plant foods are collected from the wild or produced by cultivation, many must be processed prior to consumption. Food preparation often involves techniques such as grinding, soaking or heating, all of which are fairly straightforward yet may have a significant affect on nutritional content. For example, reduction in particle size through grinding or pounding and chemical changes induced by cooking can increase digestibility, while foods roasted directly in hot ashes can contain increased levels of essential elements such as iron and calcium

Table 5.3 Plant parts exploited. Although the range of wild plant species used by different groups may be similar, these data suggest that the plant parts used may differ according to a group's relative dependence on wild resources. For example, while most of the groups shown here gather a high proportion of plants for their fruits, it is largely hunter-gatherers who exploit a high proportion of *wild* plants for storage organs such as seeds and/or roots (see [1]Felger & Moser 1985; [2]Smith & Kalotas 1985; [3]Milliken *et al.* 1992; [4]Schmeda-Hirschmann 1994; [5]Maikhuri & Gangwar 1993; [6]Densmore 1974; [7]Cane 1989). This is presumably associated with characteristic differences in the nutritional content of different plant organs (Table 5.4)

Plant part	Seri[1]	Bardi[2]	Waimiri Atroari[3]	Ayoreo[4]	Khasi/ Garo[5]	Chippewa[6]	Western Desert[7]
Mode of production	Hunting/ gathering	Hunting/ gathering	Horticulture	Horticulture	Horticulture/ livestock	Hunting/ gathering	Hunting/ gathering
Fruit	63	57	88	67	31	49	37
Seed	30	13	6	0	2	5	30
Flower	13	0	0	3	9	5	2
Root	7	22	1	18	13	15	17
Other	20	18	4	6	62	32	15
Seed + root	37	35	10	18	15	20	47

Data represent the proportion of edible wild species used for specific parts.

Table 5.4 Nutritional content of plant organs. Although rather crude, these data—taken from a study of food plants used by traditional Khoisan people—give an indication of the nutritional characteristics of different plant organs. For example, storage tissues such as kernels, tubers and seeds show consistently higher energy content than other tissues in the same plant, while organs such as leaves, and the flesh of fruits may, in some species, accumulate relatively high levels of essential micronutrients (data taken from Arnold *et al.* 1985). To some extent at least, this may explain the pattern of plant organs used by groups with different modes of production

Species	Organ	Protein (g 100 g)	Energy (kJ 100 g)	Minerals (> 20% ADR)	Vitamins > 20% ADR)
Adansonia digitata	Fruit	2.7	1292	Ca, Mg, K	thi, nic, vit.C
	Kernel	33.7	1803	Ca, Mg, Fe, K, Cu, Zn	thi
Coccinia adoensis	Tuber	1.1	289		nic
	Fruit	1.8	112		vit.C
Cucumis africanus	Leaves	1.3	90	Ca, Fe	nic, vit.C
	Fruit	2.8	163		nic, vit.C
Citrullus lanatus	Flesh	0.4	60		nic
	Seed	17.9	1274	Mg, Fe, Cu, Zn	thi, rib, nic

ADR, average daily requirement; Ca, calcium; Mg, magnesium, K, potassium; Fe, iron; Cu, copper, Zn, zinc; thi, thiamin; rib, riboflavin; nic, nicotinic acid; vit.C, vitamin C.

Table 5.5 Famine foods in northwestern North America. In a recent study Turner and Davis (1993) interviewed indigenous elders and examined reports in the existing ethnobotanical literature to accumulate information regarding the historical role of famine and survival foods in the lives of traditional peoples of northwestern North America. They demonstrated that more than 100 species of plants were used under a variety of circumstances to alleviate hunger and to assist survival in times of food scarcity. As a result, these workers were able to describe four main categories of plant foods used in times of food scarcity: group 1—traditional plants which are used regularly, but are particularly useful as survival foods during periods of stress (in winter and early spring); group 2—alternative food plants which although used to some extent during normal circumstances make a minimal contribution to traditional diet except in times of stress; group 3—true famine foods which are eaten only in times of extreme hunger; group 4—hunger suppressants and thirst quenchers

Source	Group 1	Group 2	Group 3	Group 4
Inner bark	10	3	1	1
Lichens and algae	2	1	2	1
Green vegetables	20	2	0	10
Root vegetables	29	6	4	3
Fruits	4	28	1	4

(Stahl 1989). In some cases famine foods, and even certain staples are highly toxic due to the presence of secondary compounds, and often complex processing methods are necessary to render these plants edible. For example, the famine foods *Clathrotropis macrocarpa* and *Humirianthera rupestris* of the Waimiri Atroari both require processing to remove toxins, while several important food sources of the Australian Aborigines have proved fatal when foods have been prepared inadequately (Box 5.1).

Box 5.1 Plant processing in Aboriginal Australia. Despite recent changes in Aboriginal life, which have led to increased dependence on commercially available commodities such as flour, sugar and other processed foods, communities such as the Arnhem Land Gidjingali continue to prepare traditional bush foods, several of which are treated in order to remove toxins or bitter qualities. According to the archaeological record, the technology used in the processing of cycads has been present in Queensland for almost 5000 years, and while early European explorers such as Sir Joseph Banks became violently ill after experimenting with cycad kernels, recent studies have demonstrated that all cycad foodstuffs eaten by Aborigines, are free not only of the neurotoxins which produce a condition known in cattle as 'Zamia staggers', but also of the carcinogen cycasin (see Jones & Meehan 1989; Beck 1992).

Throughout their long history the Aboriginal peoples of Australia have depended on wild foods—and in particular on wild plants—to fulfil much of their subsistence needs. While the nomadic groups of the vast central desert relied fairly heavily on the seeds of wild grasses such as wild millet (*Panicum decompositum*) others, in many parts of the Australian continent, have shown considerable dependence on the underground parts of a range of species, including the roots of bulrush and bracken, and the tubers of daisy yams (*Microseris scapigera*), parsnip yams (*Dioscorea transversa*) and lilies (*Nymphaea* spp.) However, particularly during periods of seasonal stress, Aboriginal groups have often used food plants which, although rich in nutrients are inherently toxic, and require sophisticated processing which removes toxins while maintaining nutrient quality. Two such plants include certain varieties of yam (*Dioscorea* spp), and a number of cycads (*Macrozamia* spp and *Cycas* spp), the latter of which often contain potent neurotoxins and the insidious carcinogen *cycasin*.

CYCAD PROCESSING IN PREHISTORIC QUEENSLAND

From about 4500 years ago, the technology used in the processing of *Macrozamia* nuts and other cycads has existed in Queensland, in almost the same form in which it is applied today. *Macrozamia* is a genus of cycads which produces large fruiting bodies and whose 'kernels' are both highly nutritious and extremely toxic. Their food value is exceptionally high—about 43 per cent

continued

---

continued

carbohydrate, and 5 per cent protein, and with the careful use of fire, stands of cycads could be encouraged to yield more food per hectare than many cultivated crops (Flood 1983). Consequently, cycads were used not only during periods of stress, but also to support large gatherings of people on ceremonial occasions. Removing the toxins from cycad kernels was a lengthy and complicated process, with slightly different methods used in different regions. One method involved cutting open the kernels which were then soaked to leach out the toxic chemicals; another involved fermenting the dissected kernels in grass lined pits, over a period of up to 5 months. Following the detoxification, the flesh of the cycad kernel could then be ground into flour and used to prepare cycad bread which could be stored for several months (Jones & Meehan 1989).

RECENT STUDIES IN CYCAD PROCESSING

Evidence of present day use of cycad processing techniques is found among the Aboriginal populations of the Arnhem Land Aboriginal reserve. Here the kernels of toxic varieties of cycad (*Cycas* spp) are flattened and leached for a period of 24 hours, before they are pounded and leached for a further period of anything between 7 days and 5 months (Beck 1992). After the removal of toxins, the cycads may either be eaten raw or else may be ground into loaves which are baked in ashes. The toxins present in cycads include not only a range of neurotoxins which have a profound effect on behaviour, but also a carcinogen known as *cycasin* which constitutes the main hazard in toxic forms of *Cycas*. During a recent study, chemical analysis of all forms of cycad used as traditional foodstuffs has revealed that in all cases the samples of aboriginal foods assayed contained no cycasin— even where specific kernels which were considered to be non-toxic were eaten raw. This illustrates, both the efficacy of the cycad preparation process, and the ability of local people to discriminate between toxic and non-toxic kernels.

Food processing can also play an important part in food storage, a practice which is vital to both agricultural societies and *delayed-return* hunter-gatherers, who, unlike *immediate-return* groups rely on stored foods during at least part of their annual cycle. For example, to the Aborigines of the Western Desert the desert tomato (*Solanum chippendalei*) was an important food which not only contained appreciable amounts of energy and vitamin C but which could also be stored for several months following processing (Brand & Cherikoff 1985). Like processing techniques, storage methods vary in complexity, ranging from the use of pottery vessels to store dried foods, to the application of plant toxins which inhibit potential pests. Nevertheless, despite their considerable knowledge and ingenuity, the long-term preservation of many foods presents serious problems to many traditional peoples, particularly in the humid tropics where damage due to water, insects, birds and disease remains a common threat.

Wild Plants as Animal Feed and Fuel

Although food plants dominate much of the ethnobotanical literature, wild resources are often equally important both in feeding domesticated animals and in attracting game animals. Indeed many traditional groups demonstrate a considerable knowledge of both the nutritional quality of different grazing or forage species and the ecological interactions between particular wild species of plants and animals. For example, in Ghana more than 300 species of shrubs, grasses and other herbaceous plants—most of which grow wild—have been used as animal feed (Abbiw 1990), while the Wodaabe Fulani pastoralists of Niger avoid grazing areas where the plant species present provide insufficient vitamin A (see Gupta & IDS Workshop 1989). Similarly, the *caboclo* farmers of Lake Coari in Brazilian Amazonia, recognise about 70 plant species which are associated with specific game animals (Frechione *et al.* 1989), while traditional peoples in many parts of the world are familiar with tree species which, through their association with local bee populations, provide particularly good sources of honey (Crane 1985). Wild plants also often represent the only source of fuel available to traditional societies, and although the majority of plant species can, theoretically, be used as fuel, many species are recognised for particular burning qualities. In particular, those, such as the South American species of *Ryania*, which produce toxic smoke are carefully avoided (Milliken 1992), while many communities choose certain plants for specific purposes on account of their characteristic burning properties. Examples of this have been described recently among the hunter-gatherer Seri and the horticultural Waimiri Atroari (Table 5.6).

Distribution and Behaviour of Wild Plants

While the recognition of edible species is clearly fundamental to the successful exploitation of wild foods or fuels, equally important is a detailed knowledge of the spatial and temporal availability of wild resources. For example, pastoralists and other nomadic peoples can make vital subsistence decisions only on the basis of accurate knowledge of both the geographical distribution and seasonal availability of essential natural resources. Studies into the nature of this complex area of TBK have been helped enormously by the construction of *ecological maps* and *economic calendars*—diagrams of resource availability which are produced in collaboration with local specialists. For example, work carried out with a group of Somali pastoralists in northeast Kenya has produced detailed resource maps, which indicate the location of water

Table 5.6 Burning properties and the use of plants as fuels. While most woody species could be used as a source of fuel, evidence from several indigenous groups shows that often particular species are used preferentially for specific purposes. In some cases, local preferences have already been explained empirically—for example species used as kindling are highly inflammable due to presence of high concentrations of resins; equally, woods which are strictly avoided have often been found to produce toxic fumes on burning (see [1]Felger & Moser 1985; [2]Milliken et al. 1992; [3]Abbiw 1990). Clearly, local people know much about the different properties of potential sources of fuel, a skill which has been recognised in recent years. Indeed, in one OXFAM project based in the Upper Volta, collaboration with local women (who were responsible for fuel collection) facilitated the selection of nine plant species which were regarded as particularly useful on the basis of characteristics including not only burning characteristics, but also growth rate, ease of access and grazing deterrence (Cherry 1985)

Group	Species (and characteristics)
Seri[1]	*Bursera* spp (resinous), used as kindling
Seri	*Bursera microphylla*, used for firing pottery
Seri	*Monanthochloe* sp, used for smoke signals
Waimiri Atroari[2]	*Ferdinandusa* sp (burns well even when freshly cut), used in wet weather
Waimiri Atroari	*Ryania* spp (produce toxic fumes), not used
Waimiri Atroari	*Protium* spp (resinous), used as kindling
Waimiri Atroari	*Eschweilera rhododendrifolia*, used for cooking manioc
Ghana[3]	*Celtis* spp (burns with hot flame and little smoke), is the preferred fuel for many

sources and the distribution of different types of soil and vegetation (Gupta & IDS Workshop 1989); elsewhere seasonal diagrams have been produced, covering all the major subsistence activities and ecological changes that occur within the economic year (Conway 1989). Although these techniques have been used most frequently in the analysis of traditional farming systems, studies into wild resources have recently demonstrated their value in uncovering local knowledge of the spatial distribution and seasonal behaviour of important wild plant and animal species.

Spatial Distribution of Wild Resources

Traditional knowledge of wild resource distribution is often both sophisticated and complex, and allows the recognition of characteristic *resource units* or *ecological zones*, which are normally based on specific environmental indicators. For example, the Bardi Aborigines of northwestern Australia recognise a total of 12 distinct ecological zones based on the characteristics of local plant communities, while the Amazonian *caboclos* around Lake Coari distinguish about 40 resource units according to their characteristic floral and faunal composition and to other environmental criteria including soil type and seasonal variations in water level (Frechione *et al.* 1989). Many natural resource units are also characterised by the presence of particular plant species as in the *buritizal* and *açaizal* around the Amazonian Lake Coari, which are recognised by local *caboclo* farmers on the basis of their concentrations of the plants *Mauritia flexuosa* and *Euterpe oleracea* respectively. The Lake Coari *caboclos* further divide their horizontally defined units into a range of vertical levels, which can be either terrestrial/aboreal or aquatic. Like most of the resource units recognised, many of these vertical levels are associated with particular faunal resources and are exploited very specifically in hunting and fishing (Box 5.2).

Box 5.2   Natural resource units in the Brazilian Amazon—the recognition of ecological zones by the *caboclo* peoples of Lake Coari. While the recognition of wild resources is clearly fundamental to their efficient use, a knowledge of resource availability is equally essential where societies remain dependent on wild species to fulfil their subsistence needs. In this example, we see how the *caboclo* inhabitants of Amazonia distinguish a complex mosaic of natural resource units, whose occurrence is associated with the location of economically important plant and animal species. Hence, the Lake Coari *caboclos* are able to predict the distribution of important game animals or plant resources on the basis of the distribution of at least 40 distinct types of natural resource unit.

The *caboclo* cultures of South America are characterised by rural communities and represent a combination of Amerindian and colonial elements, yet although the culture of the Amazonian *caboclos* developed as a consequence of European colonisation, they remain profoundly influenced by their indigenous ancestors, and their patterns of forest resource use resemble, very closely, those of existing indigenous cultures. In the Lake Coari region of the Brazilian Amazon, *caboclo* inhabitants prossess an extensive folk knowledge of the ecology of the Lake, recognising a complex mosaic of 40 resource units according to diagnostic floral, faunal and other environmental characteristics (Frechione *et al.* 1989).

ECOLOGY AND SUBSISTENCE IN THE LAKE COARI REGION

To the horticultural *caboclos* of Lake Coari, life is determined by the high/low water changes of the Coari River system which consists of three blackwater rivers. These rivers join to produce Lake Coari, which in turn flows into the white-

continued

— continued —

water Amazon River. Different resources are associated with each river type, and are profoundly influenced by the dynamics of the river system itself. Flooding is usually predictable, with the maximum flood level occurring in June, while the lowest water levels occur in October. However, every 7 to 10 years, unusually high flood waters occur which affect both floral and faunal distributions and human subsistence activities. For example, despite the relative fertility of the lower alluvial deposits, *caboclo* farmers do not plant all their crops in these areas, as severe, erratic flooding could prove devastating here. On the other hand, crops planted on lower lands may result in significantly increased productivity due to the soil fertility; hence successful *caboclo* farmers are those who can predict accurately, when devastating floods will arrive—allowing them to maximise productivity without the risk of crop losses.

RESOURCE UNITS AROUND LAKE COARI

In a study based on information provided by Dr Francelino da Silva, himself a *caboclo*, 40 distinct resource units (*lugares de fartura*) were recognised and named on the basis of a range of environmental characteristics. For example, *praia branca* describes the white sandy beaches of Lake Coari, which provide a habitat for birds and turtles during the dry season; *praia suja* describes another dry season habitat, where wet or muddy beaches provide a feeding ground for a great number of birds. Further examples of these units are shown in the table below:

Resource unit	Ecological characteristics	Resource unit	Ecological characteristics
Praia verde	Dry season beaches, short vegetation with birds feeding on weeds and insects	*Embaubal*	Section of *várzea* where *embaúba* trees are predominant
Restinga	Natural river levees, forested and remaining exposed during the dry season	*Buritizal*	Concentration of buriti palms (*Mauritia flexuosa*)
Chavascal	Transition area between Coari rivers and the lake itself; low vegetation which is mostly inundated during the wet season	*Jauarizal*	Concentration of *jauri* trees (*Astrocaryum jauari*) usually at critical zones during periods of medium water level
Igapó	Forest area flooded during the height of the wet season	*Tabocal*	Concentration of green-and-yellow bamboo (*Guadua* spp) in the high *várzea*
Castanhal	*Terra firme* forest where the *castanheiras*	*Mariruzal*	Floating meadow dominated by *muriru*,

— continued —

_____ continued _____

Resource unit	Ecological characteristics	Resource unit	Ecological characteristics
Castanhal (continued)	(Bertholletia excelsa) are located	Mariruzal (continued)	providing food for fish and turtles

In addition to these horizontally defined resource units, the caboclos also recognise a series of 10 vertical levels of which five are terrestrial/aboreal (T1–T5), and five are aquatic (A1–A5). For example, while the middle canopy (7–15 m above ground), which occurs in most mature forests is the principal zone for aboreal mammals and large birds, the high canopy (15+ m above ground) is regarded as difficult to hunt because of its height, and is more useful for forest products such as nuts and honey. Just as particular horizontal units are associated with certain types of animal, so too are these vertical components. For example, the sloth (Choloepus didactylus) is located primarily in the understory (T3) zone of the igapó forest area, while important fish species Geophagus sp and Lecostomus emmaginatus are both found at the river/lake bottom (A5) zone in protected pockets of the lake system.

Seasonal Behaviour of Wild Resources

While ecological mapping can provide useful information on the location of specific plants or plant communities, the construction of economic calendars can reveal extremely detailed local knowledge of both climatic factors, and of plant and animal *phenology*. For example, during interviews in Wollo, Ethiopia, two informants were able to recall the total number of days in which rain fell for each month of the previous 5 years (see Conway 1989), while according to Smith and Kalotas (1985), the Bardi Aborigines of northern Australia note that the fruiting of *Avicennia marina* (*ngurngulu*) indicates the arrival of the mosquitoes and marks the beginning of the season for bush fruits such as the nuts of *Pandanus* spp (*idul*). This type of detailed ecological knowledge is seen clearly in the economic calendars of many traditional communities including the Lake Coari *caboclos* who are able to organise very specific subsistence activities on the basis of their understanding of complex, seasonal interactions which occur between the wild species of distinct resource units (Box 5.3).

MANAGING WILD RESOURCES: ETHNOECOLOGY IN AFRICA AND AUSTRALIA

As we have seen, the development of sophisticated ethnobotanical and ecological knowledge among traditional societies, is fundamental to the

Box 5.3   Temporal changes in wild resource availability. In addition to the physical distribution of distinct resource units, the *caboclo* farmers of Lake Coari are aware also of temporal changes in resource availability. These temporal variations occur both daily, and on a more seasonal basis, the latter of which are influenced largely by changes in water level. Through understanding both these fluctuations in resource availability, and the complex interactions between the flora and fauna of distinct resource units, local people are able to organise their economic activities effectively on both a spatial and temporal basis (data modified from Frechione *et al.* 1989).

One important consideration associated with the successful exploitation of the wild resource units recognised by the Lake Coari *caboclos*, is that each of its characteristic floral or faunal species may be optimally exploited only at certain times of year. Hence, the local *caboclo* populations must recognise not only *where*, but *when* important resources are available. Like the resource units themselves, the seasonal availability of many wild species is profoundly influenced by the behaviour of the river system, and as a result, changes in water level provide an important indicator of seasonal events.

WATER LEVEL AND RESOURCE EXPLOITATION AROUND LAKE COARI

As the distinct resource units around Lake Coari are periodically altered by fluctuating water levels, so too are the species found in them. For example, while areas of *praia branca* are inundated during periods of high water level, during the dry season these areas are exposed, providing rich habitats for various bird, insect and reptile species (Frechione *et al.* 1989). Many of the other recognised resource units are similarly found only during periods of low water, as illustrated in the table below.

Resource unit	Characteristics
Praia suja	Dry season, wet or muddy beaches; habitat for birds and insects
Praia verde	Dry season beaches covered in short vegetation; habitat for mammals, birds and insects
Restinga	Natural river levees covered in forest; habitat for animals, birds, insects and reptiles
Laguinho	Small lake connected to river by narrow stream during rainy season; habitat for birds, fish, insects and reptiles
Canas seco	Navigable channel in the lake during low water; habitat for animals, birds, fish and reptiles
Pocinho	Small lake which does not dry during periods of low water, habitat for fish

Several other resource units, on the other hand, are exploitable throughout the whole sequence of changing water levels. For example, the *costa*—the bank of

continued

continued

the Amazon River—provides an all-year habitat for various species of animal, bird, fish and insect; similarly, the _castanhal_—areas of _terra firme_ forest where concentrations of Brazil nut (_Bertholletia excelsa_) are located—provides a home to animals, birds and insects throughout the annual cycle.

CHANGES IN WATER LEVEL AND ECOLOGICAL RELATIONSHIPS

In addition to understanding the effects of water level on individual resource units, local knowledge of changes in resource availability includes an understanding of the ecological relationships between the floral and faunal components of distinct resource units. For example, a close relationship exists between the _chavascal_—the transition area between rivers draining into Lake Coari and the lake itself—and the _castanhal_ (see Box 5.2).

Like the resource units themselves the relationships between distinct areas, are mediated largely by seasonal changes in water level. In the dry season, animals leave the _castanhal_ of the _terra firme_ forest, and migrate to regions of _chavascal_ (see Box 5.2) in search of water. Some small birds and mammals are attracted also to the dry season vegetation that flourishes in the damp beaches ringing the exposed _chavascal_; likewise wading birds are abundant as the shallow waters teem with small fish. However, during the rainy season, the _chavascal_ is flooded, and its resources become available to larger fish which migrate into the swollen rivers. _Caboclos_ at this time may exploit these larger fish using deep water fishing nets and other specialised technology. At this time also, _arati_ fruit are collected, while the primary activity is the collection of _castanhais_ (Brazil nuts), which are important both in subsistence and for selling in markets. In addition to these seasonal variations in the resources of the _chavascal_ and _castanhal_, local inhabitants also describe daily interactions between different resource units. For example, birds such as _massaricos_, _marrecos_ and _mergulhões_ move each day between the _chavascal_ and _castanhal_ in search of food and shelter.

successful recognition, location and utilisation of wild plants not only as food, fodder and fuel, but also as attractants for game or deterrents of pests. Yet developing patterns of resource use which are sustainable in both the short and long term is an equally important aspect of subsistence, and in recent years, several ethnobotanical studies have aimed to investigate this area of TBK. Through the examination of both the methods used in traditional ecological management, and the functional implications of these techniques, such studies have revealed a number of traditional methods which appear crucial to effective resource management, some of which are now being used in the conservation management of designated national parks in several parts of the world.

Modern Herders and Hunter-gatherers: Depending on Wild Plant Resources

Both the nomadic pastoralists of Saharan and sub-Saharan Africa and the Australian hunter-gatherers have depended on wild plant resources for millennia, their traditional subsistence strategies proving success-ful in some of the most hostile environments known. In both cases, sub-sistence is based on regular seasonal migration, which in turn is determined by the distribution and seasonal availability of wild plant foods or forage; and in both cases it is suggested that their nomadic life-styles offer a competitive advantage where environmental extremes and often poor soils might limit agricultural success (e.g. Scott & Gormley 1980; Keesing 1981; Flood 1983). However, in recent years, widespread controversy over the adverse environmental effects of traditional sub-sistence practices has prompted increasing research into the nature and consequences of the exploitation of wild plant resources.

Nomadic Pastoralists of Africa

Often adopting a mixed economy based on a combination of domesti-cated livestock and horticultural produce, the pastoral mode of sub-sistence is characterised by seasonal transhumance between winter grazing lands and summer settlements (Dyson-Hudson & Dyson-Hudson 1980). It is suggested that pastoralism represents an offshoot of early mixed agriculture in the Near East (Keesing 1981) and although increasingly marginalised by continuing agricultural expansion, this strategy still represents an important subsistence adaptation in many arid and semi-arid regions (Best & Baxter 1994). Today, pastoral soci-eties with varying degrees of dependence on supplementary crop cultivation are found in many parts of the world, some of the most widely studied being those such as the Fulani and Maasai of Saharan and sub-Saharan Africa (Table 5.7).

The success of the pastoralist life relies to a considerable extent on the conversion of low-quality grassland into dairy products, and the bulk of the pastoralist diet is derived from lifestock—predominantly from milk, but also from blood, meat and animal fat (Homewood & Rodgers 1991). Well aware of both their animals' nutritional requir-ments and the varying characteristics of pasture quality, the pastoral-ists diversify their herd to maximise productivity and to spread risk, while using their knowledge of the resources available to determine their movements. For example, both archaeological and contemporary evidence suggest that pastoralists optimise migratory routes through a

Table 5.7   Pastoralist communities of Africa. Of an estimated 24 million pastoralists throughout the world, almost half inhabit the marginal arid and semi-arid regions of the African continent. These traditional pastoralist peoples have herded their livestock for centuries, exploiting and managing the arid grasslands of Saharan and sub-Saharan rangelands. Through careful diversification of their livestock, and a detailed knowledge of wild resources, these nomadic peoples have survived in some of the harshest conditions known

Estimated population and distribution of pastoralist communities of Africa

Afar	(110 000)	Ethiopia, Djibouti
Dinka	(500 000)	Sudan
Fipa	(NS)	Tanzania
Fulani	(6 000 000)	Chad, Central African Republic, Cameroun, Guinea, Mali, Niger, Nigeria, Senegal
Maasai	(200 000)	Kenya, Tanzania
Nuer	(300 000)	Sudan
Oromo	(NS)	Ethiopia, Kenya
Somali	(2 000 000)	Ethiopia, Somalia, Kenya, Djibouti
Tuareg	(900 000)	Algeria, Libya, Niger, Burkina Faso, Mali

NS, not specified; data taken from Burger (1990).

combination of practices including visiting good quality pastures, avoiding wildebeest (which often carry diseases that are transmittable to livestock), and planting the seeds and tubers of edible plant species *en route* (Box 5.4). During dry periods, however, the herders' animals are often unproductive, and temporary food deficits are often made up through the consumption of grain, which may be either cultivated, exchanged with neighbouring agriculturalists, or else gathered from the 60 or so wild grasses which are harvested throughout Africa (Box 5.5).

Box 5.4   TBK and resource use among African pastoralist societies. Through their intimate understanding of both their domesticated livestock and the natural resources available, the traditional herders of Africa have been able to maximise nutritional returns through a combination of livestock diversification and seasonal transhumance.

Early observers of pastoral societies have often assumed that traditional herders had little control over either the composition of their herd, or the migration routes they used during seasonal transhumance to grazing pastures. For example, in 1956, Richard-Molard claimed the Fulanai peoples of Africa demonstrated an 'overwhelming passion for useless cattle' (Grayzel 1990). More recently, in the light of research which reveals the benefits of traditional herd diversification and

continued

___ continued ___

the optimisation of migratory routes, scientists are beginning to acknowledge the efficacy of the pastoralists' methods of herd and pasture management, and as early as 1977 the UN declared that traditional pastoralism represented perhaps the most efficient and sustainable method for converting rangeland vegetation into products of use to humans (see Scott & Gormley 1980).

ANIMAL NUTRITION AND HERD DIVERSIFICATION

Research in recent years has demonstrated that traditional pastoralists manipulate the composition of their herds in order to optimise the use of the often marginal resources available. For example, while animals such as goats are mainly dependent on browse species, those such as cattle and sheep are predominantly graze animals; equally, while goats drink frequently, camels can survive without access to water for several days. In addition, different animals vary in other respects such as their breeding cycles and susceptibility to pests or disease. Hence, through careful attention to herd composition, pastoralists are able to display flexibility in relation to resource use and adaptability to disaster (see Smith 1992).

Characteristic feeding requirements of domestic livestock

Camels — may constitute less than 3 per cent of domestic stock, yet will browse on undesirable species such as *Acacia* despite the presence of deterrent thorns, and can provide up to 60 per cent of the milk available during critical periods.

Sheep — depend on high energy graze species (herbs, grasses and vines) for about 74 per cent of their dietary intake, and on browse species (shrubs and trees)—which tend to have a higher crude protein content—for 21.6 per cent; they drink frequently.

Goats — depend on graze species for about 52.5 per cent, and browse species for about 47.5 per cent of their dietary intake; they drink frequently, but in small quantities.

Cattle — traditional breeds such as the indigenous zebu (*Bos indicus*) which compose most of the Maasai herd, are relatively tolerant of stress factors such as heat, low water availability, and demonstrate considerable levels of trek-hardiness, adaptability to low intake and disease resistance.

FORAGE QUALITY AND MIGRATION

In addition to their understanding of the differential feeding requirements of their livestock, traditional pastoralists of Africa—and elsewhere—possess also, an intimate knowledge of the grasslands available, avoiding areas where plants are toxic (Smith 1992) or nutritionally inadequate (Gupta & IDS Workshop 1989), or

___ continued ___

continued

where diseases are harboured, while ensuring access to salt licks and high-quality pasture. That traditional herders hold detailed TBK regarding both pasture quality and its seasonal availability has been illustrated in a study of Maasai classification of common grass species, in which Maasai rate grasses according to their palatability to ungulates. Some examples of these Maasai ratings are shown in the table below (Peterson & McGinnes 1979).

Species	Maasai name, palatability and other characteristics
Aristida spp.	_orkiriaan_—'useless' grass which cattle eat only when it is green; very susceptible to wind and trampling and cattle cannot eat it once it is lying flat.
Cenchrus ciliaris	_endiamoinwa_—'very good' for cattle, lasting through the dry season; especially common in drier areas.
Chloris pycnothrix	_engipumbu_—'useless' for cattle.
Chloris roxburghiana	_engaidosi_—medium quality grass.
Cymbopogon excavatus	_olkujita lorung'o jinia_—cattle 'hate' it.
Cynodon niemfuensis & _Cynodon plectostachys_	_emurruwa_—preferred by cattle as it remains green when other species have dried up; however, it is avoided in its early growth stages (from the beginning of the rains until January) as it is toxic at this time.
Digitaria macroblephara	_erikaru_—probably considered the best grazing grass throughout Maasailand.

Based on their knowledge of the nutritional and ecological characteristics of the grass species available, traditional pastoralists are able to optimise their seasonal migratory routes to ensure, as far as is possible, that the nutritional and other health requirements of both themselves, and their livestock, are met.

Hunter-gatherers: The First Australians

The first Australians apparently arrived from Indonesia about 40 000 years ago, and by the arrival of the European pioneers in 1788, Australia was inhabited by 500 distinct language groups, with an estimated total of about 500 000 people (Flood 1983). However, following the start of British colonisation, this population had fallen to 60 000 by the 1890s. Today there are approximately 250 000 Aborigines in Australia—about 2 per cent of the total population—most of whom now live in the country's towns and cities, yet in some regions a small number of people continue to live a nomadic or semi-nomadic life (Table 5.8).

Box 5.5 Pastoral diet and sources of wild foods. While pastoralist societies depend largely on their livestock to fulfil their dietary requirements, during seasonal periods of stress most become increasingly dependent on grain. Some groups obtain grain either through cultivation, or through exchange with neighbouring agricultural communities; others, however, remain dependent on harvesting wild grasses, and exhibit a detailed knowledge of the nature and availability of important grass species.

Among pastoral nomads such as the Tuareg of Africa, the annual transhumant cycle is determined by seasonal changes in the distribution of natural resources available—changes which are strongly reflected in the pastoralist diet. For while much of their diet is derived from milk and blood products from their livestock, during periods of seasonal stress, many pastoralists become increasingly dependent on grains, either from cultivated crops, or wild grasses.

SEASONAL CHANGES IN PASTORALIST DIET

During the wet season, domesticated livestock can graze freely on high-quality pasture, producing large quantities of milk; during drier periods, however, grain consumption increases considerably, as grasslands dry up and both cultivated cereals and wild grasses produce mature seed (see Smith 1992).

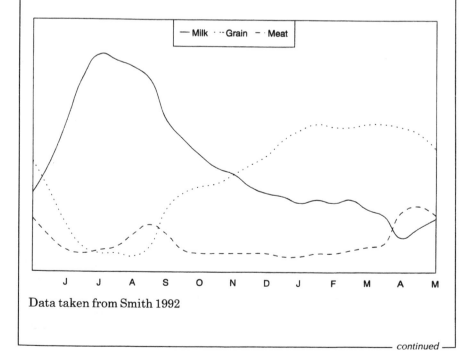

Data taken from Smith 1992

continued

___ *continued* ___

SOURCES OF WILD GRAINS

Towards the end of the rainy season, the Tuareg begin to harvest wild grains, several of which grow in large stands facilitating large scale harvesting. The characteristic of some of the grass species collected are outlined in the table below.

Grain species	Characteristics
Aristida pungens	Found in deserts occupied by northern Tuareg—often the only food available during dry periods.
Panicum turgidum	Drought resistant grass found in the central Sahara; leaves highly palatable for grazing.
Cenchrus biflorus	One of the most nutritious cereals containing 21 per cent protein; occurs in the Sahel in pure stands of up to several square kilometres.
Krebs complex:	
Panicum laetum *Eragrostis* spp *Dactyloctenium* sp *Brachiaria deflexa*	The krebs complex includes a range of grass species which are given the collective term *kreb* and extend over a considerable region in parts of Nigeria and Sudan, and have been an item of extensive trade in local markets.
Oryza barthii	The progenitor of domesticated African rice, this species is found in shallow water holes and lakes; it seeds abundantly and is still harvested on a massive scale.
Echinochloa stagnina	Extremely important source of food in Niger delta and shores of Lake Chad; may occur in massive, nearly pure stands; also a source of sugar, and provides high quality fodder as flood waters recede.

Of these wild grass species, *Panicum* is perhaps the most important to the Tuareg, being not only the most abundant, but also the first to ripen, while requiring very little preparation (Smith 1992). Significantly, some of these wild grass species can prove more nutritious than domesticated wheats (Harlan 1989), and can occur in densities which produce grain yields per hectare which are similar to those obtained by many subsistence farmers.

In spite of sustained contact with horticultural peoples in Papua New Guinea and Southeast Asia, the Aborigines of Australia have never adopted agricultural methods (Chase 1989; Yen 1989), but have demonstrated a continued dependence on wild plant foods. Indeed wild relatives of species which have been domesticated elsewhere—including *Dioscorea, Ipomoea* and *Oryza* spp—have often provided the indigenous

Table 5.8 Aboriginal communities of Australia. Hunter-gatherer societies remain in many parts of the world, particularly throughout Arctic and sub-Arctic regions, as well as in much of Asia, South America, and parts of Africa. However, the Aboriginal peoples of the Australian continent represent perhaps the most well known—and among the most widely studied—groups of hunter-gatherers. Today, relatively few Aborigines remain, their numbers having reduced drastically since the arrival of Europeans; nevertheless, in certain areas, some remote communities continue to value and retain much of their traditional cultural and environmental knowledge (see [1]Cane 1989; [2]Williams 1982; [3]Goodale 1982; [4]Smith & Kalotas 1985; [5]Jones & Meehan 1989; [6]Sutton & Rigsby 1982)

Current Aboriginal language groups and their distribution	
Pintubi and **Kukatja**	Remaining groups left the Great Sandy Desert of NW Australia in 1960s; began to return in 1980s.[1]
Yolngu	About 900 people in Yirrkala, Gove Peninsula (NE Arnhem Land, Northern Territory).[2]
Tiwi	Over 1000 Tiwi-speaking people are believed to live on the Melville and Bathurst Islands of northern Australia.[3]
Bardi and **Djawi**	Between 400 and 600 Aboriginal people live in the Dampierland Peninsula of NW Australia.[4]
Gidjingali	The Gidjingali live in the Aboriginal lands of Arnhem Land in Australia's Northern Territory.[5]
Yir Yoront	The Yir Yoront live in the Cape York Peninsula of Northern Queensland.[6]

Data taken from Burger (1990).

Australians with a major source of carbohydrate during seasonal scarcities of game animals or fish. Today, traditional foods remain important, not only as a source of nutrition, but as an integral part of the Aboriginal culture, and considerable local prestige is accrued to women who produce large quantities of well-prepared bush foods (Jones & Meehan 1989). Hence, even where commercial foods now provide a considerable proportion of the Aboriginal diet, indigenous vegetable foods are still gathered on a regular basis and play an important part in affirming traditional cultural identity. As a result, despite enormous changes imposed on many Aboriginal societies, some at least have managed to maintain their traditional knowledge regarding the nature and availability of wild resources. For example, among the Gidjingali in Australia's Northern Territory, bush foods continue to make a significant contribution to the local economy such that much of the TBK regarding wild foods has been retained (Box 5.6).

Box 5.6   Wild food use among the Arnhem Land Gidjingali. Nowadays, flesh foods contribute about 50 per cent of the Gidjingali diet, with about two-thirds of the remainder (33 per cent) provided by processed flour and sugar. The remaining 17 per cent of the diet is provided by wild plants, the gathering and processing of which are still important, and take place almost daily (Jones & Meehan 1989).

Within their Arnhem Land home, the Aboriginal Gidjingali occupy the flood-plains of a large coastal river, the Blyth, where they exploit riverine wetlands, coastline and river edge, as well as the ecualypt forests which grow on the surrounding low hills (Jones & Meehan 1989). Until the mid-1950s the Gidjingali subsisted entirely through hunting and gathering for food, and while access of social security now allows the purchase of processed commodities such as flour and sugar, much of this hunting and gathering activity continues today.

WILD PLANT FOODS OF THE GIDJINGALI

In the recent past—and to the present day—the Gidjingali have gathered four main genera of carbohydrate—*Nymphaea*, *Dioscorea*, *Eleocharis* and *Cycas* (Jones & Meehan 1989). Several of these, including certain varieties of *Cycas* and *Dioscorea* are toxic, and are treated to remove toxins or bitter qualities. At certain times of year, favoured staples become unavailable, and at such times are replaced by species such as *Pandanus* which is exploited for its 'kernels', the palm *Livistonia humilis* which is used for its pith, and the Polynesian arrowroot (*Tacca leontopetaloides*) whose tubers are eaten. In addition to these high-energy staples, a wide variety of tree fruits are eaten, including various species of *Terminalia*, *Canarium*, *Syzygium* and *Ficus*. Characteristics of some of the more commonly used foods are outlined in the table below.

Major plant foods	Nutritional and ecological characteristics
Dioscorea spp	Found in the shade of vine thickets yams are harvested to maximise regrowth in the following year; *D. bulbifera* is carefully treated to remove toxins.
Cycas media	The 'nuts' of this species require careful treatment to remove potent neurotoxins; used to make a bread called *ngatjo*, which was a staple food towards the end of the dry season.
Nymphaea spp	Waterlilies have provided a major source of carbohydrate during the later stages of the wet season; the stalks were eaten fresh, while the seeds and corms were processed prior to eating.
Eleocharis dulcis	The corms of the swamp rush were collected by women during the early dry season.

continued

— continued —

Major plant foods	Nutritional and ecological characteristics
Cyperus spp and *Vigna* sp	The roots of these less favoured foods were also collected, particularly where limited supplies of fresh water might have restricted cycad processing.

SEASONAL PATTERN OF EXPLOITATION

Like other traditional communities, much of the success of the Gidjingali is dependent on their detailed knowledge of the distribution and availability of important plant foods, with the seasonal nature of plant exploitation forming an important feature of the communities' subsistence activities. Yet while during much of the year, carbohydrate can be obtained from a range of plant species, during the late dry season, when the harvesting of *Eleocharis* corms becomes difficult, cycad nuts are collected and used to prepare *ngatjo* (cycad bread); this can be stored for some time to provide sustenance during periods of stress which precede the onset of the summer rains.

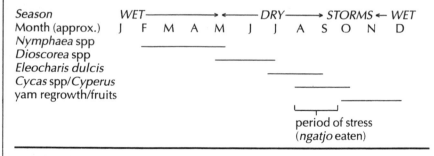

The figure above illustrates the major plant sources exploited at different times during the Gidjingali economic year, showing the dependence on species such as the toxic *Cycas*, at certain times. However, this plant diet was normally supplemented by additional foods including geese, which were hunted by the men during May and June, as well as a range of fish, reptiles, wallabies and shellfish.

Environmental Manipulation: Methods in Wild Resource Management

While early colonial observers have generally regarded nomadic peoples as essentially opportunistic with neither motivation nor strategy for environmental protection (see Lamprey 1983), more recent research suggests otherwise. Indeed, a range of botanical, archaeological, historical and ethnographic evidence demonstrates that both

Australian Aborigines and African pastoralists—as well as other nomadic peoples—use a number of management practices in the manipulation and conservation of their wild resources. In Australia, extensive evidence of traditional land management practices can be found in the historical records of the early settlers, and in many cases this is supported by additional archaeological and botanical data. For example, the intensive management of resource patches dominated by *Dioscorea hastifolia* and other tuberous plants has been reported in Western Australia (Hallam 1989). Here, both documentary and archaeological evidence suggest that traditional techniques used in the harvesting, replanting and protection of yams may have facilitated the maintenance and proliferation of this valuable food source. Among traditional pastoralists too, it is evident from more recent observations, that many have developed elaborate management practices and social institutions which protect their renewable natural resources. For example, throughout the Middle East, the traditional communal property system known as *hima* has probably been developed as a means of conserving the limited natural resources of the desert regions of the Arabian peninsula (Shoup 1990). Similarly, in Africa, traditional methods of grassland management, including practices such as controlled burning and the annual protection of emergency grazing land, seem to encourage the development of high-quality pasture (Homewood & Rodgers 1991).

Fire Technology and Resource Management

One of the most significant and widespread practices used in traditional resource management, is the use of fire. In recent years this has become one of the most widely studied aspects in the traditional management of wild resources, and this subject is now well represented in the archaeological, ecological and enthographic literature (e.g. Lewis 1982; Flood 1983; Concar 1994). Often regarded in the past as uncontrolled and destructive (see Flood 1983; Homewood & Rodgers 1991) it now seems likely that controlled burning practices may actually have an important part to play in modern ecological management, as recent scientific experiments have begun to demonstrate important functional consequences of complex fire management techniques (Concar 1994).

Fire has been used within traditional societies for many reasons, playing a vital part in communication, warfare and hunting game; it has also been used in manipulating the distribution, diversity and relative abundance of certain plants, in controlling interactions between plants and herbivorous animals, and in eradicating harmful pest and disease species (Lewis 1982, 1989; Homewood & Rodgers 1991). For example, in the Australian desert, careful firing of important seed-producing staples such as *Fimbristylis oxystachya* and *Panicum australiense* has encouraged the formation of extensive 'natural' stands, allowing the bulk harvesting of these foods; similarly in the rangelands of Tanzania, careful burning of dry, coarse grass, has proved effective in destroying dormant and free-living stages of livestock parasites while stimulating the regrowth of palatable green shoots (Homewood & Rodgers 1991).

However, while fire tolerant species may benefit from regular burning, important fire-sensitive communities must be protected from damage. For example, the protected areas of Arnhem Land in the Northern Territory of Australia continue to support residual pockets of rainforest vegetation, which represent both an important source of edible yams and a vital shelter for game animals such as kangaroos and wallabies (Lewis 1989). The need to protect these vital natural resources is well understood by local peoples, and traditionally, only burning practices which ensured the protection of these areas have been employed (Box 5.7). Clearly then, far from being haphazard or random, the successful use of fire requires both a sophisticated knowledge of the behaviour of fire itself, and a detailed understanding of the life cycles and habitats of local plants and animals.

Box 5.7   Traditional burning in Aboriginal Australia. Recent studies into traditional burning practices have revealed a considerable range of ecological functions which are fulfilled by the judicious use of fire. For example, while fire-resistant fodder or food species are encouraged and their quality improved, detrimental factors such as pest populations and the encroachment of bush can be limited. In addition, many of these studies have revealed a profound underlying knowledge of both the behaviour of fire, and its ecological consequences, which results in the effective management of many wild resources.

Growing interest in, and research into Australian Aborigines' use of fire, has produced a body of information which suggests that Australia's indigenous communities have, for millennia, used fire to influence local environments. For example, Hallam's early study *Fire and Hearth* discusses the burning practices used in Western Australia (see Lewis 1982), while more recent studies have recorded details of the environmental impact of Aboriginal fire-use (see Baker *et al.* 1992).

continued

— *continued* —
ECOLOGICAL FUNCTIONS OF FIRE

Far from the haphazard burning of vegetation, used simply in activities such as the hunting of animals, communication or warfare, Aboriginal burning practices fulfil a variety of ecological functions which are fundamental to Aboriginal subsistence.

- Stimulating the growth of forage species used in attracting game
- Extending the growing season of forage species
- Stimulating the growth of fire-resistant food staples such as *Fimbristylis oxystachya* and *Panicum australiense*
- Increasing the palatability and availability of grass species
- Synchronising the flowering and fruiting of important sources of plant foods such as cycads
- Destroying parasites and disease vectors
- Preventing the encroachment of bush

(See Lewis 1982; Flood 1983; Cane 1989; Hallam 1989; Yen 1989.)

MANAGING THE USE OF FIRE

However, while the benefits of controlled burning seem clear, it is equally important to consider factors such as the protection of fire-sensitive plant and animal species. Indeed, detailed investigations of traditional burning practices have revealed numerous factors which are considered when carrying out controlled burns.

- Protection of fire-sensitive resource areas
- Life cycles and habitats of local fauna
- Effective use of fire breaks and back fires
- Regeneration rate of local vegetation
- Vegetation structure and condition of grazing species
- Timing of burns—on both a seasonal basis and in terms of burning frequency
- Factors such as wind, temperature, dew point, relative humidity, time of day, size of area, slope and season—all of which can affect the temperature and spread of a given fire

(See Lewis 1982; Hallam 1989.)

Ecological Study of Traditional Resource Management

The grassland savannas of East Africa—the traditional rangelands of many African pastoralists—are commonly renowned for their enormous diversity of wildlife, and particularly for their wide range of large ungulates such as wildebeest (*Catoblepas gnu*) and gazelles (*Gazelle* spp). However, in recent years, these protected areas have attracted

considerable attention from conservationists who argue that the traditional herding practices of pastoralist peoples are detrimental to wild
populations—that livestock and ungulates compete for grazing and that
overgrazing and trampling are leading to ecological degradation (see
Homewood & Rodgers 1991). For example, it has been claimed that both
the grazing of livestock and the firing of pastures have caused a reduction in the area of forest and woodland, while leading to increased soil
erosion and the spread of *Eleusine* sp, an unpalatable tussock grass
which is unsuitable for grazing. In response to such claims, several
national parks including the Serengeti, Amboseli, Tarangire, Maasai
Mara, Nairobi have now excluded indigenous pastoralists from their
traditional lands (Sindiga 1984).

However, while there has been little hard evidence to support many of
these claims, more recent investigations into the ecological impact of
herding practices suggest that traditional methods of pasture management may, in fact, be *beneficial* to maintaining these important habitats
(e.g. Coughenour *et al.* 1985a,b). For example, it has been demonstrated
that high-quality grasses such as *Andropogon greenwayii* and *Themedra
triandra*, which provide an important food source to wild ungulates, are
stimulated by moderate defoliation due to livestock grazing and may
even be dependent on dry season burning for their survival (Tainton *et
al.* 1977; Coughenour *et al.* 1985b). In addition, not only have hazards
such as ticks and the incidence of severe wildfires been shown to
increase dramatically as controlled burning is abandoned, but it has
also been observed that wild animal populations actively avoid the
unpalatable grasses of unmanaged pastures (Norton-Griffiths 1979).

A similar situation has also been seen in Australia, whose unique
flora and fauna includes a large number of endemic and increasingly
rare species. With about two-thirds of its land mass currently receiving
less than 50 cm of rain annually, Australia represents the world's most
arid continent (Brand & Cherikoff 1985), and its frequent droughts, high
temperatures and wildfires have all contributed to the evolution of the
fire-tolerant grasslands and *Eucalyptus* woodlands which dominate
much of the landscape. However, equally important in influencing the
Australian environment, are the Aboriginal peoples, whose land management techniques date back several tens of thousands of years. For
example, digging sticks dating back 10 000 years have been recovered
from *Wyrie Swamp* in South Australia, while habitat manipulation
through the use of fire is apparent in the fossil record from at least 25 000
years ago (Flood 1983).

Unfortunately, as the numbers of Aboriginal people have fallen and
traditional management practices have declined, scrublands, bush and

forest have invaded extensive areas of grassland crucial to many grazing animals (Flood 1983). Meanwhile, overgrown habitats with a high buildup of plant litter have proved susceptible to massive conflagrations which are both difficult to manage and much hotter than controlled fires. Such uncontrolled burning is not only extremely dangerous to human populations, but can cause irreversible damage to fire-sensitive species which cannot tolerate intense heat. For example, the survival of the cypress pine *Callitris tropicana*—an ancient relic of a colder wetter era in Australia's prehistory—is now dependent on cool burning early in the dry season, as more intense wildfires later on can prove fatal. Similarly, the survival of a number of endangered animals, including the magpie goose (*Anseranas semipalmata*), the partridge pigeon (*Petrophassa smithii*) and pale field rat (*Rattus tunneyi*), is also increasingly jeopardised as traditional management practices are replaced by a growing number of wildfires (Concar 1994).

In the light of these and similar findings elsewhere, interest in traditional management practices has increased enormously. Since the 1980s, a large-scale experimental burning project has been underway in Australia (see Concar 1994), while earlier research in Africa has already revealed that the prevention of livestock grazing, use of fires, or other traditional activities may be highly detrimental to the long-term security of wildlife habitats such as the Serengeti grasslands (Norton-Griffiths 1979). Consequently, habitat managers in some parts of the world are beginning to look towards traditional methods to assist in the effective management of protected habitats, particularly in Australia where controlled burning has been reintroduced into national parks such as the Kakadu National Park of the Northern Territory, which has recently been designated a World Heritage Site. These potential applications of traditional management methods are discussed further in Chapter 11.

SUMMARY

Throughout the world today, millions of traditional peoples continue to depend, at least to some extent, on wild plants as sources of food, fodder and fuel. Among these communities, knowledge relating to these plants includes not only an awareness of characteristics such as their toxicity or palatability, but also details of factors such as methods used in their harvesting or preparation, their spatial distribution and seasonal availability, and their ecological relationships with important animal species. In addition, recent research into traditional modes of subsistence have revealed, contrary to previous assumptions, that wild resources are actively managed by the communities who depend on them. Indeed, studies into the TBK and ecology of both hunter-gatherer peoples of Aboriginal Australia and the nomadic pastoralists of Africa, have revealed that traditional methods of resource management may be fundamental to the conservation of some of the world's important habitats, and in some areas, traditional management methods have already been reintroduced to assist in the conservation of national parks and other protected sites.

Chapter 6

Traditional Botanical Knowledge and Subsistence: Domesticated Plants and Traditional Agriculture

Robert Booth, myself and others have received much professional credit for the work on diffused light storage of potatoes . . . Few people realise, however, that this was a technology which [scientists] first learned from Third World farmers.

Robert Rhoades 1989 in *Farmer First*

INTRODUCTION

At present, traditional, or low-input agriculture is thought to support more than one-quarter of the world's population (Wolf 1986), largely in fragile or resource-poor areas such as the tropical forests, savannas, deserts and wetlands of India, Eastern Asia, Latin America and sub-Saharan Africa (Chambers *et al.* 1989). Although encompassing a wide range of distinct horticultural systems, all traditional farming methods can be distinguished from industrial or green revolution agriculture by a number of characteristic features, including their low reliance on the use of agrochemicals and commercial seed, and their total dependence on local precipitation. The role of the traditional farmer is therefore to optimise the use of those resources which are available locally, to minimise the risk of crop failure, and to ensure the sustainability of their production methods. This chapter begins by outlining the resources and methods used by traditional farmers, before examining two important areas in greater detail—the management of crop genetic diversity, and the ecology of traditional agroforestry.

ETHNOECOLOGY AND TRADITIONAL AGRICULTURAL PRACTICE

A wide range of agricultural systems are employed in traditional farming societies throughout the world, ranging from the *shifting cultivation* of the tropical forests to the permanent *wetland agriculture* of coastal Africa (Table 6.1). As we have seen in the previous chapter, many of these farming communities have remained dependent on nutritional supplements from a wide range of wild foods and display a considerable knowledge of the nature, behaviour and distribution of the useful species available. This knowledge of the natural world—of the properties of different plants and the relationships between soils, plants and animals—is also used in manipulating ecological functions such as nutrient recycling and plant defence, the exploitation of which is fundamental to the success of traditional low-input agriculture.

Resource Availability: Environmental Factors and Crop Repertoires

Traditional agriculture knowledge (TAK) may be broadly divided into two main types: knowledge which relates to the physicochemical and biological resources available, and knowledge of the methods and technologies which are suitable for their sustainable exploitation. Over the last decade, growing evidence has demonstrated clearly that traditional farmers possess an extensive and detailed understanding of their agricultural resources, including the location and functional properties of specific soil types, the seasonal availability of water and nutrients, and the agronomic characteristics of a wide range of crop cultivars.

Physicochemical Resources

In studying traditional agricultural knowledge one research method which has been widely used is that of ecological mapping which was outlined in the previous chapter. For example, the mapping of areas under certain crop varieties has been used in determining the location of distinct agroecological niches, such as areas which favour particular crops, or those which are particularly susceptible to flooding (Conway 1989; Gupta & IDS Workshop 1989). The recognition of distinct ecological zones is fundamental to traditional farming, and just as specific

Table 6.1   Characteristics of traditional farming systems. A wide range of agricultural systems are employed in traditional societies, ranging from the wetlands of mangrove swamps, to the run-off cultivation often found in arid and semi-arid regions. In most cases, a variety of these methods are used together in order to spread risk where crop failure is common due to factors such as drought and disease. The success of these low-input systems often relies on the careful selection of crop varieties which are well suited to specific micro-environments, and to the use of cropping patterns which maximise resource use and minimise susceptibility to damage (see Richards 1985)

Traditional agricultural systems and their common characteristics

---

Rain-fed agriculture—dependent entirely on local precipitation

Irrigated agriculture—watered by any process other than direct precipitation
  estuarine swamp agriculture—unrestricted irrigation by tidal river water

  wet season floodplain agriculture—seasonal swamps used with or without some element of water control

  dry season floodplain agriculture—dry alluvial floodplains seasonally planted and supplemented by manual irrigation

  run-off (seep-zone) agriculture—silty, *hydromorphic* soils accumulate at the bottom of slopes and are watered by run-off from above

Shifting cultivation or *swidden cultivation*—common throughout the tropics, this is characterised by *rotational fallowing*; areas of forest or savanna are cleared and burned, cultivated for a short period (normally up to 3 or 4 years) and fallowed for up to several decades

Permanent cultivation—areas which are under cultivation every year, and which are fertilised through natural processes such as silt deposition or are composted artificially

Traditional gardens—include *dooryard* or *house* gardens which supplement field cultivation in many traditional farming systems

Traditional agroforestry—combines agricultural production with forest management and includes swidden agriculture and the *nomadic agriculture* of the Kayapó Indians of Brazil

---

resource units are exploited for particular wild species, only certain zones are recognised as suitable for agriculture. Indeed, while the *caboclo* farmers around Lake Coari recognise at least 40 distinct resource units, only a proportion of these are considered suitable for periodic cultivation (Frechione *et al.* 1989). The classification of ecological zones on the basis of land-use capability appears to be widespread among traditional agriculturalists, who recognise areas of agronomic potential on the basis of attributes such as soil texture and colour, hydrological patterns, and—most frequently—the plant communities they support (e.g. Box 1989; Hecht & Posey 1989; Salick 1989; Dialla 1993). On the basis of this type of knowledge, specific crops or crop varieties can be planted according to their specific needs. For example, among the Amuesha horticultural communities of the Palcazu Valley, Peru, farmers classify their land according to various factors, including natural vegetation and cropping potential, allowing them to locate their crops according to their characteristic growth requirements (Table 6.2).

In a number of cases, traditional systems of soil classification have been investigated chemically, and often a broad degree of correlation has been demonstrated between local categories and conventional soil analysis (e.g. IDS Workshop 1989; Dialla 1993). However, local classification systems frequently identify additional micro-environments which, although agriculturally significant, are not easily recognised using conventional scientific methods. For example, in Zambia, specific areas known as *dambos* are characterised by their ability to hold moisture during the dry season; yet while of crucial importance to local cultivators, these functional micro-environments were not recognised in land classification maps constructed by European researchers (Kean 1988).

In addition to the detailed recognition of specific agroecological zones, traditional farmers also exhibit a considerable knowledge of seasonal and other periodic changes. For example, around Lake Coari, the successful *caboclo* farmers are those who can predict the arrival of potentially devastating floods on the basis of ecological indicators such as the behaviour of water birds (Frechione *et al.* 1989). Equally, where shifting cultivation is practised, the timing of the burning of swidden sites has a profound influence on the nutrients available for crop growth. Hence, while a well-timed burn will normally reduce most of the trees to nutrient-rich ash suitable for planting, (at the same time eliminating potential hazards such as pathogens, pests and weeds), a burn which is less timely may leave much of the biomass unburned, or may even lead to nutrient loss and soil erosion if burning is immediately followed by heavy rainfall. Hence, the economic activities of traditional agriculturalists are strictly determined by seasonal fluctuations in a range of environmental factors, and this is manifest in the detailed nature of their complex economic calendars (Table 6.3).

Crop Repertoires

Complementing their knowledge of their agroecological environment, traditional cultivators generally demonstrate a considerable knowledge

Table 6.2   Land classification and planting patterns in Palcazu Valley, Peru. The Amuesha classify land according to a range of characteristic features including soil type, fertility and natural vegetation, and their major division between lowlands and uplands is not so much one of altitude, but one of parent material—for while the lowlands are largely alluvial, upland soils are derived essentially from weathered rock. The characteristic cropping potential of each recognised land type is known to the Amuesha farmers who plant their crops accordingly. For example, while the broad distinction between *pampa* and *altura* roughly parallels the major cropping division between maizelands and lands which cannot be used for maize, since maize is also susceptible to aluminium toxicity, low pH and flooding, its cultivation is further limited to lowlands which seldom flood, or to exceptionally rich microsites in upland regions. Equally, as beans and peanuts require light, fertile soils, these are habitually planted on beaches or low floodplains, while taro, which is extremely tolerant, can be grown in relatively poor soils (Salick 1989)

Amuesha land classification	Characteristic features		Associated crops
	Environmental	Floral	
Muepeñ	*Pampa*, high floodplains	*Astrocaryum, Cedrela, Ceiba Iriartea*	Maize
Puemetar	*Pampa*, beaches	*Phragmites*	Beans
Achpet	*Pampa*, islands	*Erythrina*	Maize
Astsets	*Pampa*, swamps	*Mauritia*	Aguaje
Tsamat	*Altura*, red soils	*Cedrela, Copaifera, Inga, Parkia Pouteria, Ocotea, Virola*	Rice, cassava
Huallatsttsen	*Altura*, white sands	*Diplotropis, Hevea*	Rubber
Carhuash	*Altura*, yellow sands	*Euterpes, Ochroma*	Barbasco
Quellhue	*Altura*, black soils	*Astrocaryum, Cedrela*	Maize

Pampa, lowlands; *altura*, highlands.

Table 6.3 Seasonal indicators and economic activities—the economic calendar of the Amazonian Desâna horticulturalists. The Desâna horticulturalists occupy the upper regions of the Rio Negro, where they subsist on a mixed economy of hunting, fishing, gathering and shifting cultivation. Their economic activities are timed according to a series of highly complex seasonal indicators, some examples of which are illustrated in this simplified table. In many cases, certain aspects of plant behaviour—such as the flowering or fruiting of particular species—are used as seasonal indicators. On the basis of these and various other characteristic biological and environmental factors, the economic activities of the Desâna can be accurately timed to optimise productivity and ecological sustainability (see Ribeiro & Kenhiri 1989)

Calendar Month	Local indicators	Economic activity
October	*Pit viper illumination* constellation; heavy rains	Cut trees, clear underbrush and collect mushrooms
November	*Pit viper tail* rains; *inga* fruits ripen; peach palm blossoms	Collect fruits of *Inga* and *Pouteria caimito* and collect mushrooms
December	First fish spawning; termites begin to fly (*Cornitermes* sp.)	Set fish traps
January	Rains diminish; star apple (*Pouteria caimito*) no longer bears fruit	Burn timber from October tree felling during 8–15 day *Inga dry season*; plant crops
February	*Armadillo rainy season* begins; fish swim upstream	Collect fruits of peach palm (*Bactris gasipaes*) and cucura (*Pourouma cecropiifolia*)
March	*Peach palm summer* dry season	Trees felled in November and underbrush cleared in January are burned
April	*Shrimp constellation* rain; third fish spawning; leaf cutter ants fly	Harvest umari fruits (*Poraqueiba sericea*)
May	Intense flood rains	Temporary and permanent fish traps set; remove underbrush for subsequent maize planting
June	3-day dry spell *adze feathered ornament, little summer*	Burn underbrush cleared in May
July	Grasshoppers take flight; intermittent rains	Collect edible larvae
August	Appearance of tree larva (*Erisma japura*) and the *inga larvae*	Collect edible larvae; hunting and fishing
September	3-day dry spell *larva old summer*	Burn underbrush cut in August

Table 6.4 Folk varieties of important crop species. In many traditional farming systems, staple crops are represented by a range of cultivars each of which fulfil a number of specific requirements. For example, among the Hopi, at least 21 different cultivars of maize are maintained on the basis of functional traits such as ecological tolerance and productivity, or on the basis of aesthetic criteria, such as taste and ceremonial significance (Soleri & Cleveland 1993). Similarly, the Ka'apor select a number of cultivars of important tuber crops—including manioc—on the basis of factors such as their colour, toxicity and size (Balée & Gély 1989). As the major limiting factor in crop production generally lies in the nature of the substrate, one of the most important factors in optimising productivity lies in choosing cultivars which are well suited to particular nutrient and moisture regimes

Group	Domesticated species	Total no. cultivars	Number of cultivars of important crop species	
Hopi (USA)	19	96	*Zea mays* (corn)	21
			Phaseolus vulgaris (beans)	9
			Cucurbita spp (squash)	9
Ka'apor (Brazil)	28*	83	*Manihot esculenta* (manioc)	19
			Ipomoea batatas (sweet potato)	5
			Dioscorea trifida (yam)	3

*Also manage many species which are not domesticated.

also of a wide range of domesticated crops. The plant species exploited in traditional horticultural systems include true *domesticates* or *cultigens*—domesticated crops which are dependent on human interference to complete their lifecycle; *semi-domesticated* plants which may be propagated and managed but are not obligately reliant on human intervention; and *protected* plants which although occurring spontaneously in managed plots are encouraged or protected by practices such as weeding and pruning. On the whole, traditional farmers spread risk both by cultivating a number of different crop species and by exploiting several *cultivars*, or varieties of important staples. For example, Harold Conklin's study of the Philippino Hanunóo (Conklin 1954b) describes more than 100 distinct cultivars of rice (*Oryza* spp), while Boster (1984) reports that the Peruvian Aguaruna recognise more than 50 varieties of their main staple manioc (*Manihot esculenta*). These *folk varieties* or *landraces* vary in a range of characteristics such as flavour, size, storage properties, time of maturation or resistance to disease, which allow them both to fulfil a range of nutritional and cultural requirements and to ensure some productivity under often unpredictable circumstances (Table 6.4).

Hence, the remarkable genetic diversity found in traditional crop repertoires—which reflects generations of experimentation and selection by local farmer-breeders—provides a means of security where environmental conditions are often variable and unpredictable, such that it is common to find low yielding varieties of staple crops maintained alongside those capable of higher yields, on account of functional characteristics such as drought-resistance, storage properties, or the ability to exploit a range of micro-environments (Wilkes 1989). The development, maintenance and efficient exploitation of this diverse *crop germplasm* involves considerable expertise, and often requires a detailed knowledge of the agronomic properties and functional characteristics of several tens—or even hundreds—of individual crop cultivars, many of which can be discriminated only by the basis of extremely subtle distinctions.

Agricultural Management: Optimising Yields

In addition to a recognition and understanding of the environmental and genetic resources available, the success of low-input agriculture is ultimately determined by their efficient management; this in turn, is largely dependent on the effective manipulation of natural ecological processes. For example, the ability of different plant species to exploit specific micro-environments—such as shaded or sunny sites—may be exploited using complex *multicropping* systems, while valuable agricultural soils can be maintained through the manipulation of natural processes such as nutrient recycling and the binding of soil by certain types of plant root. Through the exploitation of these, and other ecological functions—particularly the complex processes of plant succession—traditional cultivators have been able to cultivate fragile environments, while conserving valuable agricultural resources for future generations.

Husbandry, Pest Control and Soil Management

While traditional farming systems have long been viewed as 'primitive' by many external observers, evidence to suggest the contrary has been recognised for some time. Even as early as 1943, a report published by the West Africa Commission noted that many of the local farmers' practices, appeared to be both agronomically and ecologically sound (Sampson & Crowther 1943). For example, it was noted that Igbo farmers of eastern Nigeria planted the quick-growing *Acioa barteri* in fallow fields, thus speeding up the process of soil regeneration; similarly, a series of scientific experiments soon verified the value of spreading ashes in maintaining soil fertility. The report even suggested that perhaps the plough had been tried and abandoned in Nigeria, on the

Table 6.5   Traditional methods of crop husbandry. In traditional low-input agricultural systems, problems such as soil fertility, water stress and pest outbreaks are managed through the manipulation of natural processes. For example, the deliberate introduction of natural predators such as *Azteca* ants is used to control specific crop pests—in this case leaf-cutter ants of the genus *Atta*. Similarly, light-requiring weeds such as cogon grass (*Imperata cylindrica*) can be eradicated through the deliberate manipulation of shade (see Anderson & Posey 1989; Lightfoot *et al.* 1989)

Agronomic requirement and possible management practices

Improved growth—germination devices and whole plant treatments such as pruning, plus factors such as plant spacing and multicropping are all manipulated to improve yields

Crop protection—practices such as burning, intercropping and the eradication of alternative hosts can help to minimise levels of pest and disease organisms, while weeds may be selectively controlled through techniques such as burning, shading and weeding

Crop storage—both physical and chemical methods of protection may be employed including the use of sealed storage vessels, the application of phytochemicals, and the use of desiccating treatments such as sun-drying

Fertility management—soil fertility may be manipulated in several ways, including the exploitation of alluvial deposits and the use of practices such as fallowing, burning and mulching

Water management and soil conservation—while natural irrigation systems are frequently exploited, practices such as maintaining plant cover and mulching may help in restricting water loss, while minimising soil compaction and erosion

---

basis that its fragile soils were insufficiently stable to withstand prolonged ploughing activity (see Richards 1985), and indeed many recent reports support the notion that 'modern' technologies are often unsuited to traditional farming environments. For example, in the Western Desert of Egypt, traditional farmers regard the introduction of commercial pesticides as responsible for recent increases in postharvest losses (Parrish 1994), while in Africa the introduction of new technologies during the 'Green Revolution' has led to significant ecological deterioration (Titilola 1994).

Today, it is increasingly recognised that traditional agricultural management practices encompass a wide range of methodologies, which address a variety of problems associated with food production. They vary from methods which improve the success of plant propagation or which encourage increased yields, to those which are used in protecting stored foods or in maintaining the quality of soils (Table 6.5). Horticultural practices employed in improving plant yields include germination devices such as soaking, scarifying and shade manipulation, as well as whole-plant treatments such as pruning, grafting and injuring which promote flowering and fruiting in some species (Fernandez 1994). Equally, increased yields due to reducing plant loss may be achieved using various approaches to pest control. Many of these involve the exploitation of plants' natural defence mechanisms; for example, among the Ka'apor of Brazil, slow growing, toxic cultivars of manioc are planted at the swidden edge to deter attacks from leaf cutter ants (*Atta* sp). Other methods of pest control involve the use of additives such as lime and crushed peppers (*Capsicum* spp) which act as insect irritants, while the *caboclo* farmers of the Amazon deliberately encourage birds such as the predatory *caboré* hawk, and the *bem-ti-vi* which eats large quantities of insect pests (Frechione *et al.* 1989); Parrish 1994). Finally, in maintaining the fertility, stability and moisture content of soils, again various methods are employed. Where possible, natural processes such as regular alluvial deposition are exploited, while artificial treatments such as burning, mulching and bucket irrigation are of widespread importance.

Perhaps one of the most important methods in traditional agricultural management, is the use of multicropping systems which exploit the diverse characteristics of different crop species in order to optimise productivity (Table 6.6). Although rare in temperate zone agriculture, the practice of multicropping—the planting of a range of crops in the same field, during a single growing season—is common throughout the tropics, and in recent years has started to attract considerable scientific attention. Consequently, there now exists a body of evidence which supports the benefits of multicropping, especially in soils of low or indifferent fertility. For example, in recent farmer-driven experiments in Guatemala, it was found that through intercropping groundnuts with beans, production was increased by 50 per cent compared with separate plantings (Bunch 1989).

Germplasm Management

Like the effective propagation and protection of cultivated plants, the effective management of crop germplasm is also central to agricultural success and requires not only the maintenance of existing genotypes, but also the development of improved varieties and the selection of semi-domesticates for transplantation. While the management of semi-domesticates may depend largely on practical factors such as the abundance, ease of access, palatability and preparation requirements of a particular species (Rocheleau *et al.* 1989), the development of novel vari-

Table 6.6 Multicropping techniques used in traditional farming systems. The term multicropping covers a range of cropping patterns which are based on spatial or temporal associations of complementary species or varieties. These patterns are recognised in contrast to sole cropping or *monocultural* systems which are common to industrial agriculture (see Richards 1985). The use of multicropping systems has been found to confer a number of benefits, including the measurable reduction of the risk of crop failure, the reduction of outbreaks of pests or disease, and the increase of crop yields by up to 80%. Such benefits are based on the differential requirements and behaviour of particular species thus maximising resource-use efficiency. For example, where cereals and legumes are intercropped, the shallow root systems of cereals exploit the moisture available in soil surface layers, while the deep roots of legumes are able to tap water resources at lower levels. Similarly, where fast-growing cultivars are intercropped with slow-growing crops, major demands are placed on resources at different times throughout the growing season (see Richards 1985)

Traditional systems of multicropping

Intercropping (mixed cropping)—generally refers to mixed *polycultures* where complementary crops are physically planted together

Sequential cropping—refers to rotational patterns where crops are planted and harvested sequentially within the same site during a single growing season

Relay cropping—refers to rotational patterns where sequential crop cycles overlap in time

Patchwork cropping—specific microhabitats are exploited as particular species are planted according to microvariation in soil fertility, light intensity and other agronomic factors

Multistorey cropping—combinations of crops which grow to different vertical levels are manipulated to form a range of microsites suitable for specific crops or cultivars

---

eties involves a combination of spontaneous and artificial cross-fertilisation, the selection of promising progeny on the basis of certain desirable traits, and the subsequent testing and propagation of improved genotypes (Fernandez 1994). Spontaneous hybridisation can be encouraged using a number of management techniques. For example, periodic habitat disturbance can encourage the growth of wild or weedy species thus facilitating genetic exchange between domesticated crops and their wild relatives. Similarly, the deliberate planting of cross-fertile domesticates or the manipulation of natural pollinators can also promote continued *introgression* or *gene flow* between different genotypes. These methods, and the implications of conserving traditional folk varieties are discussed more fully in later sections.

Productivity and Experimentation in Traditional Agricultural Systems

Although the efficacy of traditional farming systems can be assessed only on the basis of both their annual productivity and their long-term sustainability, such studies are hampered by practical difficulties in assessing overall yields in complex multicropped systems and in measuring sustainability over relatively short periods of time. Nevertheless, some insight into traditional agricultural productivity has been provided in a recent study of Yanomami (also known as Yanoama or Yanomamö) horticulture (Smole 1989). Using calculations based on estimated plant densities and mean yields from a sample of harvested plants, this study suggests that, even under marginal conditions, the production of staple crops such as manioc can yield sufficient energy per hectare to feed about 17 people each year—more than a third as much as is produced by a hectare of wheat under industrial agricultural conditions over the same period (Table 6.7). Other studies have further revealed that when the costs of agricultural inputs are considered, low-input agriculture can prove to be more than three times more *efficient* than industrial agriculture (Table 6.8). Moreover, there exists

Table 6.7 Productivity in traditional farming systems. Estimates of the edible yields from staple crops such as banana, plantain (*Musa* spp) and manioc (*Manihot esculenta*) demonstrate that traditional agricultural systems can produce considerable crop yields. The figures presented here, show that 1 ha of land can yield sufficient energy to fulfil the requirements of up to about 17 people *per annum*. In addition, where *Musa* spp are grown the area is normally intercropped with additional species such as yam (*Dioscorea* spp) and sweet potato (*Ipomoea batatas*), which increase considerably the overall energy yields per hectare. While the protein levels of these plant staples is rather low—less than 1.5%—nutrient supplements are often derived from additional subsistence activities such as hunting, fishing and gathering wild species

Crop	Annual edible yield (kg ha–1)	Energy content (kJ kg–1)	Total energy (kJ ha–1)	Individuals fed (ha–1)*
Cowata plantain[1]	14158	5055	71.5 million	15
Manioc[1]	5293	15286	80.9 million	17
Banana (plantation)[1]	11578	3885	45.0 million	10
Wheat[2,3]	15000	14020	210.3 million	45

ha, hectare; *assuming the average daily energy requirement per person is 13000 kJ (Arnold *et al.* 1985); [1]data taken from Smole (1989) have been converted to standard units;[2] mean US wheat yield taken from Khanna and Shukla (1991);[3] mean energy content of wheat taken from Simpson & Conner-Ogorzaly (1986).

Table 6.8 Efficiency of traditional farming systems. In a series of trials in Nigeria, the returns (value/cost ratio) produced using commercial fertiliser on sole crop and intercropped farms were compared for different management practices. Where sole cropping occurred, net returns were considerably better for 'improved' practices, yet where mixed crops were planted, net returns were up to threefold higher using local management techniques. In this example, it is clearly demonstrated that through combining new innovations with traditional practices, West African small-holders, like other traditional farmers, are able to maximise production efficiency in marginal environments (see Richards 1985)

Cropping pattern	Value/cost ratio	
	Local practices	'Improved' practices
Monocrop		
Sorghum	5.6	10.2
Maize	4.1	12.2
Mixed crop		
Maize/sorghum	20.2	17.7
Yam/maize	77.3	24.6
Sorghum/cowpea	13.5	8.4

the additional possibility of further yield improvements as traditional farming communities continually experiment with a wide range of variables including new planting materials, the effects of varying factors such as crop mixtures, plant spacing and shade, the development of non-toxic pest controls, and the improvement of tools and labour saving devices. For example, Roland Bunch (1989) has described how crop yields in various parts of Central America have increased yields as a consequence of local experiments with intercropping, pest control and the development of native species.

FARMER-BREEDERS: THE HOPI INDIANS OF THE AMERICAN SOUTHWEST

Throughout the USA, the dietary requirements of many Native Americans have been fulfilled largely by wild resources. In areas such as the southern regions of the Great Basin, and in most of California, wild plants such as piñon and acorns were important; elsewhere animals ranging from buffalo and deer to small rodents and invertebrates provided much of the staple diet. Only in the American Southwest has agriculture formed the dominant subsistence strategy from prehistoric times

(Driver 1969), a long-lasting tradition which is strongly reflected in the rich ceremonial life of the region's traditional farming communities.

The American Southwest is a largely arid region which encompasses most of Arizona and New Mexico, as well as the northernmost parts of several states of Mexico—namely Sonora, Sinaloa, Chihuahua and Durango. The earliest evidence of farming here comes from the Bat Cave, New Mexico, where a small popcorn maize has been dated from about 2500 BC (see Driver 1969), and by 1000 BC both the common bean (*Phaseolus vulgaris*) and improved maize varieties were present. By the beginning of the Christian era, these domesticated plants had spread throughout much of the Southwest, and were cultivated by the farmers of the early Mogollon, Hohokam and Anasazi traditions. The latter of these ultimately formed the basis from which the modern agricultural-ists—the Hopi, Zuni and other Pueblo peoples—ultimately emerged. Today, this region is recognised as part of the *Aridoamerican* centre of crop diversity (Nabhan 1985) and harbours an important pool of native germplasm, the maintenance of which may be entirely dependent on the continued practice of traditional dryland agriculture (Brush 1991; Soleri & Cleveland 1993).

Dryland Farmers: The Hopi of Arizona

While almost all the Southwest peoples have farmed to some extent, the Pueblos of Arizona and northern New Mexico have been most closely associated with agriculture throughout the last millenium. Developing from earlier farming cultures who lived on a diet of cultivated maize supplemented with wild plants and small animals, the modern Pueblo peoples remain predominantly maize farmers, the most well known of whom are perhaps the Hopi of the Colorado Plateau in northern Arizona and New Mexico. One of the best known—and most persistent—of the indigenous groups in the USA, the Hopi and their ancestors have prac-tised intensive dryland agriculture for millenia. Throughout this time they have developed a diverse range of crops and farming practices which are particularly well adapted to local environmental conditions and have formed the basis of a number of ethnobotanical studies.

Hopi Indians

In spite of recent tourism and the influx of commercial high-yielding crop varieties, the Hopi, like many of the Pueblo Indians still retain many of their traditional values and practices. The reasons for this are partly his-torical, dating from about the time of the first European contact in 1539,

Table 6.9 Hopi agriculture—physicochemical constraints. Low precipitation levels, strong winds and high summer temperatures combine to make water the most limiting factor in Hopi agriculture. However, after several thousand years of agricultural heritage, the modern Hopi have developed both agronomic practices and drought-adapted cultivars which are successful under these extreme conditions. Unfortunately, while the Hopi agriculturalists are still considered as superior dryland farmers, much of their knowledge and germplasm may be lost due to rapid acculturation and socio-economic change (see Soleri 1989; Soleri & Cleveland 1993)

Physicochemical limitations in Hopi agriculture

Short growing season
Time between freezing temperatures is limited to between 120 and 160 days (depending on location)

Limited moisture
The Hopi reservation has little in the way of surface water and is subject to variable annual precipitation (150–230 mm)

Frequent drying winds are common, especially at the beginning of the growing season, leading to high rates of water loss through evapotranspiration

High summer temperatures also contribute to evapotranspiration

Soil quality
The soils of the Hopi reservation have been classified as unsuitable for agriculture in the USDA Land Capability Classification (Brady 1974)

when, despite European pressure, the Pueblo peoples secretly preserved their own culture, resisting change and largely avoiding contact with Spanish settlers. Later, when Mexico achieved independence from Spain in 1800, the Pueblo settlements were formally acknowledged as self-governing towns and the people's right to their land was guaranteed. Today about 7000 Hopi Indians inhabit the Hopi Reservation in northeastern Arizona (Arizona State Data Center 1992 cited in Soleri & Cleveland 1993) while a further 2000 live in Upper and Lower Moenkopi, which are situated outside the boundary lines of the reservation as currently recognised by the Federal Government. Here, despite the difficulties imposed by the arid and infertile environment, the Hopi have developed an agricultural system which has sustained them for well over 1000 years.

Hopi Agriculture: Dryland Farming

The Hopi Indian reservation is located in the high desert of northeastern Arizona where the prevailing physicochemical conditions would challenge any farmer (Table 6.9). Most of the agricultural activity takes place in the southern, lower part of the reservation where water availability is determined by local topographical and geological features. In areas where water from snow or rain seeps from the sides of the *mesas* or tablelands, moisture can accumulate in the heavier soils which are buried beneath a layer of dry sand. Here, the Hopi locate their orchards, planting fruit trees and a number of crops such as melons, squash, gourds and beans (Soleri & Cleveland 1993). Elsewhere, occasional springs provide fresh water for drinking and for irrigating terraced gardens where a range of vegetables including chilis, carrots and lettuce can be cultivated (Soleri 1989). In the dry fields, where much of the Hopi maize is grown, water is provided by moisture which is stored in the soils over the winter, and by both direct rainfall and run-off from mesas during the growing season itself (Bradfield 1971).

The continued cultural significance of agriculture is demonstrated by its strong ties to the mythology and ceremonial life of the Hopi who trace their farming traditions to their origins as a people (Nequatewa 1967). As one of the most important crops in their repertoire, maize has a major role in Hopi folklore and each child is dedicated an ear of maize during early infancy (Brown *et al.* 1952). In addition, sweet corn varieties feature extensively in one of the most important Hopi ceremonies—the *Niman* or Home Dance, which marks the beginning of the harvest in July. These ceremonial aspects of Hopi agriculture have been vital in its survival and have played a significant part in shaping the development of some important Hopi crops (Soleri 1989).

Hopi Germplasm Management: Selecting Crops for Arid Lands

The continued success of Hopi farmers has been greatly facilitated by the development of specific landraces or folk varieties (FVs) which are well adapted both to local environmental conditions and to specific socio-cultural requirements. Unlike the modern crop varieties (MVs) of industrial agriculture which are normally grown each year from commercial seed, the FVs of traditional agricultural systems are continually selected by—and dependent on—the influence of both local farming practices and prevailing environmental conditions. Hence, although many of the Hopi crops, such as maize (*Zea mays*) and beans (*Phaseolus* spp) were originally domesticated elsewhere, most have developed under long-term selection by the Hopi and on the basis of their distinct morphological, biochemical or physiological adaptations, they are considered to be *native crops*.

Table 6.10 Hopi agriculture—folk varieties of native crops. Of the 19 crop species regarded as native crops in Arizona, nine were reported in the Hopi crop repertoire in 1989. Most of these species are represented by two or more folk varieties, as in many cases, continued introgression with wild relatives has maintained high levels of intraspecific genetic diversity. Although several of these crops were originally domesticated elsewhere, many of these Hopi varieties have been modified through continued gene exchange and selection— both natural and artificial—for more than 500 years, and are genetically distinct from folk varieties elsewhere (see Brown *et al.* 1952; Nabhan 1985; Soleri & Cleveland 1993)

No. Hopi varieties of crops native to Arizona	Introgression observed in the American Southwest
Amaranthus cruentus	Gene exchange observed between wild amaranths and domesticated species grown by the Hopi and other southwestern farmers
1 *Capsicum annuum*	Gene exchange has been observed between domesticated chilis (*C. annuum*) and wild chiltepines in the Sonoran region
3 *Cucurbita* spp	Introgression reported between wild populations and traditional landraces in the Sonoran region
1 *Helianthus annuus*	The domesticated sunflower is cross-compatible both with *H. anomalus*—a species which is protected by the Hopi, and with wild *H. annuus*
5 *Lagenaria siceraria*	No gene exchange with wild populations in the southwestern region
2 *Phaseolus acutifolius*	No data found
4 *Phaseolus lunatus*	No wild populations occur sufficiently far north to introgress with Hopi varieties
10 *Phaseolus vulgaris*	Wild populations in northern Mexico are cross-compatible with domesticates and with wild populations of *P. coccineus*
17 *Zea mays*	Three separate waves of domesticated maize introduction have resulted in the subsequent mixing of these ancient genotypes

*Occurs as a volunteer in irrigated gardens. Data modified from Nabhan 1985.

Folk Varieties and Genetic Diversity

Within Arizona, at least 19 domesticated species are regarded as native crops each of which is represented throughout the region by a great diversity of varieties (Nabhan 1985). Several of these are grown by the Hopi who alone cultivate up to 17 FVs of maize, which are either recognised as performing well under certain conditions, or are maintained on the basis of factors such as their culinary qualities (Soleri & Cleveland 1993). These traditional Hopi varieties do not represent a static pool of genes (Table 6.10), but are continually developed in response to environmental or socio-economic change. Indeed, a recent comparison of the Hopi repertoire in 1989 with that of an earlier report has revealed a number of important modifications due to factors such as increased seed availability and changes in selection criteria (Soleri & Cleveland 1993). For example, recent increases in the use of commercially produced foods has reduced the need for maize varieties which can be stored for months in case of a poor harvest, while the advent of mechanised seed grinding has limited the demand for soft-seeded varieties. On one hand then, it is clear that genetic variation plays an important part in traditional crop management. However, it is also vital that desirable genotypes are maintained in a stable form, and the Hopi must therefore manage their crop germplasm in a manner which fulfils two conflicting criteria: genetic diversity must be encouraged to meet unpredictable changes in demand, yet where desirable characteristics are to be maintained, further genetic exchange must be minimised.

Diversification and Stabilisation: the Role of Traditional Agricultural Management

The fine balance required between promoting genetic diversity on the one hand, and maintaining genetic stability on the other, can be achieved through the implementation of a range of crop management practices. Diversification is largely dependent on two major processes: the introduction of new genotypes from elsewhere, and genetic *introgression* between different varieties, species or genera. Hopi farmers are particularly well known for their experiments with new germplasm, and are regarded as being prepared to 'try anything once' when it comes to seed (Whiting 1939). This type of experimentation has played an important part in Hopi agriculture for centuries—indeed, these seasoned agriculturalists have acquired seed from a range of different sources since they began farming, and while some of their crop plants are endemic to the region such as the garden vegetable *nanakopsi* (*Monarda menthaefolia*), others, like the tepary bean (*Phaseolus acutifolia*), have been borrowed either from nearby peoples or from further afield (Whiting 1939; Soleri & Cleveland 1993). In addition, the Hopi farmer breeders continue to experiment with new genotypes which arise as a consequence of both deliberate and controlled introgression with related species or cultivars, thus facilitating the development of novel landraces with specific, desirable traits.

Yet, while the introduction of external germplasm and the selection of novel hybrids play a significant part in Hopi agriculture, many established FVs also remain vital to total crop productivity—particularly

Table 6.11 Number and distribution of local varieties used by Hopi farmers. In 1989, the Hopi crop repertoire included a total of 55 Hopi varieties and 40 which were non-Hopi, these included both a number of commercial varieties, and several folk varieties from neighbouring farmers. These data clearly demonstrate the relative importance of both different crop species and of distinct crop varieties. For example, while all of the farmers interviewed grew maize, only 16% were cultivating sunflower. Equally, the continued importance of traditional Hopi varieties is reflected in the high proportion of farmers who were growing them. Nevertheless, the dynamic nature of Hopi repertoires is also illustrated here, with 42% of the total crop germplasm having been introduced from elsewhere. It is significant to note that the greatest proportion of non-Hopi crops are grown in irrigated gardens, suggesting that the hardy field crops developed *in situ* are particularly well-suited to the harsh conditions of the dry fields and orchards (Soleri & Cleveland 1993)

	No. of varieties		% Farmers growing cultivars			
Crop	Hopi	Non-hopi	Hopi only	Non-Hopi only	Both	Total
Fields						
Maize	17	4*	48	0	52	100
Watermelons	2	2	46	6	34	86
Lima beans	4	1	72	0	12	84
String beans	3	1	38	6	36	80
Squash	3	6	38	6	34	78
Melons	4	4	40	18	14	72
Field beans	7	2	36	8	18	62
Gourds	5	1†	60	0	0	60
Tepary beans	2	0	34	2	0	36
Sunflower	1	1	8	8	0	16
Orchards						
Fruit trees	6	8	46	4	22	72
Gardens						
Vegetables	1	10	0	48	4	52

*Two introduced varieties cultivated on experimental basis only; †introduced variety cultivated on experimental basis only.

where cultivars which are able to tolerate local environmental extremes have been developed. For example, while introduced varieties account for about 90 per cent of the crops grown in irrigated gardens, established Hopi FVs account for as much as 64 per cent of the crop varieties grown in dry fields and orchards (Table 6.11). This distinction in crop distribution illustrates the central role of local cultivars where water stress is most common, and these data are supported by interviews with local farmers, 70 per cent of whom regard commercial sweet corn as less drought-hardy than Hopi varieties (Soleri & Cleveland 1993).

Both the development of new cultivars, and the maintenance of established genotypes are dependent on the selection of suitable seed via the processes of both human and environmental selection. For while conscious selection is necessary in establishing aesthetic traits such as colour and taste, unconscious and natural selection also can play a significant part in selecting for tolerance to environmental stress (see Table 6.12). For example, while the blue maize variety *kokoma* is selected artificially on the basis of its deep blue colour, Hopi maize seedlings exhibit unusually elongated radicals and coleoptiles, which are apparently selected by traditional planting strategies rather than through deliberate choice (see Soleri & Cleveland 1993). Other factors involved in maintaining specific genotypes can include the spatial or temporal segregation of cross-compatible populations, and the Hopi farmers' fields are often highly fragmented with distinct interfertile varieties assigned to separate fields (Brown *et al.* 1952).

It is clear from this discussion, that traditional crop varieties such as those of the Hopi represent an important reservoir of genetic material which may prove valuable in future crop development—particularly in the production of improved stress-tolerant varieties; it is equally clear that both the continued development and conservation of this diverse germplasm is dependent on maintaining traditional farming practices. Indeed in recent years there has been increasing concern over the conservation of these diverse landraces in *ex situ* seed banks, and a number of reports have highlighted the critical importance of traditional management techniques in the conservation of this valuable native germplasm—an important concept which is discussed further in Chapter 12.

TRADITIONAL AGROFORESTRY: CROP PRODUCTION IN AMAZONIA

By definition, agroforestry is characterised by management techniques which combine agriculture and/or livestock production with tree crops and/or forest species on the same unit of land (Denevan & Padoch 1988). Throughout Amazonia, and other tropical forest regions of the world, agroforestry plays a vital part in traditional subsistence, as the majority of agricultural productivity is based on *swidden agriculture* or *shifting cultivation*, a practice which is characterised by the short-term cultivation of cleared fields, followed by an extended period of managed forest (or savanna) regeneration. Dating back to at least the late Pleistocene (Groube 1989), the traditional practice of shifting cultivation has recently attracted considerable attention from

Table 6.12 Methods of germplasm management used in traditional farming systems. Successful germplasm management requires a fine balance between promoting the diversity necessary to meet changing conditions and demands, and maintaining desirable cultivars in a stable form. While the conscious manipulation of crop genotypes occurs through practices such as artificial pollination and deliberate selection of seed, agricultural practices which encourage the growth of weedy crop relatives, or which create specific micro-environments also play a significant part in traditional germplasm management. Hence, both environmental and agronomic influences function together in traditional crop development. (See: [1]Brush 1991; [2]Nabhan 1985; [3]Fernandez 1994; [4]Bellon & Brush 1994; [5]Soleri & Cleveland 1993; [6]Boster 1984; [7]Maurya 1989; [8]Bellon & Taylor 1993)

Techniques in germplasm management

Promoting diversity—spontaneous hybridisation
The growth of weedy species in response to periodic disturbances (associated with agricultural activity) facilitates continued genetic introgression between domesticated crops and their wild, or weedy relatives[1,2]

Close planting of cross-fertile varieties may lead to selectable genetic improvements[1]

Planting of species which encourage natural pollinators may promote spontaneous out-crossing[3]

Promoting diversity—artificial hybridisation
Methods such as hand pollination may be used where spontaneous out-crossing does not readily occur[3]

Maintaining desirable genotypes
Land-fragmentation and temporal segregation can function to restrict undesirable out-crossing[1,4]

Rigorous selection of seed plays a central part in maintaining existing cultivars[1,5]

Certain traits may also be selected unconsciously, either due to genetic linkage of certain characteristics, or where practices such as mixed cropping lead to the indirect selection of individuals which are particularly well suited to certain agricultural practices[6,7]

Natural selection also plays an important part, particularly in marginal environments where selection for tolerance of disease, pests and other environmental stress factors is able to continue[8]

researchers who are interested in its potential as a non-destructive alternative in the economic development of tropical forests. While such research has been carried out in many regions, the swidden cultivators of Amazonia have provided a particular focus of attention for many ethnobotanists (Table 6.13).

Table 6.13 Shifting cultivators of Amazonia. The indigenous peoples of the Amazonian rainforests represent some of the least assimilated—yet perhaps some of the most widely studied—traditional peoples in the world (Burger 1990). For centuries, these forest-dwelling communities have managed the biological wealth of the regions, forests and waters, developing sophisticated agricultural practices which appear to be ecologically sustainable on a long-term basis

Group	Region	Source
Aguaruna	Peru	Boster 1984
Amuesha	Peru	Salick 1989; Salick & Lundberg 1990
Chácobo	Bolivia	Boom 1989
Desâna	Brazil	Ribeiro & Kenhiri 1989
Ka'apor	Brazil	Balée & Gély 1989
Kayapó	Brazil	Anderson & Posey 1989; Hecht & Posey 1989; Posey 1984
Runa	Ecuador	Irvine 1989
Yanomami	Venezuela, Brazil	Smole 1989

Ethnoecology and Swidden Cultivation

Following early misconceptions of tropical agroforestry as 'simple' and 'haphazard' (see Conklin 1954a) it is now clear that swidden agriculture in fact represents a closely integrated system which combines the production of domesticated crops with the manipulation of wild plant resources. The *swidden cycle* is initiated when a suitable site is selected, cleared and planted with a range of domesticated plants; after 1–5 years of crop production, regeneration of the site is then managed for many years to encourage the growth of useful plant species. These *fallows,* or areas of *secondary forest* regeneration, may be harvested for between 15 and 40 years before the site is considered suitable to be cleared and planted again, by which time soil nutrient levels have recovered from their previous exploitation. Using this strategy of long-term field rotation, tropical agriculturalists are thus able to produce domesticated plants on a sustainable basis, while effecting the concentration of useful wild plants into readily located resource islands.

Although details, such as planting times and specific cropping strategies, can vary enormously according to the local environment, cultural tradition or individual preference, the overall cycle has certain features in common throughout Amazonia. The initial phase of swidden cultivation involves the intensive management of agricultural plots and the harvest of both domesticated and wild species, while the *ecological management* of ageing swiddens is based on the deliberate manipulation of normal ecological succession. Cultivated fields and managed fallows both produce a wide range of useful plants, yet the species composition of different stages in the cycle is quite different and each may be recognised as a distinct type of resource unit. For example, the Ka'apor of Brazil recognise at least eight forest zones, six of which are recognised as specific stages in forest regeneration (Balée & Gély 1989). Hence, swidden cultivation is clearly much more than simply a temporary opening up of forest followed by opportunistic harvesting, but rather represents a complex agroforestry system where human intervention alters natural ecological succession to ensure the success of useful plant species (Irvine 1989).

Initiating the Swidden Cycle

Choosing the location of a new swidden site represents an important process in both agronomic and social terms, often requiring consultation with shamans or other spiritual leaders in addition to the consideration of practical factors such as soil type and gradient. For example, when a Yanomami gardener chooses his site, he takes into account a range of factors, including the locations of his existing gardens, the risk of enemy raids, and the suitability for growing plantains (Smole 1989). Once chosen, the site is cleared using machetes, although a number of large trees or rootstocks are normally left *in situ* to maintain soil stability—for where thin forest soils are exposed to tropical rainstorms, accelerated erosion can soon lead to loss of nutrient-rich topsoils. The clearing process requires a considerable input of labour, and is generally carried out by men, after which the slashed vegetation is burned to produce a nutrient-rich ash, while eradicating a range of pest and disease organisms. Where possible, the felled vegetation is left until dry before burning begins; however, where humidity is high, or rains begin early, poor burning may leave much of the timber unburnt (Smole 1989). As differential burning can affect both the production of ash and the removal of pests, the newly burned swidden site represents a complex range of micro-environments which differ in terms of nutrient composition, light intensity and the presence of pests.

Crop Management and Soil Conservation

In most cases the planting of crops takes place following burning, and the swidden plots are characteristically planted with a range of species. These include both domesticated staples such as manioc

(*Manihot esculenta*), maize (*Zea mays*), yam (*Dioscorea* sp), sweet potato (*Ipomoea batatas*) and bananas or plantains (*Musa* spp), as well as a number of useful semi-domesticates which are transplanted from the surrounding forest. In addition, certain species may be grown in specific areas which are managed as *house* or *yard gardens*. For example, the Amuesha, Ka'apor and Chácobo cultivate a variety of fruit trees and medicinal plants in gardens which they continually weed and replant even after the rest of the swidden site has been allowed to revert to secondary forest (Balée & Gély 1989; Boom 1989).

In both fields and gardens, cultivated plants are generally propagated using clonal materials derived either from plants in old swidden plots, or from the surrounding forest. For example, a single manioc root may be split to produce several planting stems, while cuttings taken from forest species may also be used successfully. In planting their crops, farmers take advantage of plot microdiversity as a means of maximising productivity and minimising crop damage. For example the Kayapó plant drought-resistant sweet potatoes in the open swidden centre where levels of light and heat are highest, while nutrient-rich micro-sites are often used for fast-growing species with high nutrient demands (see Box 6.1).

Box 6.1   Plot microdiversity and crop distribution in swidden cultivation. In recent years, several hypotheses have been put forward to explain the practice of what has been named *concentric 'ring' agriculture*. Stocks (1983) has suggested that it can minimise shading and reduce plant disease; that it can maximise the dispersal of a given crop thus reducing problems of pest and disease; that it locates those plants which are vulnerable to insect predation away from the forest; that it places nitrogen-requiring species near to nutrients from forest leaf fall. Hecht and Posey, on the other hand, have recently suggested that the manipulation of soil fertility gradients may explain this type of agronomic structure (Hecht & Posey 1989). In each case, it is evident that specific micro-environments within the swidden site are both created and exploited to optimise agricultural productivity.

In addition to considering the cropping potential of a given agroecological niche, swidden horticulturalists optimise crop productivity through the careful manipulation and exploitation of environmental microvariation occurring within a given site. For example, particularly nutrient rich pockets may be used as nursery sites (Richards 1989), while the positioning of certain species may have a profound influence on plant–animal interactions such as attracting game, or deterring pests (Balée & Gély 1989).

continued

continued

THE CREATION AND EXPLOITATION OF MICRO-ENVIRONMENTAL VARIATION

Several studies into crop distribution in swidden sites have revealed patterns of cropping which reflect the deliberate manipulation of environmental micro-diversity within a given plot. For example, the Brazilian Gorotire Kayapó clear new swiddens such that the crowns of trees fall towards the outer edges of the swidden site, thus creating areas of high nutrient concentration on burning. Through this and other soil management practices, a range of micro-environments of differential fertility are created, as illustrated in the table below.

Region of swidden site		Mean fertility element				
	pH	P (p.p.m.)	K (p.p.m.)	Ca (meq.100 g^{-1})	Mg (meq.100 g^{-1})	OM (%)
Centre zone	5.50	11.00	225.0	3.97	0.80	3.15
Intermediate zone	5.43	2.64	133.0	0.97	0.70	2.76
Edge zone	5.31	3.44	182.0	2.13	1.12	3.69
Forest	4.50	1.17	68.2	0.27	0.68	1.95

P, phosphorus; K, potassium; Ca, calcium; Mg, magnesium; OM, organic matter.

The high concentrations of calcium and organic matter found in the centre of the site are due to more complete burning and external inputs from mulches and ashes from cooking fires, which together create a fertile micro-environment suitable for the growth of sweet potatoes. The second, intermediate ring supports the growth of a range of crops including manioc, yams and beans, the planting of which is linked to soil microdiversity within the ring itself. Finally the outer ring, or edge zone is primarily devoted to *Musa* and yams which are less nutrient requiring than many other crops (Hecht & Posey 1989).

Elsewhere, planting patterns in swidden sites are thought to be related to factors other than fertility, including pest concentrations and light intensity. For example, the Ka'apor of Brazil plant fast-growing varieties of manioc (which are susceptible to leaf-cutter ants) only in the centre of recently cleared fields (where management practices ensure low ant concentrations); in contrast, slow-growing, non-susceptible varieties are planted around the swidden edge where pest levels are often much higher (Balée & Gély 1989). In addition to planting crops in areas which are most favourable to their particular needs, cropping patterns may also be manipulated to fulfil additional practical functions. This is illustrated again among the Ka'apor who create angular fields which maximise the proportion of field/forest edges. Here, spontaneous species including 'wild' manioc (*Manihot quinuiparila*) flourish, acting an effective attractant for game such as Brocket deer (*Mazama americana*) (Balée & Gély 1989).

Although manipulation of the swidden plot is most intensive during the initial period of cultivation, both weeding and soil management remain important throughout the swidden cycle. Three particularly important techniques used in soil manipulation include *in-field burning*, *mulching*, and the direct application of nutrients from various sources (Hecht & Posey 1989). The first of these—the practice of in-field burning—is carried out largely during the first few years of the swidden cycle, providing nutrients for later crops and reducing populations of pests and unwanted competitors. As in all burning practices the timing and temperature of these fires can have a profound influence on both nutrient release and the risk of crop damage and is always avoided at times when valuable fruit trees such as *Caryocar brasiliense* are in flower. Mulching, on the other hand, involves the application of leaves or crop residues and functions not only to maintain nutrient levels, but also provides protection against raindrop compaction and high soil temperatures, and can help to conserve soil moisture during early-season periods of drought. Among the Kayapó, favoured mulches include the leaves of the palm *Maximiliana maripa* and banana (*Musa* spp), the husks of *Bixa orellana*, as well as the residues of crops such as rice, beans and sweet potato. Finally, direct nutrient additions can involve the application of a variety of materials including the ashes of particular plant species, and the nests of termites (*Macrotermes bellicosus*) or ants (*Azteca* spp). These fertilisers are applied largely to crops which are long-lived such as *Musa* spp, certain varieties of yam and pineapples, or are used in preparing planting mixtures which promote the establishment of seeds or cuttings. Later in the swidden cycle, nutrient addition to fallow fields may also be achieved through the planting or protection of nitrogen-fixing trees such as the leguminous *Inga* or non-leguminous *Trema micanthra* (Hecht & Posey 1989).

Managing Ecological Succession

Following the initial period of cultivation, the level of human interference falls, yet active management of swidden fallows can continue for several decades. During this time, useful plants and species are protected or transplanted, while others are removed or pruned to limit competition for resources such as light and nutrients (Irvine 1989). Even when the understorey eventually opens up as the expanding canopy cuts out much of the available light, the harvesting of useful products from both wild and semi-domesticated plants continues.

Table 6.14   Changes in plant communities throughout the swidden cycle.
In the swidden cycle of the Ka'apor of Brazil, the young swidden is characterised by a high proportion (50%) of domesticated plants, including staples such as manioc, sweet potato and yam. As regeneration of secondary forest progresses, the total number of useful plant species increases, although the number of domesticated plants is considerably lower. At the beginning of the old swidden phase, a few staples are still harvested, yet much of the food available now comes from fruit trees which were either planted or protected at the beginning of the cycle. This productive period exhibits the greatest variety of both food plants and game attractants, many of which are encouraged by the use of specific management practices. By the beginning of the fallow period, management is minimal and domesticated species are no longer present, yet certain fruit trees are still harvested, while the characteristic vegetation of old fallow sites often continues to attract a considerable variety of game (see Balée & Gély 1989)

Swidden phase	Number of species					Characteristic plants
	Crops	Wild	Total	Food	Game	
Young swidden (<2 years)	28	28	56	21	23	*Tikuwi*—fast-growing manioc (*Manihot esculenta*)
Old swidden (*c.* 2–40 years)	19	59	78	30	46	Slow-growing manioc and various fruits
Fallow (*c.* 40–100 years)	0	23	23	14	18	Mature *Hymenaea courbaril*, *Spondias speciosum* and *Theobroma speciosum*

Fallow Management and Food Production

Contrary to earlier beliefs, it is now clear that the swidden fallows of tropical horticulturalists are anything but abandoned, but are actively managed over a period of several years to produce a series of semi-natural *resource islands*. During the early stages of regeneration, many of the tree species present such as plantains and cacao (*Theobroma cacao*) may have been deliberately introduced into swidden sites; however, as regeneration proceeds, an increasing proportion of plants are spontaneous volunteers which are deliberately protected during maintenance weeding and selective cutting. Thus, even as the number of domesticates and semi-domesticates decreases, the total number of plants used in subsistence remains high. For example, in the old swiddens of the Ka'apor, while less than a third of the species present are planted, more than half are used as foods, while almost 80 per cent function as game attractants (Table 6.14). In addition to supplying many sources of food, these old swiddens are also exploited for fuel, the fallen timbers of trees such as *Protium* spp, *Dodecastigma integrifolium* and *Sagotia racemosa* being considered as excellent sources of firewood (Balée & Gély 1989).

The notion that the species composition of managed fallows represents the result of careful management has been clearly illustrated in a recent report on succession management among the Runa of Ecuador. Here, both the structure and composition of plant communities were found to be strikingly different in managed and unmanaged sites, the unmanaged plots being largely composed of a uniform canopy of *Cecropia*—a genus of fast-growing pioneers which accounted for 90 per cent of the stems greater than 10 cm *diameter at breast height* (d.b.h.). In contrast, the managed sites were much more open and considerably more diverse than those which were unmanaged, containing 20 per cent fewer stems per hectare, while exhibiting an overall modified *diversity index* of 9.19 compared with a value of 4.36 in unmanaged plots (see Table 6.15).

Additional Methods in Forest Management: The Kayapó of Brazil

While swidden cultivation represents probably the most widespread form of traditional agroforestry, additional methods of forest management in Amazonia have also been identified, including the *nomadic agriculture, hill gardens* and *apêtê*, or *forest island* management of the Kayapó. About 2500 Kayapó currently live in a 2 million ha reserve located in the Brazilian states of southern Pará and northern Mato Grosso, where they subsist on a diet of cultivated sweet potato and manioc, supplemented through fishing, hunting and gathering. Their diverse methods of forest management have been studied in some detail (Posey 1984; Anderson & Posey 1989), and in each case secondary succession is manipulated to provide a range of resource-rich areas which play a vital part in subsistence. For example, *forest fields* and *trailside plantings* are essential to human survival during annual migrations which can last up to several months; *apêtês* in *cerrado* or savanna zones also provide a range of important plant and animal resources (Table 6.16). While relatively little is yet known about the ecological repercussions of these additional management strategies, it is clear that these practices enhance habitat diversity and encourage the formation of biologically diverse communities—an important factor which may prove significant in future efforts to promote biodiversity conservation alongside continued economic activities in many of the fragile, yet species-rich regions of the world.

Table 6.15 Levels of diversity in managed and unmanaged fallows. Like other swidden cultivators of the Amazonian rainforests, the Runa horticulturalists of Napo Province, Ecuador use a variety of methods in managing ecological succession in their swidden fallows. These methods include the planting of domesticated and semi-domesticated tree crops, the planting and protection of forest species, and selective cutting and weeding. As a consequence, managed fallows demonstrate high canopy diversity compared with unmanaged areas. Using a modified *Simpson diversity index* based on numbers of individuals within each *family*, it can be demonstrated that the floral diversity of the managed fallow is probably considerably higher than that found within unmanaged sites (see Irvine 1989)

Family	Number of individuals		Family	Number of individuals	
	Managed	Unmanaged		Managed	Unmanaged
Annonaceae	3	3	Meliaceae	4	6
Apocynaceae	3	0	Menispermaceae	1	0
Araliaceae	9	0	Monimiaceae	2	2
Bignoniaceae	6	8	Moraceae	107	205
Bombacaceae	2	3	Myristicaceae	3	0
Boraginaceae	3	0	Myrtaceae	24	6
Burseraceae	1	4	Nyctaginaceae	0	7
Compositae	12	3	Olacaceae	1	0
Dichapetalaceae	0	1	Palmae	14	4
Elaeocarpaceae	0	1	Piperaceae	20	9
Erythroxylaceae	0	1	Rubiaceae	72	13
Euphorbiaceae	11	40	Rutaceae	16	1
Flacourtiaceae	3	12	Sapindaceae	1	1
Guttiferae	5	6	Sapotaceae	6	2
Lacistemataceae	0	1	Solanaceae	24	21
Lauraceae	0	6	Stericulaceae	1	2
Lecythidaceae	3	6	Tiliaceae	7	2
Leguminosae	30	28	Urticaceae	1	3
Malpighiaceae	0	1	Verbenaceae	11	5
Melastomataceae	71	39	Vochysiaceae	1	0
Modified diversity index*				9.19	4.36

Data taken from Irvine (1989); * $D = N(N-1)/\Sigma(n_1(n_1-1)$, where N=total number of individuals and n_1=number of individuals in the n_1th *family*).

Table 6.16   Diversified forest management among the Kayapó. In addition to the clearing and cultivation of swidden fields, and their subsequent management during regeneration, the Kayapó also manage a range of additional sites, each of which have particular practical functions. For example, trailside plantings provide sustenance and medicinal plants for travellers during their frequent month-long treks, while hillside gardens provide an important food reserve in times of crop failure (Posey 1984; Anderson & Posey 1989)

Forest management techniques of the Brazilian Kayapó

Nomadic agriculture—Posey (1983) has described a broad system of 'nomadic agriculture' which includes:

   Primary and secondary forest transplants—on many occasions trips are made into primary and secondary forest specifically to gather certain plants for transplanting into old regenerating swidden sites.

   Forest fields—during their extensive travels—often for up to months at a time—the Kayapó transplant useful plants to concentrated spots near trails and campsites to form so-called *forest fields*.

   Trailside plantings—in addition to forest fields, the sides of trails themselves are used as planting zones—often of up to 4 m in width; these are planted with numerous species including yams, sweet potatoes, fruit trees and medicinal plants.

Plantations in forest openings—openings in primary forest are seen as natural prototypes for gardens. These openings, which may be either natural or man-made, create new microhabitats and planting zones.

Hill gardens—tuberous plants such as Zingiberaceae and Araceae are planted in well-drained hillside plots which are principally reserved for food resources in case of floods or crop disasters.

Cerrado plantings (apêtês)—within the savanna regions of their traditional lands, the Kayapó manage their resources with the controlled use of fire, and the construction of *forest islands* (apêtês) which are initiated in mounds of compost and are planted with useful tree species to produce 'islands' of woody vegetation. These islands consequently contain unusually high concentrations of important plant resources.

SUMMARY

Today, a significant proportion of the world's population remains dependent on small-scale low-input agriculture to fulfil their subsistence needs. These traditional agricultural systems vary from the dryland production of the American Southwest, to the swidden cultivation of the tropical forests; yet each system is characterised by its dependence on local resources and locally developed technologies. In order to optimise food production in these low-input farming systems, farmers must possess a considerable knowledge both of the nature and characteristics of the resources available, and of the methods suitable for sustainable crop production under conditions which are often marginal for agricultural activity. This chapter discusses the traditional botanical knowledge regarding the cropping potential of different agroecological niches as well as methods of husbandry used to optimise long-term crop productivity; it also examines the role of local botanical knowledge in maintaining stocks of high-quality germplasm which are able to meet changing environmental and cultural demands.

Chapter 7

Plants in Material Culture

Fibre plants are second only to food plants in terms of their usefulness to humans and their influence on the advancement of civilisation. Tropical people use plant fibres for housing, clothing, hammocks, nets, baskets, fishing lines and bowstrings. Even in our industrialised society, we use a variety of natural plant fibres: . . . In fact the so-called synthetic fibres now providing much of our clothing are only reconstituted cellulose of plant origin.

Mark Plotkin 1988 in *Biodiversity*

INTRODUCTION

While early studies into *material culture* often concentrated on items collected from archaeological investigations, current studies demonstrate an increasing interest in the *artefacts*, or man-made objects of existing traditional societies. For the material culture of a given society refers to the total range of objects produced by that society, including functional items such as tools, shelter and clothing as well as more decorative arts and crafts. Traditionally, many of these items have been made from plant materials, and even as commercial alternatives—such as aluminium pans and modern clothing—have been widely adopted, the use of plants in traditional art and technology remains an important aspect of traditional botanical knowledge (TBK). This chapter discusses the current roles of plants in the manufacture of traditional goods, and outlines some of the specialist skills which are involved in the production of such items.

MATERIALS AND METHODS IN TRADITIONAL ART AND TECHNOLOGY

For millennia, plants have provided human societies with an enormous range of useful materials, ranging from wood for construction and fibres

Table 7.1 Proportion of useful plants used in material culture. This table presents data from recent ethnobotanical inventories of five distinct cultural groups. These data are not intended to facilitate a direct comparison between cultures, as methods of data collection differ in each case. However, they do indicate that construction and technology account for between 18 and 61 per cent of the total plant uses reported, suggesting that, on the whole, plants remain an extremely important resource in traditional material culture. (See: [1]Boom 1989; [2]Appasamy 1993; [3]Phillips & Gentry 1993a; [4]Medley 1993; [5]Milliken et al. 1992)

Group	Total no. plant uses	% of total uses				
		Com	Fuel	Med	C&T	Food
Chácobo (Bolivia)[1]	366	0.5	6	46	18	27
Kadavakurichi reserve (India)[2]	97	0	22	26	22	31
Mestizo (Peru)[3]	1067	17	4	15	51	13
Pokomo (Kenya)[4]	126	1.5	7	18	61	12
Waimiri Atroari* (Brazil)[5]	225	0	6	15	52	27

*Data presented for 1 ha plot.
Com, commercial; Med, medicinal; C&T, construction and technology (material culture).

for textile manufacture, to pigments used as dyes and resins which provide a range of sealants and adhesives. In industrialised societies, many plant products have now been replaced with synthetic alternatives, and in some cases these have also been adopted by traditional peoples. For example, the introduction of metal and plastic containers has led to a decrease in the production of domestic baskets among the Seri of Baja, California, while the use of aluminium canoes by the Amazonian Waimiri Atroari has resulted in the loss of many skills required for traditional canoe construction (Felger & Moser 1985; Milliken et al. 1992). Nevertheless, recent reports on the existing material culture of traditional societies suggest that both wild and cultivated plants remain vital to many aspects of traditional life, including the construction of shelters and boats, the manufacture of hunting implements and other tools, and in the production of small items such as containers, toys and ornaments.

Timber and Non-timber Plant Products

The continued significance of plants in traditional material culture is illustrated in Table 7.1, which indicates that for groups such as the Pokomo of Kenya, the Waimiri Atroari of Brazil, and the mestizos of Tambopata, Peru, material culture can account for more than half of the

Table 7.2 Traditional uses of timber and non-timber plant products.
While timber remains important in the construction of homes, boats and other
items, a vast range of non-timber plant products are also important in traditional material culture. All parts of plants may be used—ranging from leaves
and stems, to seeds and exudates—to produce a variety of tools, weapons and
aesthetic items

Products	Uses
Timber products	
Round wood	House posts, beams, dug-out canoes, fencing, footbridges
Sawn timber	Flooring, walls, canoes, furniture
Split trunks	Walls, doors, flooring, roof gutters, paths
Non-timber products	
Leaves	Thatch, fans, cigarette papers
Stems	Furniture (rattan), tying materials, fish traps, spears, basketry
Branches	Bows, spears, toothbrushes
Petioles	Rafts, blow-gun darts (petiolar spines)
Bark	Walls, roofing, canoes, soot used in strengthening pottery
Roots	Tying materials, canoe paddles (buttress roots)
Inflorescence	Arrow shafts
Seeds	Personal adornment, games
Fibres	Textiles, ropes, paper, hammocks, basketry
Exudates	Glazes for pottery, waterproofing (rubber), adhesives, illumination
Pigments	Dyes and paints, body painting, tattoos

Data taken from sources throughout this chapter.

total plant uses reported. Many of the species used in material culture
are exploited as timber which remains essential to the construction of
shelters, fences and other large items; yet a number of *non-timber plant
products* (NTPPs) derived from leaves, resins, pigments and fibres are
also of considerable importance. For example, the leaves of large palms
or grasses are commonly used in thatching for traditional dwellings,
while plant fibres are used in products ranging from paper and textiles to
cordage and baskets (Table 7.2). As a result of extensive ethnographic
and ethnobotanical research there now exists a considerable body of
data concerning the uses of both timber and non-timber forest goods.

Plants Used as Timber

As in most societies, timber remains fundamental to traditional life—
particularly in the construction of both temporary shelters and more
permanent homesteads. Indeed, the construction of dwellings can

Table 7.3 Proportion of timber species used in the construction of dwellings. Most types of traditional dwelling rely to a large extent on a supply of suitable timber for basic components such as load-bearing house posts and roof supports, as well as for walls, doors and flooring. Hence it is not surprising that house construction accounts for much of the timber used in many traditional societies. Timber is also used for other items such as canoes, paths and furniture, although these additional uses account for a much smaller proportion of the total uses of timber (see [1]Phillips & Gentry 1993a; [2]Medley 1993; [3]Milliken *et al.* 1992)

Group	Total uses for timber	Percentage of total uses			
		House	Trans	Furn	P&B
Mestizo (Peru)[1]	451	92.5	3	4	0.5
Pokomo (Kenya)[2]	32	56.0	31	13	0
Waimiri Atroari (Brazil)[3]	92	85.0	12	3	0

Trans, transport; Furn, furniture; P&B, paths and bridges.

account for more than 90 per cent of the uses of timber reported within a given community (Table 7.3). The characteristic dwellings built by different peoples vary enormously in their design and construction, according to both the materials available, and the prevailing environmental conditions. Yet with the exception of some traditional shelters of the Arctic, which were generally built entirely from blocks of snow, most traditional dwellings have depended on timber to some extent. For example, in North America, while the portable *tipis* of the Plains Indians consisted of a frame of poles covered with a tailored buffalo hide, the large plank houses of the Northwest Coast were built entirely of wooden planks attached to a framework of poles (Driver 1969). Similarly, the permanent mud huts still common to West Africa are supported by wooden frames which reinforce the walls and help to prevent cracking (Abbiw 1990). However, the use of timber is not restricted simply to the construction of homesteads, but remains important in building other commodities such as boats, bridges and furniture.

The number of species considered suitable for a particular construction purpose can vary enormously. For example, certain *mestizo* populations of Amazonian Peru, recognise 19 separate use-categories for timber, ranging from *roundwood* which is used for house posts to *sawnwood* used for purposes such as canoe construction; yet while 42 species are considered as suitable for the former, only four are used in the latter (Table 7.4). For more complex structures such as permanent housing, or large vessels, a range of different species may be required, each fulfilling specific functions within the total structure. For example, the Waimiri Atroari use more than 16 species as timber in the construction

Table 7.4   Use-categories for timber species as defined by *mestizo* populations in Peru. The *mestizo* peoples of Tambopata in northeastern Peru recognise at least 19 distinct use-categories for timber. The differences in the numbers of species considered suitable for each category reflect differences in both the amount of timber required for particular tasks, and the types of timber which are suited to specific uses. For example, the relatively large amount of timber required for house posts may necessitate the use of a range of species, while specific needs of buoyancy, water resistance and workability may severely restrict the number of species which are suitable for building canoes (see Phillips & Gentry 1993a)

Use-category	No. of species	Use-category	No. of species
Sawnwood for floors and walls	223	Split trunks for floors	3
Roundwood for beams	122	Split trunks for roof and floor supports	3
Roundwood for house posts	42	Split trunks for roof gutters	2
Sawnwood for furniture	16	Trunks for footbridge construction	1
Sawnwood for house posts	12	Temporary canoes	1
Hollowed trunks for canoes	9	Raft construction	1
Roundwood/split trunk for fence posts	7	Split trunks for shelving	1
Split trunks for partition walls	5	Split trunks for paths	1
Split trunks for outside house walls	4	Split trunks for temporary buildings	1
Sawnwood for canoes	4		

of *malocas*—large round houses which provide permanent accommodation. However, while some of these species may be used fairly generally, others are used only very specifically on the basis of their characteristic physical or chemical properties. This is true of species such as the Annonaceous *Guatteria olivaceae*, which has traditionally been the preferred species for the construction of the *malocas'* outer walls, on account of its reputed resistant to lances—presumably an important consideration before the cessation of intertribal wars. Similarly, *Minquartia guianensis*, a member of the Olacaceae which, renowned locally for its strength and resistance to rot, is used almost exclusively for the main uprights of the *malocas* by the Waimiri Atroari and a number of other Amazonian groups. Indeed this species is so highly respected by the nearby Tembé Indians of Pará (Brazil), that the breaking of a taboo against its use as firewood, is said to result in death (Milliken *et al.* 1992).

Although the specific plants available to a given community will vary

enormously in different parts of the world, it is evident from the literature that functional criteria such as shape, strength and durability
commonly play an important part in the selection of suitable timber.
For example, in Ghana, species such as *Afzelia africana* and
Craterispermum laurinum are recognised as particularly valuable on
the basis of their resistance to termites and fire, respectively, while
Canarium schweinfurthii—and again *Afzelia africana*—are both used
in the construction of flooring and stairs due to their considerable resistance to abrasion (Abbiw 1990). Similarly, in Brazil, the wood from
various species of *Mezilaurus* is used in the construction of dugout
canoes and larger boats on account of its considerable resistance to
rotting (Milliken *et al.* 1992). These differences in the functional properties of wood are ultimately based on species-determined characteristics
of woody tissues, including factors such as the presence, size and distribution of vessels and fibres, and the accumulation of lignins, tannins
and other secondary compounds (see Chapter 2). For example, plants
such as *Diospyros* spp which have very thick cell walls and contain many
lignified fibre cells have relatively heavy wood—and therefore a high
specific gravity as outlined in Box 7.1. Those species with thin cell walls,
on the other hand, tend to be much lighter. Owing to their high proportion of strengthening fibre cells, heavier woods also tend to be stronger
and are commonly used in construction.

**Box 7.1   Factors affecting the functional characteristics of wood—physical charac-
ters.** While the strength and density of wood play a fundamental part in determining
the timber quality of a given species, other physical wood properties such as flexibility, size and shape are also important in influencing the potential uses of a given tree
species.

Species-specific characteristics of woody tissues play a fundamental part in
determining the potential uses of a given tree species. For example, the density or
specific gravity of woody tissues can have a profound effect on the strength of
timber, while additional features of wood anatomy can have significant influences on other important physical characteristics such as flexibility.

THE SPECIFIC GRAVITY OF WOOD

The specific gravity, or density, of wood is equal to the mass of 1 cm³ of oven-dried
material. The specific gravity (sp.gr.) of the wall substance of secondary xylem is
about the same for any plant species (sp.gr.=1.53); hence differences in the sp.gr.
of different types of wood depend on the amount of wall material present within a

_ *continued* _|

continued

given volume of tissue. Wood density is a species-specific characteristic, such that wood from different species will vary fairly predictably in their behaviour. For example, as the mass of 1 cm^3 of water is 1 g, then any wood with a sp.gr. of less than 1 will float. Very heavy woods such as _Krugiodendron_ are generally too dense to work, and commerical timber normally has a specific gravity of between 0.35 and 0.65. For example, pine, which is commonly used in house construction, has a specific gravity of about 0.5.

Plant species	Specific gravity (g.cm^{-3})	Use
Natural range of wood density[1]		
Aeschynomene sp (Leguminosae)	0.04	
Ochroma sp (balsa wood)	0.10–0.16	
Pinus halepensis (Pinaceae)	0.48	
Eucalyptus camaldulensis (Myrtaceae)	0.52–0.68	
Quercus calliprinos (Fagaceae)	0.80	
Krugiodendron sp (Rhamnaceae)	1.40	
Plants used by the Pokomo of Kenya[2]		
Ficus sycomorus (Moraceae)	0.47	Temporary canoes (1 year)
Populus ilicifolia (Salicaceae)	0.50	Temporary canoes (2 years)
Alangium salviifolium (Alangiaceae)	0.56	House posts, furniture
Polysphaeria multiflora (Rubiaceae)	0.63	House posts
Diospyros mespiliformis (Ebenaceae)	0.66	Canoes, furniture (very durable)

Data taken from [1]Fahn (1990); [2]Medley (1993); also see Simpson & Conner-Ogorzaly (1986).

It is clear from these data, that most of the woods used in construction by the Pokomo of Kenya, have a sp.gr. of between 0.471 and 0.663, which closely coincides with the range used for commercial timber (Medley 1993). However, additional wood characteristics can also have a considerable influence on the usefulness of a given timber species. For example, although timber from _Diospyros mespiliformis_ is very durable, it is rather susceptible to cracking in the sun, due to additional anatomical characteristics which affect its hygroscopic nature.

ANATOMICAL AND MORPHOLOGICAL FACTORS AFFECTING WOOD CHARACTERISTICS

While the strength and density of wood is determined largely by the amount of wall material per unit volume; flexibility, on the other hand, is influenced by both

continued

continued

the length of individual fibre cells with the secondary xylem, and the extent to which they overlap. Long, overlapping fibres result in wood such as that from the trunks of _Brosimum utile_ (Moraceae) and _Ficus nymphaefolia_ (Moraceae) which provide elastic slats used in the bedframes of the Chácobo Indians (Boom 1989). The suitability of a species to a particular task can also be influenced by its _hygroscopic_ properties, for the ready absorption of moisture from the atmosphere is likely to limit the dimensional stability of the wood, often due to problems such as warping and cracking. Growth habit is another feature likely to exert a strong influence on the applications of any given species. For example, among the Pokomo, trees such as _Pavetta sphaerobotrys_ (Rubiaceae) and _Drypetes natalensis_ (Euphorbiaceae), while of suitable size and strength for house posts, rarely produce a straight pole; in contrast, many Anonaceous trees are characterised by their long straight trunks, and their use in construction is widespread (Milliken _et al._ 1992).

However, specific gravity is not the only criterion which determines the timber quality of a given species. This is demonstrated among the Pokomo of Kenya who frequently use the trunks of _Polysphaeria multiflora_ (Rubiaceae—sp.gr. 0.629) as house poles—indeed this species alone accounts for more than 50 per cent of the wood cut by local peoples. In contrast, another species, _Oncoba spinosa_ (Flacourtiaceae), which has a higher specific gravity (sp.gr. 0.645) is rarely used for poles due to its pronounced susceptibility to rotting—a factor which is often related more to a species' chemical characteristic than to its physical density alone. Indeed, as rotting represents the result of bacterial or fungal degradation, it is the presence of antimicrobial substances such as resins, tannins and essential oils which generally determines the durability of wood. Such considerations can help to explain patterns of traditional plant use such as the widespread use of timber from members of the Lauraceae and Myristicaceae both of which contain high concentrations of essential oils (Box 7.2).

Box 7.2 Patterns of plant use—secondary products and phylogenetics. Several plant families are characterised by their role in providing commercial timber on the basis of their members' functional and/or aesthetic qualities. Many of these are also important in providing timber for traditional construction purposes, including several of those illustrated here. Clearly, the successful use of plants in material culture requires a considerable understanding of the species available, and individuals often demonstrate a comprehensive knowledge of the wood characteristics of local species.

In addition to the important physical characteristics of different timber species, factors such as durability, or the ability to deter pests are strongly influenced by

continued

_____ *continued* _____

the chemical make-up of the wood. For example, the accumulation of antimicrobial phytochemicals can have a significant effect on a species' ability to withstand rotting. In many cases, species within a given plant family exhibit similar physical and chemical characteristics, such that patterns of plant use may be phylogenetically determined.

CHEMICAL FACTORS AFFECTING WOOD CHARACTERISTICS

The durability of wood is dependent on its ability to minimise both microbial degradation and the destructive activities of termites and other pests. This in turn is dependent largely on the presence of secondary chemicals such as resins, tannins and oils, which are thought to protect the living tree from a range of pest and disease organisms. For example, the bark of several timber species used by the Waimiri Atroari apparently contain bioactive principles which are used in activities such as traditional medicine or pest control. These include *Duguetia caulifora* and *D. flagellaris* (Annonaceae) whose trunks are used in house construction, and whose bark is used for medicinal purposes; the wood of *Guarea scabra* (Meliaceae) is used in house construction, while its inner bark is used in pain relief; the trunks of *Irianthera juruensis* and *I. paraensis* are used in house construction while their resinous bark is used in the treatment of fungal infections.

PHYLOGENETICS AND PLANT USE

In many cases, the members of particular plant families accumulate characteristic chemicals which are able to limit degradation, such as the Burseraceae which commonly accumulate bioactive resins, and the Lauraceae whose tissues often contain aromatic oils. Physical characteristics such as growth habit and wood density may also be similar in related species. Hence, where timber species used by groups living within similar environments are compared, certain plant families can often appear especially important. This is clearly demonstrated in the tropical rainforests of South America, where, throughout Amazonia, five plant families—the Annonaceae, Arecaceae (Palmae), Fabaceae (Leguminosae), Lauraceae and Myristicaceae are commonly used in construction. Although this is partly due to their strong representation in local vegetation, it is also due to attributes such as shape, strength, workability and the presence of secondary compounds which are common among particular families (see Boom 1989; Milliken *et al.* 1992; Phillips & Gentry 1993a).

Characteristic features and distribution of plant families providing timber

---

Annonaceae—A family of plants which have aromatic tissues, and whose tree species are often characterised by their long, straight trunks; provide almost 10% of the timber species used by the Waimiri Atroari. (Tropical, especially Old World)

_____ *continued* _____

— *continued* —

Bombacaceae—A family of tree species, many of which exhibit stout trunks suitable for canoe construction; e.g. *Scleronema micranthum* used in canoes in Amazonia. (Tropics)

Chrysobalanaceae—Several yield siliceous timber which can be hard to work, but are resistant to pest damage. (Pantropical)

Lauraceae—Tissues contain aromatic oils and many species are valuable as timber; provides about 12% of timber species used by the Waimiri Atroari. (Amazonia and SE Asia)

Leguminosae—Many species are valuable as timber and/or produce bioactive chemicals; provide up to 17% of the timber species used by the Pokomo. The Waimiri Atroari claim that the wood of *Swartzia panacoco* can last for 60 years even if buried underground. (Ubiquitous)

Moraceae—Most species produce bioactive latex, and many are characterised by the flexible nature of their wood. (Tropics and sub-tropics)

Myrtaceae—Many provide valuable timber and accumulate aromatic oils; several species are used in house construction by the Waimiri Atroari and Bardi. (Tropical and warm regions, especially Australia)

Sapotaceae—Many provide valuable timbers and some produce latex; provide about 11% of timber species for the Waimiri Atroari, and 7% of those used by the Pokomo. (Tropics and sub-tropics)

Data taken from Allaby (1992); Milliken *et al.* (1992); Medley (1993); Smith & Kalotas (1985).

Other Plant Materials Used in Construction

Of course timber is not the only plant material which is used in construction, as additional materials are also required for applications such as roof construction. In some societies, these materials may be provided by animal products. For example, the tents of the nomadic Tuareg pastoralists of West Africa are usually covered with goatskins, while the Seri of Baja, California often used pelts from various animals to waterproof the roofs and walls of their traditional brush houses or *haaco hahéemza* (Felger & Moser 1985; Smith 1992). Yet frequently, roofing materials are obtained from a range of plant parts, including leaves, stems, sheets of bark, or—less frequently—split wood which is used for shingles. As with other construction materials, different roofing materials are chosen on the basis of their different functional characteristics, and while some—such as the leaves of *Carapa procera*—are resistant to damage from termites, others are easily worked or are particularly water resistant (Table 7.5).

For some traditional societies, bark provides not only a roofing

Table 7.5 Plants used as roofing materials in traditional dwellings.
Various plant parts—particularly the leaves, bark and stems—are used in roof
construction by traditional societies. For example, in Ghana at least 15 different
species have been recorded as useful for making roof shingles, while the stems of
Agelaea trifolia and *Calotropis procera* are sometimes used as thatching (Abbiw
1990). However, in many cases, roofing materials are provided by leaves; indeed
the Waimiri Atroari use the fronds of at least six species of palm for roofing,
while almost 100 different species of grasses and sedges are considered suitable
for thatching in Ghana (Abbiw 1990; Milliken *et al.* 1992). The Bardi Aborigines
also use a number of grasses in constructing their shelters, yet prefer to use the
bark from two myrtaceous trees—*Melaleuca dealbata* and *Eucalyptus miniata*,
whose bark can be stripped in large sheets following the first rains of the wet
season (Smith & Kalotas 1985). As with timber, suitable plant materials are
chosen on the basis of functional characteristics, such as their availability,
durability and water-proofing properties (see: Densmore 1974; Conner &
Simpson-Ogorzaly 1986; Harborne 1988; Abbiw 1990; Milliken *et al.* 1992)

Use and characteristics of traditional roofing materials

Betula papyrifera (**Betulaceae**)—The ease of bark removal and the
predominance of this species play a considerable part in determining its use
among the Chippewa of North America. In addition, many species of birch
contain protective chemicals, including *B. resinifera* whose toxic resin repels a
number of mammalian predators, while the phenolics of *B. pendula* and *B.
pubescens* are toxic to various invertebrates.

Bridelia ferrugina—A decoction of the bark is mixed with clay to produce a
cement used in flat roofs in Ghana.

Calotropis procera—Stems are used for some roofing in Ghana.

Carapa procera (**Meliaceae**)—The termite-proof leaves of this species are
sometimes used in Ghana.

Eucalyptus miniata (**Myrtaceae**)—The bark of this, and many other species of
Eucalyptus is characterised by the ease of its removal and its high
concentrations of condensed tannins which may provide protection against
both herbivory and microbial degradation.

Palmae—The leaves of many palms species are used as thatch in many
tropical regions. In Amazonia the leaves of *Genoma deversa* are sometimes
used but are not a material of choice, lasting only a few years, compared with
those of *Mauritia carana* which last between 8 and 10 years.

Poaceae—Almost 100 different species of grass are used for thatching in
Ghana.

Terminalia ivorensis (**Combretaceae**)—As its wood is split easily and can last
for up to 15 years, this species is often used to prepare shingles for roofing in
Ghana.

material, but represents one of the most important materials used overall in construction. Indeed, in regions such as the northern USA and British Columbia (which are dominated by coniferous forests or by mixed broadleaf and coniferous woodlands) indigenous peoples have often relied on bark for the construction of both traditional dwellings and canoes. Like wood, the functional properties of bark can vary enormously, its physical and chemical properties depending largely on species-determined factors, and in particular the specific nature of the *periderm* (Figure 7.1). In species which produce only a thin cork layer outside the cork cambium, the outer surface normally looks smooth as in white birch (*Betula papyrifera*), while a thicker layer of cork has a surface which is cracked and ridged. Equally, the way in which subsequent cork cambiums arise, has a considerable influence on the appearance of the outer bark, for where these developing cambiums arise as a series of overlapping scales, the outer layers are sloughed off accordingly, and a *scaly bark* is formed. However, where subsequent periderms are formed as entire cylinders, a *ring bark* is formed, where the dead outer tissues are sloughed off as hollow cylinders. Hence, in some plants such as certain species of *Eucalyptus* and *Melaleuca* whose bark is intermediate between these two extremes, the outer layers can be peeled off in large sheets suitable for roofing or other purposes. For example, the bark of the white or paper birch (*Betula papyrifera*) and the black ash (*Fraxinus nigra*) have been used in covering the traditional domed wigwams of the Great Lakes Indians, while the heaviest bark from larger birches has been used in making large canoes, which could carry several people (Driver 1969; Densmore 1974).

A final source of construction material, is provided by the hollow, flexible stems of two types of plant—the bamboos (Bambusaceae) and the climbing palms or *rattans* (subfamily Calamoideae), which are widely distributed throughout the tropics. Both taxa include a large number of species, and both are characterised by mature stems which are made up of a series of hollow segments, each of which consists essentially of a cylinder of vascular tissues and their associated fibres. The stems of these plants can vary from a few millimetres to over 10 cm in diameter, and can grow at rates of up to 5 m per year (Myers 1992). By virtue of their unusual structure, these tough, flexible *canes* offer an effective combination of flexibility, light weight and strength which has proved invaluable to many traditional peoples—particularly in African and Asian communities, where furniture, scaffolding, bridges and houses have all been constructed using these materials (Simpson & Conner-Ogorzaly 1986; Morakinyo 1994).

In addition to these major building materials, the construction of

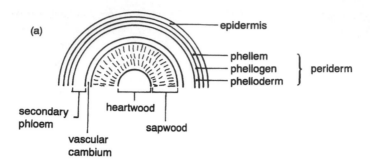

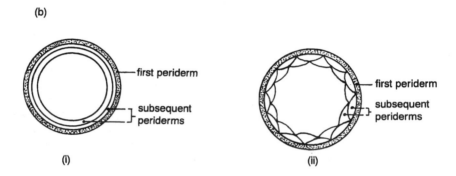

Figure 7.1   Bark formation in woody plants. (a) The periderm of a plant is a compound secondary tissue comprising the cork cambium (*phellogen*), the cork layer (*phellem*) and in some species a living, parenchymatous tissue known as the *phelloderm*. Together these tissues replace the function of the epidermis during secondary growth. The initial phellogen—the lateral cambium responsible for producing the cells of the first periderm—develops in living tissues in or near the epidermal tissues and, like the vascular cambium, it exhibits periodic meristematic activity; (b) With continued lateral expansion, the protective periderm itself is periodically replaced by new periderms, each forming at new cork cambiums which arise sequentially in the living tissues of the inner bark. As this occurs, any tissues exterior to the innermost cork cambium are cut off from the nutrient and water supply and die, thus becoming part of the outer bark which is eventually sloughed off with continued growth. (i) Where these subsequent phellogens form a continuous cylinder as in species of *Vitis* and *Clematis* a ring bark is formed which may be sloughed off as a hollow cylinder; (ii) where they arise as a series of overlapping scales, a scaly bark forms (Fahn 1990). As the constituent cells of the periderm vary, so too do the functional characteristics of the bark. For example, in *Eucalyptus* spp the cells of the cork layer are often thick-walled and filled with tannins, affording protection against a variety of pests and pathogens; in *Betula* spp layers of these tannin-containing cells alternate with layers of thin-walled cells, causing the outer bark to peel like sheets of paper

dwellings, boats and other items requires a range of smaller, yet equally important components. Strong tying materials are needed for wooden frameworks, while sealants are necessary for caulking boats. Again plants can provide a range of fibres, resins and gums which fulfil these purposes. For example, in Amazonia, the aerial roots of certain members of the Araceae are used as tying materials, while in Baja, California, a mixture of animal fat and the gum from the elephant tree (*Bursera microphylla*—Burseraceae) was commonly used as pitch for sealing boats earlier this century (Felger & Moser 1985; Milliken *et al.* 1992). Useful plant fibres and exudates are considered in greater detail in later sections.

Plants in Art and Technology

Almost every type of plant organ has found some application in traditional art and technology. Fibrous stems, roots and leaves have provided materials for basketry, cordage and textiles; specific types of wood have been used in the manufacture of a wide range of tools, toys and small utensils; seeds and flowers have been used in making necklaces and ceremonial attire. Meanwhile, a range of plant extracts and exudates have provided pigments, dyes, resins and adhesives which have fulfilled a range of functional and aesthetic requirements.

Traditional societies often possess a very complex technology, as tools used for different purposes can fulfil very specific functional requirements. For example, the Seri Indians use at least five different types of harpoon for catching marine animals: two types of turtle harpoon for winter and summer fishing; a double-pronged fish harpoon for fishing from boats; a single-pronged fish harpoon used for catching large fish from the shore; various smaller harpoons are used for catching swimming crabs (Felger & Moser 1985). Similarly, the Amazonian *caboclos* of Lake Coari use at least 13 different tools for catching fish and turtles in specific habitats, while the Waimiri Atroari employ a whole range of traps and spears, each of which is used specifically for catching fish, turtles or caiman (Frechione *et al.* 1989; Milliken *et al.* 1992). Tools used in hunting also include items such as arrows, blow guns, clubs and boomerangs (Table 7.6); other utensils range from small items such as needles, nails and sandpaper, to much larger commodities such as looms and canoe paddles (Table 7.7). Plants are also important in less utilitarian aspects of traditional life, providing the raw materials for a range of decorative and ceremonial items, as well as for various games and toys. For example, the Chippewa use a number of grasses and bark materials for making dolls, while the dried stems of wild onion (*Allium stellatum*) have been used to

Table 7.6   Traditional plant-based tools used in hunting and warfare. In many cultures, traditional technology remains important in both subsistence activities and in some cases, in defence. These tools range from relatively simple items such as the one-piece blow guns of Amazonia which are made from the hollowed stems of *Iriartella setigera* (Palmae), to the complex turtle harpoons of the Mexican Seri. These traditional harpoons consisted of a mainshaft of several pieces of flexible wood such as white mangrove (*Laguncularia racemosa*—Combretaceae) joined with cord made from mesquite roots (*Prosopis glandulosa*—Leguminosae). For winter fishing these shafts would be about 6–10 m in length, while those used during the summer (when turtles remain nearer the surface) were approximately half this length. The barbed metal harpoon point would be embedded in a short, detachable foreshaft of mesquite wood using the sticky *lac* from creosote bush (*Larrea divaricata*—Zygophyllaceae) or from the brittlebrush (*Encelia farinosa*—Compositae), and bound with mesquite cord. Mesquite cord also provided the traditional toggle line which was used to pull in the harpooned prey (Balick 1984; Felger & Moser 1985)

Implements used in hunting and warfare (not fibres)	
Arrowheads and shafts	Fishing reels and traps
Axe handles	Harpoon points and shafts
Blow guns and blow-gun darts	Shields
Boomerangs	Spears
Bows	Wrist braces
Hunting clubs	

Table 7.7   Traditional plant-based tools. A vast range of utensils are traditionally made from non-fibrous plant parts ranging from simple digging sticks to complex agricultural equipment such as the wooden *lithaos* of traditional Philippino farmers (Fujisaka *et al.* 1993). A number of smaller items are also made from plants. For example, in Kenya animal skins are attached to hollowed logs to make drums, using the recurved spines of *Combretum paniculatum* (Combretaceae) as nails; in Australia the Bardi Aborigines use the leaves of *Ficus opposita* (Moraceae) as sandpaper, and for scouring anything from sparkplugs to pots (Smith & Kalotas 1985; Medley 1993). While many of these utensils are of widespread use, others such as the cradle boards and pry bars of the Seri are more restricted in distribution (Felger & Moser 1985)

Some traditional utensils based on plants (not fibres)	
Carrying sticks	Looms
Combs	Mortar and pestles
Containers	Nails
Cooking utensils	Needles
Cradle boards	Paddles
Digging sticks	Pry bars (agave chisel)
Drills	Sandpaper
Firesticks	Small utensils (spoons, tongs, etc.)
Hives	Tool handles
Lithaos	Walking sticks

Table 7.8   Proportion of useful plants used in art and technology. These data illustrate both the total range of plant species and families used in traditional art and technology, and the number of different plant taxa which are used specifically in making tools, utensils, or items used in recreation (including games and musical instruments) (see [1]Smith & Kalotas 1985; [2]Felger & Moser 1985; [3]Milliken *et al*. 1992; [4]Medley 1993)

Group	Total no. plants used	Plants used in art and technology	No. used for specific purposes*		
			Hunting	Utensils	Play
Bardi[1]					
Species	150	54	33	10	4
Families	56	24	18	7	4
Seri[2]					
Species	384	94	15	46	45
Families	80	42	12	26	25
Waimiri Atroari[3]					
Species	319	118	32	31	10
Families	64	39	12	19	9
Pokomo[1]					
Species	97	38	5	15	10
Families	38	18	3	10	2

*Excluding those plants used as sources of fibre.

make toy whistles (Densmore 1974). Elsewhere the Bardi Aborigines use the pods of the legume *Caesalpinia globulorum* to play marbles, while the Seri Indians use the seeds of *Agave subsimplex* as beading (Felger & Moser 1985; Smith & Kalotas 1985; Milliken *et al*. 1992).

The great importance of plant products in traditional art and technology, is reflected in the range of species and families which are used by different groups, with up to one-third of all useful plant species being used in making tools and craft items (Table 7.8). The majority of these provide materials such as wood for tool handles, weapons and utensils such as mortars and pestles; canes for arrow shafts and scaffolding; leaves for cooking and storing foods; bark for containers and cooking utensils; even beads for decorative items such as necklaces and the beaded aprons (*kwyie*) traditionally worn by the Waimiri Atroari women. Again, some items demand the use of particular plant species, while others have less specific requirements. For example, both the Seri and Chippewa Indians use a considerable variety of plant species for items such as toys and games (Densmore 1974; Felger & Moser 1985). In contrast the Seri and Waimiri Atroari exploit the characteristic properties of much more specific plant taxa in the manufacture of their fishing equipment (Felger & Moser 1985; Milliken *et al.* 1992). As with the raw materials used in construction, the choice of plants used in art and technology is determined on the basis of a combination of functional criteria such as shape, ease of working, strength and flexibility.

Plants as Sources of Fibres

Plant fibres, as described in Chapter 2, are vital to traditional material culture, providing lashing materials for both the construction of buildings and the manufacture of wooden tools. They also provide materials for additional objects such as mats, baskets, brushes, hats and paper, while softer fibres are used for weaving textiles. Fibres from most types of plant organ have found some use in traditional material culture. For example, the flexible roots of tamarack (*Larix laricina*) have been used by the Chippewa for weaving bags and sewing birchbark canoes; fibres from the leaves of the palm *Phoenix reclinata* have been used for making brooms among the Pokomo; wood fibres from the yellow cedar (*Chamaecyparis hootkatensis*) have been woven into soft textiles of high quality by the aboriginal peoples of British Columbia. Yet while both fibre cells themselves, and their potential applications are extremely diverse, the range of plants exploited for fibres within a particular community is comparatively low (Box 7.3). Indeed, throughout Amazonia, four species of a single genus—*Ischnosiphon*—are used widely in basketry, while the Seri of Baja, California use only one species—*Jatropha cuneata* (Euphorbiacae)—in weaving baskets. Similarly, many peoples throughout the world rely largely on cultivated cotton (*Gossypium* spp) for the production of textiles. The use of such a restricted number of species reflects the small number of plants producing fibres which are sufficiently strong, yet flexible enough for weaving baskets and producing cloth.

Box 7.3   Plants as sources of fibre. As we have seen in Chapter 2, useful fibres can be derived from a number of anatomically distinct plant structures, ranging from the seed hairs of cotton (*Gossypium* spp), to the fibre cells which lend support in woody tissues. Moreover, while fibres found in the inner bark of tree species are largely *xylem fibres*, those found in the stems and roots of dicotyledenous species (*bast fibres*) and those obtained from the leaves of monocotyledenous species (*hard fibres*) are both essentially *phloem fibres*. The restricted nature of the plant species used as sources of fibres, reflects the fact that relatively few taxa produce fibres of sufficient length, strength and flexibility for use in weaving baskets or spinning twine. Indeed of the 56 species of plant described in the table below, more than half belong to one of four families: the Arecaceae (palms), Leguminosae, Moraceae and Sapindaceae.

TRADITIONAL USES OF PLANT FIBRES

Plant fibres have a wide range of applications, providing raw materials for items such as baskets and small utensils such as sieves and cordage used as lashing material or in making hammocks. In addition, soft fibres used in spinning twine used in textile manufacture, although in some cases traditional clothing or blankets are made from bark cloth—sheets of fibrous inner bark which are washed and pounded to produce soft, pliable sheets of fibre cells which are held together by natural plant *pectins* and *gums*.

Use of plant fibres in material culture

Basketry

Amazonia—split stems of several species of *Ischnosiphon* (Marantaceae)[1,2,3]

Baja, California—split stems of *Jatropha cuneata* (Euphorbiaceae)[4]

Kenya—leaf fibres from *Phoenix reclinata* (Arecaceae)[5]

Textiles, yarn and paper

Tahiti—inner bark of *Broussonetia papyrifera* (Moraceae) and *Artocarpus altilis* (Moraceae) used in bark cloth[6]

Philippines—leaf fibres of pineapple (*Ananas comosus*—Bromeliaceae) woven into extremely fine textiles[6]

South America—fibres from fruits of *Eriotheca globosa* (Bombacaceae) and from leaves of *Mauritia flexuosa* (Arecaceae) used to produce yarns[3,7]

Worldwide—seed hairs of *Gossypium* sp (Malvaceae) used in weaving textiles

Mesoamerica—bark of *Ficus* sp (Moraceae) used in making paper[9]

Egypt—leaf fibres of *Cyperus papyrus* used in ancient Egypt for making paper[8]

Cordage

Amazonia—leaf fibres of *Bromelia* sp (Bromeliaceae), petioles of *Heliconia* sp (Heliconiaceae) and aerial roots of *Heteropsis* sp (Arecaceae)[1]

Australia—aerial roots of *Ficus* spp (Moraceae), inner bark of *Brachychiton diversifolius* (Stericulaceae) and climbing stems of *Tinospora smilacina* (Menispermaceae)[10]

See [1]Milliken *et al.* (1992); [2]van den Berg (1984); [3]Boom (1989); [4]Felger and Moser (1985); [5]Medley (1993); [6]Lewington (1990); [7]Balick (1984); [8]Simpson & Conner-Ogorzaly (1986); [9]Miller & Taube (1993); [10]Smith & Kalotas (1985).

PLANT TAXA AS SOURCES OF PLANT FIBRES

However, while the uses of fibres are often very varied, their sources are often fairly restricted. For example, only a very small proportion of the plants used by

continued

_____ *continued* _____

groups such as the Aboriginal Bardi, and the Amazonian Waimiri Atroari are used as sources of fibre:

Group	No. plants used	Sources of fibres	Plants used for specific purposes	
			Mats/baskets	Hammocks/rope
Bardi[1]	150 (56)	5 (5)	0 (0)	5 (5)
Seri[2]	384 (80)	5 (4)	3 (2)	3 (3)
Waimiri Atroari[3]	319 (64)	35 (13)	12 (4)	26 (11)
Pokomo[4]	97 (38)	11 (7)	2 (2)	10 (7)

First number=no. of species; number in brackets=no. of families; data taken from [1]Smith and Kalotas (1985); [2]Felger and Moser (1985); [3]Milliken *et al.* (1992), Medley (1993).

Plant Extracts and Exudates

Of equal importance in the manufacture of buildings, tools and other items are a number of plant chemicals which are either exuded or extracted from various plants. These chemicals include a wide range of pigments, resins, latexes, waxes, oils and gums which are used to fulfil a number of functional and aesthetic roles—from waterproofing canoes to painting beads. For example, the Waimiri Atroari, manufacture blocks of a substance known as *manji* that provides both a strong adhesive and an effective sealant which is used for caulking canoes and boats. The *manji* is prepared from exudates of *Symphonia globulifera* (Guttiferae) which are boiled with the latex of either *Couma macrocarpa* (Apocynaceae) or *Brosimum parinarioides* (Moraceae) and moulded to form solid blocks which can be stored and used as required (Milliken *et al.* 1992). Other plant products which are used in material culture include a variety of dyes and pigments, strengthening agents and sealants used in pottery, and a range of products used in curing leather. These, as well as plant chemicals used in other aspects of traditional life, are discussed in greater detail in the following chapter.

The Production and Processing of Useful Plant Materials

The manufacture of many traditional artefacts is dependent, either entirely or substantially, on a supply of plant-based raw materials. Some of these, such as cotton and certain species of rattan, are cultivated. However, the majority of these resources are normally. harvested from wild or—at least among swidden cultivators—from managed, or semi-domesticated populations. For example, of the 30 or so plant species cultivated by the Waimiri Atroari, only four—*Bixa orellana, Bromelia* sp, *Gynerium sagittatum* and *Crescentia crujete*—are used in construction or crafts. These four species (which are used respectively for their pigments, or as twine, arrow cane or storage containers) represent only about 2 per cent of the total number of species employed in material culture; the remainder are collected from forest sites (see Milliken *et al.* 1992). Once harvested, plants are processed using a variety of techniques. Fibres may be dried and either bleached or dyed before they are woven into baskets, or rolled into yarn for weaving textiles; the processing of wood, bark and other materials also involves a range of skilful techniques, from the harvesting of large sheets of inner bark, to the delicate fletching of hunting arrows. This section outlines just a few examples of the diverse array of methods which are used in traditional technology throughout the world.

Managing Resources for Material Culture: The Bora Agroforestry Project

Among traditional swidden horticulturalists, many of the plants used in material culture are found within their swidden plots and regenerating of fallows, where often they are cultivated or at least deliberately managed. For example, between 1981 and 1983, researchers involved in the *Bora Agroforestry Project* carried out a detailed investigation of the swidden-fallow management in Brillo Nuevo, a Bora Indian village in the northeastern region of Amazonian Peru. The Bora arrived at this site about 50 years ago, and almost all the land close to the village is now in some stage of secondary regeneration following earlier swidden cultivation. During the course of the study, it was found that together, the swidden and fallow fields of the Peruvian Bora contain at least 78 species which are used in their material culture, several of which are wild species which are actively managed. For example, species such as tropical cedar (*Cedrela odorata*) and *Jacaranda copaia*, which provide valuable construction materials, are commonly protected and/or transplanted into swidden sites, while the high proportion of useful plants in older fallows indicates the deliberate manipulation of ecological succession to favour the growth of useful species (Table 7.9).

Table 7.9 Swidden fallow management of plant species used in material culture. In the traditional agroforestry of the Bora Indians of the Peruvian Amazon, changes in species composition during managed regeneration of fallow fields suggest the manipulation of ecological succession to increase proportions of plants used in material culture. For example, while young swidden sites of up to 4 years old contain mostly cultivated food species such as corn, rice, manioc, peppers and young fruit trees, the older fallow fields exhibit an increasing proportion of species which are used in construction or which provide paints and resins used in material culture. Many of these are perennial species which are actively transplanted and/or protected by the Bora, including a number of species used for timber (*Astrocaryum chambira, A. hicungo, Cedrela odorata, Iriartea* spp and *Schelea* sp), roofing *Mauritia flexuosa* and *Phytelephas macrocarpa*), fibres (*Euterpe oleracea, Gossypium barbadense* and *Ficus* spp), storage vessels (*Theobroma bicolor* and *Crescentia cujete*) or exudates such as pigments (*Genipa americana* and *Bactris gasipaes*) and resins (*Dacryodes* sp); see Denevan and Padoch (1988)

Fallow	Abundance of useful plants (%)					
	Cultivated	Construction	Canoes	Craft	Extracts*	Total[†]
10 years	8.32	30.24	0.32	6.08	1.28	37.92
15 years	0.31	27.53	0.31	9.09	2.10	39.03
20 years	2.28	50.13	0.76	0.00	4.80	55.69
35 years	6.77	14.16	0.43	0.43	5.92	20.94

*Plants used as sources of paints and resins which are used in art and technology; [†]total number of plants used in material culture; data taken from Unruh and Flores Paitán (1987).

Wood Technology Among the Waimiri Atroari

Like other traditional societies throughout the world, the Waimiri Atroari of the Brazilian Amazon display considerable botanical knowledge and technical skill in their material culture. For example, among their traditional hunting equipment, the principal weapon is the longbow or *warpa* which may be up to 2.2 m in length (Milliken *et al.* 1992). This weapon is fashioned specifically from the heartwood of the Moraceous *Brosimum guianense* which possesses the strength and flexibility required for the manufacture of these long hunting bows. The bow itself is rectangular in section and is strung with a strong three-ply cord, known as *karwa*, which is made from the leaf fibres of a cultivated species of *Bromelia*. The leaves are cut at the base and split, and the silky fibres extracted by pulling the leaf sections through a special string noose. The fibres are then washed, dried and rolled on the thigh into a twine of appropriate strength and thickness, before being attached to the wooden bow. The arrows (*pyrwa*) are constructed from the long, straight inflorescences of *Gynerium sagittatum* (Poaceae), which is also cultivated in order to maintain a regular supply. These inflorescences, which make up the arrow shaft, are straightened carefully by heating irregularities over a low fire and gently flexing, before they are dried in the sun. The arrow shaft is weighted at each end by a small plug of a dense wood such as *Myrcia minutiflora* (Myrtaceae) or *Peltogyne* sp (Leguminosae), to improve the accuracy of flight and to prevent the arrow from breaking off once embedded in the quarry. The junctions between the hollow shaft and these hardwood sections are secured using a length of *karwa*, which is looped round the joint and pulled tight by means of two toggles, thereby compressing the shaft at the point of insertion. These joints are then bound using *karwa* cord and *manji* resin. Although traditionally fitted with some type of wooden point carved from *Brosimum guianense* (Moraceae), many arrows are now fitted with a barbed steel point, whose size and shape varies according to specific game. For example, wooden arrow heads used for hunting birds are about 30–60 cm in length and are ridged; those used in hunting fishes are about the same size yet are smooth; steel arrowheads used for fishing are between 12 and 15 cm long, and are often barbed.

Bark Products in Northern USA

As we have seen, tree bark can provide an important source of raw material to many aboriginal peoples, and like other aspects of plant use, the effective harvesting and processing of the bark requires considerable botanical knowledge and technical skill. Among the Chippewa of Minnesota, the bark of the paper birch (*Betula papyrifera*) and of cedar (*Juniperus virginiana*) have been used in numerous items ranging from canoes to small utensils. For example, birch bark has a reputation for protecting even edible substances such as fish and maple syrup from decay, and has commonly been used in making a range of storage boxes or *makuks*. These lidded containers were sewn using the split roots of *Larix laricina* and varied in size from those capable of holding about 50 g of sugar, to those with a capacity of 20 or 30 times greater; larger *makuks* could even be used as buckets, once the seams were covered in pitch, and a carrying handle attached. Other items made from birch bark include cooking vessels, which are made from green bark and filled with water, and smaller items such as dishes, spoons and trays used for winnowing wild rice.

The bark itself is normally collected during June and July, when it is relatively easy to remove. Bark harvesting is carried out very carefully, ensuring the selection of a suitable tree, and carrying out ceremonial activities regarded as necessary in obtaining permission to collect bark.

The tree is then felled by axe, in such a way that the tree rests on the stump as it falls, thus preventing the bark from being soiled by falling on the ground (Densmore 1974). Following felling, a vertical cut is then made along the length of the trunk and the bark is unrolled from the trunk to produce strips of bark whose width varies according to its intended use. Sheets of bark are then tied into thick packs using strips of basswood fibres and are stored until use. In order to make *makuks* and other items, the stored bark is unrolled by exposing to the heat of a fire, which renders the sheets pliable. Once softened in this way, the bark can be manipulated as required, and will retain the desired shape on cooling, to provide bark products which can be used for up to 10 years (Densmore 1974).

Fibre Processing: Basketry in Baja, California

Like the manufacture of items from wood or bark, the successful processing of plant fibres is dependent on both the choice of plant material used, and the skill of the craftsmen involved. For the most part, basket making among the Seri Indians is largely the domain of women, some of whom have achieved such excellence in their craft that their baskets now rank among the most expensive in the commercial world (Felger & Moser 1985). The raw material is collected from wild populations of *Jatropha cuneata*, with suitable individuals being carefully selected on the basis of their pliability. Following the selection of straight branches of the appropriate size (1 m in length and 1.5 cm in diameter at the base), these are then cut at the base and taken to the village for processing.

The first stage of preparation requires the removal of any remaining twigs and leaves, after which the long stems are treated in a brushwood fire to loosen the bark thus aiding its removal. Once peeled and cooled, the branches are split into a series of splint types, each of which have specific roles in making baskets (Figure 7.2). The baskets themselves are coiled rather than woven, using a bundle of fine splints as a *foundation bundle* which is coiled and secured using flexible *sewing splints*; today baskets are made in various of shapes, ranging from bow-shaped baskets with sloping sides to deep, vase-shaped baskets which are often lidded, and which may be up to several feet in height. Many of these are decorated with complex geometric designs, that are created using dyed stitching splints, and which, at the beginning of this century, were coloured using only a reddish brown dye normally prepared from the roots of *Krameria grayi* (Krameriaceae). Today, however, the modern, commercial baskets now include patterns in black, yellow, or yellow-orange which are created using a range of plant-based dyes.

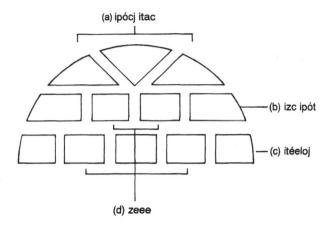

Figure 7.2 Basketry splints of the Seri Indians. Once peeled, the flexible stems of *Jatropha cuneata* are processed to form a range of splint types, each of which have specific roles: (a) the flexible rounded strips from the outermost part of the stem (*ipócj itac*) form the source of top grade stitching splints; (b) the second layer of strips (*izc ipót*) also provide stitching splints, but their woody nature renders them of inferior quality to *ipócj itac*; (c) the flat inner strips are divided to produce splints from the edges (*itéeloj*) which are used as stitching splints only in poor quality baskets; (d) the coarse, less flexible, internal splints from the inner strips (*zeee*) are used to make up the *foundation bundle*

Adapting to Change

Material culture—like any other aspect of traditional life—is in a constant state of change, as new materials and technologies are introduced, or as traditional resources are lost. In some cases, such changes have resulted from the commercialisation of plant products, which has frequently led to the overharvesting of traditional raw materials. For while local marketing of minor forest products has long played a part in traditional economic life (Padoch 1988; Dove 1994), recent intensification of demand—due to both the rapid expansion of rural populations and the increased size of urban and foreign markets—has led to unsustainable harvesting of many useful species. For example, while plant extraction for subsistence purposes by the Kenyan Pokomo does not lead to a loss of forest area, outside the protected TRNPR (Tana River National Primate Reserve), increased demand from growing populations of migrant workers has led to a measurable increase in forest degradation (Medley 1993). Similarly, the escalating demand for rattan products in Nigeria has resulted in a significant depletion of the rattan supplies available to fulfil the subsistence needs of forest dwellers (Morakinyo 1994).

In response to these changes, many communities have developed new methods of managing the supply of raw materials. As we have seen in the Bora Agroforestry Project, this may involve a move to the planting and protection of useful species—an approach which has similarly been adopted towards rattan supplies in East Kalimantan, Indonesia. Here, swidden farmers have incorporated rattan cultivation into their swidden cycle, allowing a reliable harvest after 7–10 years of fallow management (Morakinyo 1994). Elsewhere, non-destructive methods of harvesting have been developed where commercial demand for plant products have rendered earlier harvesting methods unsustainable. For example, in the Esmeraldas province of coastal Ecuador, the continued harvesting of palm leaves by tree-felling has led to a depletion of useful palms—particularly of *Astorcaryum standleyanum* whose leaves provide fibres for a wide range of marketable commodities such as hammocks, baskets, hats and even furniture. In the neighbouring Manabí province, however, the development of a special harvesting tool—essentially a metal chisel attached to a bamboo handle of up to 10 m in length—has allowed the non-destructive harvesting of leaves from the ground (Borgtoft Pedersen 1994).

The production of ironwood sculptures among the Seri Indians provides a good example of adaptation to both economic and ecological changes. For during the early 1960s, decreased availability of certain subsistence resources and the increasing influence of the tourist industry, combined to produce a significant influence on the traditional Seri life-style. In need of a source of income to supplement a diminishing resource base, certain individuals began to adapt traditional wood-carvings for a more commercial market. Originally carved from soft-woods such as *Bursera hindsiana*, the advent of modern tools made working with hardwood a possibility, allowing the sculpting of the leguminous ironwood (*Olneya tesota*), a rich dark brown hardwood which can produce a smooth, lustrous finish. The first ironwood carvings were produced by local craftsman José Astorga, and soon led to a successful trade in characteristic animal sculptures whose subtle details clearly demonstrated the intimate Seri knowledge of local wildlife (Felger & Moser 1985). At first, irregularities in the wooden pieces were filled using a resinous substance known as *lac*, which was obtained from surrounding creosote bushes (*Larrrea divaricata*); later the introduction of high-quality sandpaper allowed the production of a more even finish. Similarly, while early carvings were polished using vegetable or turtle oils, more recent pieces have been treated with shoe polish or wax to produce a high-gloss finish. Gradually then, the Seri have adopted new technologies, and developed new carving techniques, to produce high-quality carvings which, while remaining distinctly Seri, have become both artistically, and economically viable in the world art market.

SUMMARY

Despite many profound changes in traditional societies—both through increased contact with the industrialised world, and through the erosion of their natural resource base—plants and plant products continue to play a fundamental part in the material culture of many of the world's indigenous communities. They provide timber and other materials for the construction of dwellings and items of transport; they also provide a range of materials used in the manufacture of tools, crafts and decorative items. In many cases, the collection and processing of these plant materials requires considerable botanical knowledge and technical skill. For example, the crafting of the longbow used by the Waimiri Atroari of Brazil exploits the specific properties of several plant species, while requiring skills in wood technology and fibre processing. In addition to knowledge relating to the properties of specific plants and the skills required for their processing, traditional peoples must also ensure a continued supply of useful materials—often through the management of useful species throughout the swidden cycle. In some cases, however, supplies of traditional raw materials have been significantly reduced, due to factors such as an increase in commercial demand or habitat erosion in 'developing' regions. In such cases, local peoples have responded through experimentation with other materials and technologies to develop suitable alternatives.

Chapter 8

Traditional Phytochemistry

As about three-quarters of the biologically active plant-derived compounds presently in use globally have been discovered through follow-up research on folk and ethnomedicinal uses, it is imperative that ethnobotanically directed research to find active pharmaceuticals be instituted on a broad scale...

Walter Lewis and Memory Elvin-Lewis (1994) in *Ethnobotany and the Search for New Drugs*

INTRODUCTION

Just as the physical characters of plants—their fibres, wood and other useful structures—have formed the basis of the material culture in many human societies, their chemical compounds also have long been exploited by human populations. For example, archaeological investigations at the Coppergate site, York (UK) have revealed that extracts of several plant species were commonly used in textile processing during AD ninth century (Hall 1993), while in Neolithic sites in Switzerland, the recovery of seeds from the opium poppy (*Papaver somniferum*) have been interpreted as indicating its early use as an intoxicant (Rudgely 1993). Today, plant chemicals continue to provide an important range of products to both traditional and industrialised communities, including vegetable oils which have been used not only for cooking, but as industrial lubricants and in the manufacture of materials such as paints, varnishes and soaps; plant latexes, which, with their elastic properties, have provided an invaluable source of rubber; other plant chemicals are used as pigments, dyes and as tanning agents used in curing leather. Plants also provide a source of bioactive compounds, many of which can be used as poisons, stimulants or therapeutic agents (Farnsworth 1990).

As a consequence, some of the world's most economically important plant chemicals have been discovered through early ethnobotanical observation. For example, reports dating from the first explorations of Amazonia outlined a curious custom whereby native peoples would dip their feet into a certain plant exudate before holding them into the smoke of a fire to produce a flexible protective coating around the foot (see Simpson & Conner-Ogorzaly 1986; Boom 1989). In this case the tree was *Hevea brasiliensis* (Euphorbiaceae), the species from whose latex all natural rubber is now obtained. In a similar way, quinine, the antimalarial drug obtained from Peruvian *Cinchona officinalis* (Rubiaceae) was discovered as a consequence of its traditional use in treating fever (see Lewington 1990). Now, as the search for useful natural compounds has intensified, several researchers have adopted an *ethno-directed approach* in the identification of useful plants, and a considerable literature on traditional phytochemistry has now accumulated. This chapter introduces the applications of plant chemicals used by traditional peoples, discussing, where appropriate, the nature and biological activity of the chemicals involved.

PHYTOCHEMICALS USED IN NON-MEDICAL APPLICATIONS

Many plant species produce chemicals which are used in traditional daily life. The sticky *exudates* of many resiniferous plants provide adhesives used in tool-making as well as caulking materials for sealing boats, buckets and other items; pigments extracted from leaves, bark and other plant tissues are used as cosmetics and in the decoration of traditional arts and crafts; oils are used in cooking and as cleansing agents; volatile oils and pungent compounds are used in preserving or flavouring foods; toxic plants can provide a source of poisons used in subsistence activities such as hunting and fishing. In some cases, these useful chemicals are provided by primary compounds. For example, *storage products* such as seed oils and starch, or primary wall components such as pectin have provided important raw materials used in a variety of applications (Table 8.1). More frequently, however, the useful functions of plant extracts and exudates are determined by the presence of various plant secondary compounds—the terpenoids, phenolics, alkaloids and other chemical groups which are outlined in Chapter 2, and which are distributed discontinuously throughout the plant world.

Table 8.1 The nature and uses of plant chemicals. Many plant species accumulate a wide range of chemicals which fulfil many ecological functions. Primary compounds such as starches and oils store energy for subsequent growth, while secondary compounds are apparently involved in activities such as attracting pollinators and protecting plants against biological or other forms of environmental stress. In many cases, the physical and/or chemical characteristics of these substances can also be exploited by humans, providing many of the raw materials used in material culture, personal grooming and traditional medicine (see Simpson & Conner-Ogorzaly 1986; Harborne 1988; Hendry 1993)

Phytochemicals	Probable functions in Planta	Common uses
Primary compounds		
Glycerides (fatty oils)	Storage products in seeds, nuts and fruits	Lubricants; cosmetics; cooking; metal salts used as soaps
Starches	Storage products in seeds, tubers and other storage organs	*Sizing* of textiles; gelling agents and weak adhesives
Pectins	Cement cells together; form part of the primary cell wall	Gelling agents
Secondary compounds		
Waxes	Protection against water loss; defence	Waterproofing; illumination
Pigments	Attraction of pollinators; protection against UV damage; photosynthesis	Dyes; body paints; pigments; medicine
Tannins	Waterproofing; defence	Dyes; tanning agents
Resins	Wound protection; defence	Sealants, illumination, adhesives, preservatives
Gums	Wound sealing	Gelling agents; sealants and adhesives
Latexes	Wound sealing; defence	Sealants; adhesives
Volatile oils	Attraction of pollinators; defence; protection against fire	Flavours and fragrances; food preservatives; medicine
Saponins	Defence	Fish poisons; soaps
Hormones	Attraction of pollinators; defence	Birth control; pest control
Alkaloids	Defence	Pest control; medicine

Nature and Sources of Useful Chemicals

Useful phytochemicals can be isolated from various tissues and organs and from a wide range of species. Some types of chemical—including storage compounds such as starch—are widely distributed throughout the plant kingdom; others are much more restricted in their occurrence. For example, *betacyanin*, the deep red pigment of beetroot (*Beta vulgaris*) is restricted to members of the *order* Caryophyllales, which includes families such as the Amaranthaceae, Cactaceae and Chenopodiaceae (Hendry 1993), while the bitter tasting *cardenolides*—a group of toxic terpenoid compounds—are found predominantly among the Apocynaceae, Asclepiadaceae and Scrophulariaceae. As a consequence of these distribution patterns, it is often possible to identify particular plant taxa as sources of certain chemicals, as outlined in Table 8.2.

Plant Chemicals Used in Subsistence Activities

During the course of their evolution, many plant species have been protected by their ability to accumulate toxic compounds. In some cases, these toxins are effective against large herbivores; for example the oestrogen-like compounds found in plants such as the date palm *Phoenix dactylifera*, the seeds of pomegranate (*Punica granatum*) and the seeds of apple (*Malus pumila*) (Harborne 1988). Others, including pyrethrin from *Chrysanthemum cinearifolium* and azadirachtin from the neem tree (*Azadiarachta indica*), are specifically active against certain species of insect (Ley 1990). Others still, may be toxic towards micro-organisms or even other plants. On the basis of the specific nature of their biological activity, some of these toxic compounds have found a role in the subsistence activities of traditional societies, providing a source of *arrow poisons* and *fish poisons* (Box 8.1) many of which are widely used today.

Arrow poisons prepared from extracts of poisonous plants have been used throughout the world for millennia. The earliest reliable written evidence for the use of plant-based arrow poisons, appears in the Rig Veda (which emerged about 1200 BC) while in Europe, both Homer (eighth century BC) and Virgil (70–19 BC) mention the use of toxic plant extracts as arrow poisons (Mann 1994). Among many traditional communities today, plant toxins continue to provide important hunting poisons, some of the most well documented of which are the *curare* preparations of South America. Here, traditional *curare* is commonly prepared from the bark of the rainforest vine *Chondrodendron tomentosum*, the most potent constituent of which is

Table 8.2 Phylogenetic distribution of traditional sources of plant chemicals. Just as the morphological or other physical characteristics of related plant species are often similar, so too the chemicals produced by related taxa are often closely related. For example, popular flavouring herbs, such as thyme (*Thymus vulgaris*), oregano (*Oreganum vulgare*), basil (*Ocimum basilicum*) and sage (*Salvia officinalis*), all produce volatile oils containing related monoterpenes—and all belong to the family, Lamiaceae; similarly, the pungent principles of the closely related chili peppers and sweet peppers (*Capsicum* spp; Solanaceae) are provided in both by the alkaloid capsaicin. Where such chemicals have a restricted distribution throughout the plant kingdom, these natural patterns are often reflected in the ways in which particular plant taxa are utilised within traditional communities (see Smith 1976; Simpson & Conner-Ogorzaly 1986; Harborne 1988; Hay & Svoboda 1993)

Phytochemicals	Characteristic plant sources
Primary compounds	
Glycerides (fatty oils)	Widespread in seeds; however, important sources include olive (*Olea europeae*; Oleaceae) rape (*Brassica* spp; Brassicaceae) and coconut (*Cocos nucifera*; Arecaceae)
Starches	Ubiquitous; however, important sources include corn (*Zea mays*), wheat (*Triticum* spp) and sorghum (*Sorghum bicolor*) all of which are members of the Poaceae
Pectins	Ubiquitous; however, important sources include apples (*Malus pumila*; Rosaceae) and the peels of citrus fruits (*Citrus* spp; Rutaceae)
Secondary compounds	
Waxes	Ubiquitous; however, accumulated at high levels in the carnauba wax palm (*Copernicia cerifera*; Arecaceae) and candellila plants (*Euphorbia antisyphilitica*; Euphorbiaceae)
Pigments	Many are widespread (or ubiquitous) including chlorophylls, carotenoids and flavonoids; however, sources of particular pigments are generally restricted to certain species
Tannins	Like other pigments, tannins are widespread; however, important sources include various members of the Leguminosae (e.g. *Caesalpinia brevifolia*; and *Acacia* spp) and of the Fagaceae (e.g. *Quercus suber*)
Resins	Resins are common among the members of several plant taxa, including the families, Myrtaceae (e.g. *Eucalyptus* sp) and Burseraceae (e.g. myrrh; *Commiphora myrrha*) and the order Coniferales (e.g. pines; *Pinus* spp)

Table 8.2 (*continued*)

Phytochemicals	Characteristic plant sources
Gums	Widespread; however, important sources are found among the Leguminosae including gum arabic (*Acacia senegal*) and several species of *Astragalus*
Latexes	The most important sources of latex are members of the Euphorbiaceae, including para rubber (*Hevea brasiliensis*) and ceara rubber (*Manihot glaziovii*); other sources include certain members of the Sapotaceae
Volatile oils	Volatile oils are widely distributed in tropical and temperate angiosperms, particularly in families such as the Lamiaceae (mints), Myrtaceae (eucalypts), Piperaceae (pepper), Rutaceae (citrus) and Lauraceae (bays)
Saponins	Found in more than 70 plant families, particularly among the Leguminosae. Euphorbiaceae and Sapindaceae
Hormones	Include human sex hormones (particularly among the Leguminosae) and insect moulting hormones (e.g. *Taxus baccata*; Taxaceae)
Alkaloids	Chemically diverse and widely distributed; e.g. *tropane alkaloids* are common among the Solanaceae; *pyrrolizidine alkaloids* among the Compositae; *lupin alkaloids* among the Leguminosae

the alkaloid *d*-tubocurarine. This toxic alkaloid binds to the *receptors* of nerve cells (*neurons*) thus inhibiting the action of the *neurotransmitter* acetylcholine. As this transmitter is involved in causing muscle contractions and speeding up heart rate, the blocking of its normal activity leads to relaxation at low doses, and to total muscle paralysis at higher levels. In poisoned prey, this ultimately leads to death through paralysis of the heart and respiratory muscles, yet as the toxin is poorly absorbed via the gastrointestinal tract, consumption of the prey is not hazardous to humans (Mann 1994).

Like arrow poisons, plant-derived fish poisons (or *ichthyotoxins*) have also played an important part in traditional economies throughout the world. These toxins, which may be used to either kill or simply to stupefy fish, are usually employed in small rivers or pools with slow-moving water where the poisons can remain at sufficient concentrations to affect the fish. However, the art of fish poisoning is a delicate one, as the indiscriminate use of toxins may lead to the serious depletion of fish stocks in a given area. Hence, various fishing techniques have been developed in order to regulate fish supplies, including the construction of dams which allow the escape of smaller fish, while the cyclic poisoning of different areas, facilitates the periodic recovery of local fish populations (Acevedo-Rodríguez 1990). The chemical principles responsible for poisoning fish include a range of isoflavonoids, saponins, cardiac glycosides, alkaloids, tannins and cyanogenic compounds, as well as the polyacetylenic alcohol, *ichthyoctthereol*. Many of these act through interfering with some aspect of respiration; others act on the central nervous system causing a range of effects including heart failure and respiratory failure due to muscle paralysis.

Box 8.1 Traditional sources of plant chemicals used as arrow poisons and fish poisons. Plant toxins are diverse, and some are more specific in their toxicity than others. For example, while some toxins are effective only in fish, and represent no health risk to humans, others are potentially dangerous in a range of organisms and indeed poisoning due to the ingestion of fish stupefied by certain alkaloid-containing plants has been reported among several South American communities (see Acevedo-Rodríguez 1990). The plant sources used may therefore reflect a balance between the efficacy and safety of the poison obtained (see Felger & Moser 1985; Deneven & Padoch 1988; Acevedo-Rodríguez 1990; Mann 1994; Milliken *et al.* 1992).

Toxic plant products have been used in hunting, fishing, warfare, and even execution for millennia. For example, the use of certain plant poisons for stunning fish is mentioned in the writings of the Greek philosopher Aristotle (384–322 BC), and may date back to Palaeolithic times (Acevedo-Rodríguez 1990), while extracts of plants such as hemlock (*Conium maculatum*) and monkshood (*Aconitum napellus*) were widely used in executions throughout much of Ancient Eurasia (Mann 1994). Today, plant poisons continue to provide an important tool used in hunting in many traditional societies.

PLANT TOXINS USED AS POISONS

Like many other useful plant chemicals, these toxins are chemically extremely diverse, their mode of action varying considerably from plant to plant.

Chemical types	Biological activity
Arrow poisons	
Toxic alkaloids	Acetylcholine inhibitors, leading to muscle paralysis; e.g. *d*-tubocurarine
Cardiac glycosides	Inhibit sodium/potassium ion pumps in heart muscle cells, leading to abnormal heart activity (common among the Apocynaceae, Asclepiadaceae and Moraceae)

continued

— continued —

Chemical types	Biological activity
Fish poisons	
Isoflavonoids	Interfere with mitochondrial activity leading to asphyxiation; e.g. rotenone, tephrosine, lonchocarpin (restricted to the Papilionoideae tribe of the Leguminosae)
Saponins	Modify water surface tension, blocking respiration at the gills
Cardiac glycosides	Cardiac poisons as outlined above
Alkaloids	Wide range of chemical structures, leading to death through processes such as respiratory failure and muscle paralysis (particularly common among the Solanaceae, Loganiaceae and Menispermaceae)
Tannins	May act through cross-linking with gill proteins leading to asphyxiation
Cyanogenic glycosides	Release hydrogen cyanide to inhibit cytochrome oxidase, leading to asphyxiation at the cellular level (common among the Rosaceae, Flacourtiaceae and Euphorbiaceae)
Ichthyocthereol	Polyacetylenic alcohol interfering with mitochondria; leads to asphyxiation (restricted to members of the Compositae)

SOURCES OF PLANT TOXINS

Useful phytotoxins are clearly based on a range of chemical types, each of which exhibit a specific mode of action. Not surprisingly, these different chemicals are produced by a range of plant families—for example, throughout various countries in South America, poisons used in hunting and fishing are derived from at least 14 different plant families.

Country	Plant families used
Arrow poisons	
Columbia	Annonaceae, Menispermace
Brazil	Leguminosae, Loganiaceae, Myristicaeae, Palmae, Simaroubaceae
Fish poisons	
Peru	Apocynaceae, Compositae, Leguminosae
Brazil	Convolvulaceae, Euphorbiaceae, Graminae, Icacinaceae, Lecythidaceae, Leguminosae, Loganiaceae

Plant Chemicals in Material Culture

As we have seen in Table 8.1, plant chemicals provide a range of important materials which are used in traditional material culture. These include *dyes* and *pigments*, which are used both in the decoration of textiles and other artefacts, and in body painting; *tanning agents*, are used in the curing of animal skins; *adhesives* and *sealants* are important in the manufacture or waterproofing of items ranging from pottery vessels to boats; inflammable oils are used in illuminating homes. Useful phytochemicals are chemically extremely diverse and may be isolated from a wide range of plant sources. This is illustrated in Ghana, where more than 100 sources of plant pigments are used in the production of traditional dyes (Box 8.2); similarly, sealants are provided by a diverse range of resins, latexes, wax and the resinous substance, *lac* (Box 8.3). In general, the functional characteristics of these chemicals relate to their ecological role *in planta*. For example, the immiscibility of resins with water, their tendency to harden on exposure to air, and their antimicrobial activity appears invaluable in their wound-healing functions in the plant; clearly these very properties are exploited in the sealing and protection of boats, pottery and other items. Equally, the protective role of tannins, through their ability to complex with digestive enzymes and other proteins, is fundamental to their role in human culture, where their function is to stabilise the *collagen* (protein) molecules of animal skins, thus rendering the cured leather resistant to degradation (Simpson & Conner-Ogorzaly 1986).

Box 8.2   Traditional sources of plant pigments and tannins. Plants have provided an important source of pigments and tannins for millennia. While tannins have been used in preserving animal skins, plant pigments have been used in dyeing textiles or fibres used in basketry, in the decoration of ceramics or wooden artefacts, and as body paints, hair dyes and tattooing agents (Harborne 1984; Simpson & Conner-Ogorzaly 1986).

Plant chemicals have provided an important source of agents used in curing leather and in dyeing textiles for millennia. For example, chemical analysis of preserved fabrics dating back to the Predynastic period of Ancient Egypt (5000–3150 BC) has demonstrated that many were dyed using plants such as woad (*Isatis tinctoria*), madder (*Rubia tinctorum*) and safflower (*Carthamus tinctorius*) (Hall 1986). Similarly, the use of plant tannins in curing animal skins dates back to at least 1500 BC (Simpson & Conner-Ogorzaly 1986). Both tannins and pigments can be found in various plant parts, and are isolated from a wide range of plant species; both also are in use throughout many traditional societies today.

continued

continued

PLANT PIGMENTS

So far, more than 4000 plant pigments have been purified and characterised, yet despite their chemical complexity, most of these pigments fall into one of only five major structural classes: _porphyrins_ such as chlorophylls; the yellow and orange _carotenoids_; _flavonoids_ including the blue and red _anthocyanins_; red and brown _phenolic pigments_; finally, _indole derivatives_ include the deep blue pigment indigo which is isolated from _Indigofera tinctoria_ (Hendry 1993). Useful plant pigments have been isolated from many different plant families, and from almost all plant parts—including bark, leaves, seeds, fruits, stems and roots—to yield a wide range of colours from yellow through to crimson and purple to black. For example, throughout Ghana, useful pigments have been isolated from more than 100 plant species, as illustrated in the table below (see Abbiw 1990).

Colour	Plant species used
Indigo	8 including _Lonchocarpus_ sp and _Indigofera tinctoria_ (leaves)
Saffron	2: _Rhodognaphalon brevicuspe_ and _Lannea welwitschii_ (bark)
Blue	5 including _Saba florida_ (leaves, twigs, flowers) and _Cremaspora triflora_ (fruits)
Blue-black	6 including _Jatropha curcas_ (bark)
Brown	9 including _Maesa lanceolata_ (sap)
Red	21 including _Acacia nilotica_ (pods) and _Trilepisium madagascariense_ (latex, wood)
Red-brown	16 including _Lannea kerstingii_ (bark) and _Bridelia ferruginea_ (leaves)
Crimson	1: _Mirabilis jalapa_ (flowers)
Yellow	20 including _Anacardium occidentale_ (leaves) and _Borassus aethiopium_ (fruit rinds)
Orange	3 including _Bixa orellana_ and _Lonchocarpus sericeus_ (seeds)
Purple	2: _Bridelia ferruginea_ (bark) and _Pterocarpus erinaceus_ (bark, wood, roots)
Green	(combination of indigo plus yellow)
Grey	3: _Jatropha curcas_ (leaves) mango fruits (_Mangiflora indica_) and _Ziziphus mauritiana_ (bark)
Cinnamon	1: _Ziziphus mauritiana_ (bark)
Black	28 including _Acacia nilotica_ (seed), _Alchornea cordifolia_ (fruit) and _Syzygium rowlandii_ (bark)

continued

---

continued

PLANT TANNINS

Of the plant tannins, only two major groups are recognised—the large _condensed tannins_, which occur widely among higher plants, and the smaller _hydrolysable tannins_, which are restricted to a few dicotyledenous families, including the Fagaceae, Anacardiaceae and Leguminosae. The phenolic tannins are characterised by their ability to form strong cross-linkages with protein molecules, a process which helps to stabilise animal skins thus protecting them against water, heat and microbial degradation. Like other plant pigments, tannins can be extracted from many sources, including the leaves, unripe fruits, seed coats and bark of many species, where these compounds function in protecting the living plant.

Box 8.3   Traditional plant sources of sealants and adhesives. Throughout the plant world, species producing sticky exudates, or acting as high yielding hosts of the _lac_ insects, are often characteristic of particular plant families. For example, families including the Cactaceae, Stericulaceae and Ulmaceae are known to secrete mucilaginous gums; resin secretion is common among the Anacardiaceae, Burseraceae and the coniferous gymnosperms; latex secretion is particularly common among members of the Apocynaceae, Euphorbiaceae, Moraceae and Sapotaceae; _lac_ is commonly collected from members of the Leguminosae and Moraceae (see Simpson & Conner-Ogorzaly 1986).

The plant chemicals which are used most frequently in preparing adhesives or waterproof sealants are those provided by the _waxes, gums, resins_ or _latexes_ of a range of plant taxa. In addition, among several traditional societies throughout the world, additional sealants are provided by _lac_—a resinous substance of insect origin, yet which is found only on certain species of plant.

PLANT WAXES, GUMS, RESINS AND LATEXES

The waxes, gums, resins and latexes of plants constitute a range of diverse chemical types.

- Waxes, which are chemically similar to many plant seed oils, are made from simple alcohols and fatty acids, and are found covering the aerial surfaces of all higher plants. Among traditional communities, they provide sealants and polishes for both ceramic and wooden items
- True gums, which are polysaccharides of the acid salts of various sugars, are plant exudates which are produced only at specific times in response to wounding or injury. Owing to their ability to form stable bonds with water molecules, these substances can be used as water-based adhesives and gelling agents

continued

continued

- Resins are made up largely of diterpenoids, although they also contain substances such as monoterpenes, phenylpropenes and alkaloids. Owing to their antimicrobial activity and waterproofing qualities, these exudates are commonly used in sealing canoes as they prevent both water seepage and rotting due to microbial degradation or pest attack. They are also widely used as adhesives, and —due to their inflammable nature—as a means of illuminating homes
- Latexes are composed largely of polyterpenoids—polymeric compounds of up to 300 000 isoprene units. In addition, like plant resins, they often contain additional chemical substances. With their elastic properties and waterproof qualities, these substances are also in common use as adhesives and sealants

THE NATURE AND DISTRIBUTION OF *LAC*

Unlike other traditional sealants, *lac* is not directly synthesised by plants, but is produced by certain species of *scale insect* which occur naturally as parasites on a number of host plant species. The insects characteristically produce a layer of red resin on the branches of host trees, some of which are associated with greater yields than others. For example, in the Yunnan Province of China, *lac* insects (*Kerria yunnanensis*) are hosted by more than 300 local plant species, yet of these, only about 10 characteristically produce high yields of the resin (Sainte-Pierre & Bingrong 1994).

Reported in Chinese literature since at least the third century AD, *lac* is still used in many traditional societies, both as an adhesive and as a varnish. In some cases the hardened resin globules are simply collected from wild populations of host plants, as in the creosotebush scrublands of Baja, California (Felger & Moser 1985). Elsewhere, traditional farming systems have incorporated *lac* production, as in the southern region of the Yunnan Province, China. Here, traditional swidden cultivation practices encourage the growth of *lac* host trees, while farmers supplement spontaneous populations of the *lac* insects, by tying up bundles of small branches covered with live larvae. In addition, as these insects tend to settle preferentially on branches which are 3 years old, necessary pruning of trees is carefully managed to maintain sufficient branches of this age at any one time (Saint-Pierre & Bingrong 1994).

Plant Chemicals as Body Products

The range of body products provided by plant chemicals includes cleansing agents such as soaps and shampoos, and oils used in the preparation of pigments used in body painting (Table 8.3). Among the most widely investigated of these, are the water soluble *saponins* which often accumulate in the leaves of the living plant, and which, like many other secondary chemicals, afford protection against insect herbivory. Owing to the surface-active nature of these terpenoid derivatives, saponins exhibit soap-like properties which are readily detected by their

Table 8.3   Plant chemicals used as body products. While many cleansing agents which are used traditionally are based on saponins, other body products such as cosmetics and body paints involve various other chemical groups. For example, the hair conditioning shampoo which the Mexican Seri obtain from the ground fruit of jojoba (*Simmonsia chinensis*) is essentially a liquid wax, while body paints consist of various plant pigments, often mixed with a vegetable oil to facilitate their application to the skin. As this table illustrates, these chemicals can be obtained from a wide range of plant sources (see Felger & Moser 1985; Deneven & Padoch 1988; Milliken *et al.* 1992)

Group	Plant sources and uses
Cleansing agents	
Seri (Mexico)	Agavaceae (*Agave schottii*); soap
	Chenopodiaceae (*Atriplex polycarpa*); shampoo and washing clothes
	Curcurbitaceae (*Vaseyanthus insularis*) and *Brandegea bigelovii*); shampoo
	Rhamnaceae (*Zizyphus obtusifolia*); shampoo
	Simmonsiaceae (*Simmondsia chinensis*); shampoo
	Zygophyllaceae (*Guaiacum coulteri*); shampoo
Caribs (Dominica)	Graminae (*Gynerium sagittatum*); shampoo
Ecuador	Phytolaccaceae (*Phytolacca bogotensis*); shampoo and washing clothes
South America	Lecythidaceae (*Bertholletia excelsa*); soap
	Rhamnaceae (*Citrus aurantifolia*); washing clothes; (*C. sinensis*); soap
Africa	Caricaceae (*Carica papaya*); soap
	Meliaceae (*Trichilia dregeana*); soap and anointing hair and body
Europe	Caryophyllaceae (*Saponaria officinalis*); soap and delicate textiles
Cosmetics	
French Guiana	Caricaceae (*Carica papaya*); added to *anatto* body paint
	Caryocaraceae (*Caryocar villosum*); oil used in face creams and added to *anatto* body paint
	Nyctaginaceae (*Neea* spp); body paint
Bora (Peru)	Apocynaceae (*Rauvolfia macrantha*); black body paint
	Flacourtiaceae (*Casearia pitumba*); hair dye
	Hypericaceae (*Vismia* sp); cosmetic lotion
Kayapó (Brazil)	Leguminosae (*Enterolobium ellipticum*); body paint
Yanomami (Brazil)	Palmae (*Jessenia bataua*); blue body paint
South America	Bixaceae (*Bixa orellana*); red body paint
	Palmae (*Jessenia bataua*); fixes dye from *Genipa* to skin
	Rubiaceae (*Genipa* sp); black body paint

ability to cause foaming when added to water; hence their traditional exploitation as detergents and soaps.

Plant Chemicals as Fragrances and Food Additives

Two further groups of plant chemicals which, for millenia, have played an important part in human culture are the *volatile oils* of *aromatic plants*, and the *pungent principles* of spices such as ginger (*Zingiber officinale*) and chili (*Capsicum annuum*). Like many of the useful chemicals outlined in this section, these compounds are chemically diverse and can be isolated from a range of plant sources. For example, while phenolic compounds form the predominant essential oil components of the aromatic herb tarragon (*Artemisia dracunculus*), many other oils—including those of mint (*Mentha* spp) and thyme (*Thymus vulgaris*) are largely composed of terpenoids (Cotton *et al.* 1991a; Waterman 1993). Equally, while the phenolic compounds gingerol and paradol are responsible for the hot, pungent flavour of root ginger, other flavours, such as the pungent properties of garlic (*Allium sativum*) and mustard (*Brassica* spp), are based on various sulphur-containing compounds. On the basis of their pleasant fragrance and flavour, many of these chemicals are used as perfumes and/or in flavouring foods; others are exploited also for their preservative properties, the biocidal activity of many of these compounds often playing an important part in traditional methods of food storage and pest control (Box 8.4).

Box 8.4   Plants as fragrances and food additives. By virtue of their biological activity many of the plant chemicals providing flavour and fragrance compounds have also been used as protectants of stored foods and of other degradable items (see Deneven & Padoch 1988; Milliken *et al.* 1992; Medley 1993; Alkire *et al.* 1994).

Plants often contain chemicals which have a distinctive fragrance and/or flavour. For example, in many species, flower petals contain *essential* or *volatile oils* which are involved in attracting pollinators; in others, leaves contain essential oils where they are believed to play a part in plant protection (Deans & Waterman 1993). Equally, the *pungent principles* of, for example, root ginger (*Zingiber officinale*) are thought to function in protecting living plants against pests and disease. By virtue of both their aesthetic appeal and their preservative properties, many of these compounds have found a role in traditional societies— both as sources of flavours and fragrances, and as agents used in the preservation of stored foods and other items.

continued

_____ *continued* _____

VOLATILE OILS AND PUNGENT PRINCIPLES

In many cases, the essential oils of aromatic plants are composed largely of mono- and sesquiterpenes. For example, the major oil components of species such as peppermint (*Mentha piperita*), coriander (*Coriandrum sativum*) and rose (*Rosa* spp) are terpenoid in nature. Other species, such as tarragon (*Artemisia dracunculus*) and anise (*Pimpinella anisum*) contain oils which are predominantly phenylpropenoid in composition (Hay & Svoboda 1993). Additional flavour principles, contributing to the four main aspects of taste—sweetness, bitterness, sourness and saltiness—are provided by a range of chemical types. For example, the phenolic compounds gingerol and paradol are responsible for the hot, pungent flavour of root ginger, the pungent properties of garlic (*Allium sativum*) and mustard (*Brassica* spp) are based on various sulphur-containing compounds, while the alkaloids capsaicin and piperine provide the characteristic flavours of chili peppers (*Capsicum annuum*) and black pepper (*Piper nigrum*), respectively.

FLAVOUR, FRAGRANCE AND PEST CONTROL

By virtue of their diverse bioactive or aesthetic properties, many of the chemicals contributing to a plant's fragrance or flavour have been exploited in traditional communities as condiments, insect repellents, preservatives and fragrances. For example, aromatic resins may be burned as fumigants (Milliken *et al.* 1992), while tree ash may provide a common source of salt (Schmeda-Hirschmann 1994). Additional examples are outlined in the table below.

Group	Plant sources and uses
Flavours and fragrances	
Widespread	*Acorus calamus* (Araceae); flavour and fragrance
Bora (Peru)	*Quararibea obliquifolia* (Bombacaceae); salt
	Ocimum micranthum (Lamiaceae); flavour
	Geonoma acaulis and *Maximiliana sp* (Palmae); salt
Waimiri Atroari (Brazil)	*Parinari excelsa* (Chrysobalanacaeae); perfume
	Cymbopogon citrarus (Graminae); spice
Preservatives and insecticides	
Zanzibar	*Anacardium occidentale* (Anacardiaceae); preservative (fishing nets)
Kenya	*Salacia stuhlmanniana* (Capparidaceae); mosquito repellent

_____ *continued* _____

continued

Group	Plant sources and uses
widespread	*Acorus calamus* (Araceae); insecticidal
Bora (Peru)	*Lindackeria paludosa* (Flacourtiaceae); insecticide
Ecuador	*Minthostachys* sp and *Bystropogon* sp (Lamiaceae); insect repellents
Cuba	*Anacardium occidentale* (Anacardiaceae); insect repellent and fungicide
French Guiana	*Protium decandrum* and *P. sagotianum* (Buseraceae); fumigants
South America	*Bixa oreliana* (Bixaceae); insect repellent

Many flavour and fragrance compounds are widely distributed throughout the plant kingdom. In particular, volatile terpenoids are found not only among gymnosperms and angiosperms, but are also accumulated in a range of lower plants (Hay & Svoboda 1993). In contrast, phenylpropanoid volatiles are distributed more sporadically throughout higher plants, occurring only relatively infrequently among the gymnosperms, while the sulphur-containing mustard oils (isothiocyanates) and disulphides are particularly common among the Cruciferae and Liliaceae, respectively.

Production and Processing of Plant Chemicals

While certain phytochemicals, such as resins, gums and latexes are *exuded* from plants, either naturally, or in response to treatments such as wounding, others—such as the pigments and tannins—must be *extracted* from plant tissues using a range of complex processing techniques. For example, the latex of the rubber tree (*Hevea brasiliensis*) is skilfully *tapped* to ensure yields which are both safe and sustainable, while processing techniques used in the production of traditional pigments such as *Seri Blue* may facilitate complex chemical reactions which are necessary to produce the desired colour.

Collection and Processing of Plant Exudates

Plant exudates constitute any chemicals which either are naturally secreted on to the surface of the plant, or whose secretion can be induced artificially, and they include many of the latexes, gums and resins which are used in traditional societies. Substances which can be exuded naturally in significant quantities, include certain gums and resins, such as those of the brittlebush *Encelia farinosa* and of the resinous shrub *Frankenia palmeri*—both of which are collected by the Seri of Baja, California and used in sealing pottery (Felger & Moser 1985). More often, however, it appears that the production of natural resins can be greatly increased using artificial means. For example, in Southeast Asia, local resin collectors have found that only where resin-producing members of the Dipterocarpaceae are subject to a complex preparation process are significant yields achieved (Messer 1990). During this preparation process—which is based on prolonged, low-level wounding for between 6 and 12 months—small holes incised in the trunk are subsequently restripped at monthly intervals. On each occasion, the incisions are protected from disease attack by replacing the bark, and finally larger holes are made and resin collection begins. Significantly, recent investigations have demonstrated that this resin-inducing response can be mimicked through the application of close chemical relatives of *ethylene*, a naturally occurring plant hormone which is believed to play an important part in wound response. Hence, it seems likely that traditional resin-inducing methods are based on the artificial manipulation of the plant hormone system.

One of the most well documented methods of collecting plant exudates is that of *rubber tapping*, which also involves making an incision into the bark of productive trees in order to release the latex from the plants' secretory tissues. Like the process of resin induction described above, rubber tapping is an extremely skilled practice, for in many cases, high-yielding tissues are located in the inner bark nearest the cork cambium, where any damage sustained can lead to impaired bark regeneration, the introduction of disease, and often to the decreased lifespan of the affected tree. In the case of *Hevea brasiliensis*, trees are normally tapped early in the morning, when low rates of transpiration promote hydrostatic pressure in the laticifers, thus increasing the rate of latex flow. The rubber trees are milked by making a shallow, diagonal cut in the bark, at the lower end of which a small metal channel is inserted, allowing the latex to drip into a container. In traditional rubber producing areas of Brazil, the native trees are tapped every 3 days in order to maximise yields; elsewhere, domesticated rubber trees may be tapped more frequently. For example, among the Ibanic-speaking Kantu' of western Borneo, maximum latex flow occurs where trees are milked for between 15 and 19 days each month (Dove 1993). Collected in this way, continued latex production can allow mean daily yields of up to 25 g of dried rubber per tree, while production may be sustained over a period of up to 50 years or more (MacDonald 1991).

Once collected, these plants exudates may be processed in a variety of ways. The preparation of the *manji* resin of the Brazilian Waimiri Atroari has been described in the previous chapter, and many other useful resins are similarly manipulated on the basis of heating and reshaping blocks of resin, which harden on cooling and may be stored for later use. Rubber, in contrast, is *coagulated* prior to storage, using either smoke or, more recently, the addition of dilute acetic or formic acids.

Extraction and Processing of Other Plant Chemicals

Extraction procedures used in obtaining those chemicals which are not exuded vary considerably according to the chemical nature of the substance in question. For while water-soluble substances such as isoflavonoids and glycosidic compounds may be readily obtained through grinding tissues in water, others may require additional treatments such as heating or fermentation. In general, the preparation and use of plant dyes rank among the more complex methods, for the colour and fastness of the pigments used is strongly influenced by a number of factors, including pH and the presence of sugar residues. For example, while anthocyanin pigments are generally reddish in colour under acidic conditions, an increase in pH produces a shift to purple and blue; similarly, while extracts of the leaves of *Indigofera* spp are essentially colourless, treatments which promote certain chemical changes in the extract give rise to the deep blue coloration characteristic of the indigo pigment. The dyeing process is further complicated by the need for the use of *mordants*, which increase the ability of the dyes to adhere to fabrics. These mordanting agents—usually salts of metals—form a chemical link between the pigment molecule and the fibre molecules, producing both strong dye-fabric bonding, and in some cases, affecting the final colour of the dye (Simpson & Conner-Ogorzaly 1986).

Traditional Chemicals in the Commercial Market

While much of the discussion in this chapter has related specifically to the uses of plant chemicals among traditional cultures, many of these useful plant products are equally exploited in today's industrial societies throughout the world. For example, while natural plant pigments have largely been replaced by synthetic alternatives, natural tannins have retained much of their original importance in the commercial leather industry (Simpson & Conner-Ogorzaly 1986). Similarly, products such as natural rubber and carnauba wax are of considerable economic importance on a global scale (see Chapter 1), while natural volatile oils used widely in the flavour, fragrance and aromatherapy industries have an estimated value in excess of US$700 million *per annum* (Verlet 1993). In recent years too, a number of new plant products have gained significant economic value as a result of ethnobotanical research, notably the unusual wax-like oil of jojoba (*Simmondsia chinensis*) which has found a significant role in the commercial production of shampoos and cosmetics (Felger & Moser 1985; Simpson & Conner-Ogorzaly 1986).

ETHNOPHARMACOLOGY: THE PHYTOCHEMISTRY AND PHARMACOLOGY OF TRADITIONAL MEDICINE

So far, this chapter has discussed some of the non-medical applications of plant chemicals used in traditional societies, including the exploitation of their biological activity as toxins, pest deterrents and preservatives. However, one area of traditional phytochemistry which has received more attention than any other, is the use of plants in *traditional medicine*—indeed between 1993 and 1994, almost 20 per cent of the papers published in the journal *Economic Botany* were specifically concerned with the ethnobotany of traditional medical systems. Like other areas of ethnobotanical research, these studies involve a wide range of expertise, drawing on the skills of *medical anthropologists*, ethnotaxonomists and taxonomic botanists, phytochemists, pharmacologists and medical doctors.

The Nature and Study of Traditional Medical Systems

According to the World Health Organisation (WHO), 'traditional medicine' constitutes a rather vague term, used to distinguish any ancient, or culturally based health care practice from orthodox scientific medicine or *allopathy* (see Farnsworth 1994). This includes those systems which are currently regarded as indigenous, unorthodox, alternative, folk, fringe and 'unofficial', and includes both the major Asian systems such as Chinese, Ayurvedic, Unani and Unani Tibb medical systems—all of which have been well documented since ancient times—and less widespread, largely orally transmitted practices used by traditional communities elsewhere. Today, it is estimated that about 64 per cent of the total global population remain dependent on traditional medicine for their health care needs (Farnsworth 1994; Sindiga 1994).

Yet despite the worldwide importance of traditional medicine, orthodox attitudes to these diverse, culturally based systems of medicine have
been largely sceptical, and it was not until the 1970s that the systematic
study of traditional medicine and herbal remedies began to receive widespread academic attention. At this time, the WHO, encouraged by the
success of Chinese *heterox* healthcare programmes, began to encourage
the study and application of traditional medicine, with the aim of combining traditional and Western resources to create an acceptable and
effective solution to historical problems of sanitation, nutrition and
primary health care (see Lozoya 1994). Since this time, considerable attention has focused on the analysis of traditional plant remedies, and
together, workers from a number of disparate disciplines have produced a
considerable literature regarding both the nature of traditional medical
systems and the uses and efficacy of many traditional herbal remedies.

The Nature of Traditional Medical Systems: Shamanism and Herbalism

One important aspect of many traditional medical systems is that they
are characterised by the coexistence of several diverse and often com-
peting healing traditions within a single community. For example, in
Latin America, a range of folk healers—*herbalistas, curanderos, espiritistas, brujos, sobadores* and *comadronas*—are found. While *curanderos*
and *espiritistas* cure both physical and spiritual conditions using both
plants and rituals, the *herbalistas* specialise specifically in the selection
and preparation of herbal remedies; *brujos* (witches) also use herbs,
although their intentions are generally thought to be malevolent. In
contrast, *sobadores* or masseurs use massage as a major part of their
therapy, and finally *comdronas* (midwives) specialise in birthing and
maternal and infant health matters (Weller *et al.* 1992).

In Western terms, these various systems are often broadly divided into
two categories: *shamanism* and *herbalism*; the former dealing with diseases which are regarded as spiritual in origin, while the latter involves
conditions whose cause is regarded as 'natural'. Diseases of a spiritual
nature are generally the more severe, and are often symptomised by inex-
plicable internal pain. They may be considered the result of either evil
spirits entering the body, of loss of soul, or else a consequence of breaking taboos (Arenas 1987), and must be diagnosed and treated by the
shaman. In contrast, common ailments such as wounds, skin infections
and protozoan or helminthic infestations may be treated by herbalists,
many of whom have considerable knowledge of the properties of medicinal plants.

The Study of Traditional Medicine: Medical Anthropology, Medical Ethnobotany and Ethnopharmacology

Whether a given medical system is characterised by detailed documentation from ancient times, or whether it has been transmitted only through non-written media, any effective ethnomedical study must consider not only the nature of the remedies employed, but must also examine factors such as traditional concepts of the origins of disease, and local perceptions of the efficacy of particular treatments. Consequently, the study of traditional medicine is broadly divided into three main approaches: *medical anthropology* (including *clinical anthropology*) which examines cultural aspects of human health and disease; *medical ethnobotany* which studies the nature and applications of plants used within traditional medical systems; finally, *ethnopharmacology* involves the conventional chemical and pharmacological analysis of traditional remedies, including those based on plants.

Medical anthroplogy—or the anthropology of health—is a field of anthropological study which has grown enormously over the last 25 years. During this time various studies have focused on all aspects of culturally conditioned perceptions and experiences of ill health, including: local attitudes to disease and the roles of *traditional medical practitioners* (TMPs); symbolic aspects of illness; the social systems which influence local concepts of illness; the generation and dissemination of local medical knowledge (Seymour-Smith 1986). In addition, the specific sub-discipline of clinical anthropology examines the relative efficacy of conventional allopathic medicine wisdom compared with non-Western practices within a given community, in order to assist in maximising the clinical efficacy of traditional healing. Unfortunately, however, many questions regarding the efficacy of traditional medicine may remain unanswered for some time, as practical and ethical restrictions limit the types of data which can be collected in the field (Anderson 1992).

While medical anthropological studies into ethnomedicine are largely concerned with culturally based concepts of disease and the nature of local healing systems, *medical ethnobotany* involves the identification of botanical species used in traditional remedies and sometimes includes investigations into folk systems of classifying medicinal plants. Often, ethnobotanical inventories of plant species used in healing are constructed using interviews or formal surveys in combination with the use of voucher specimens or the discussion of plants found within a given area; other studies have concentrated on the identification of medicinal plants sold in local or urban markets. For example, in

Monterrey, Mexico, details were collected for 66 higher plants used in traditional medicine, based on specimens which were offered by vendors, and recording notes on paper during essentially opportunistic interviews (Nicholson & Arzeni 1993).

Finally, ethnopharmacological analyses essentially involve the extraction, pharmacological evaluation and identification of any bio-active chemicals present in traditional remedies, and within the field of ethnobotany this pertains specifically to phytochemicals obtained from plant-based treatments. In some cases, where ethnobotanical data have emphasised the need to use fresh material, simple plant extracts have been prepared in the field (Cox 1990); alternatively, fresh material can be kept reasonably well for a few days if kept free of moisture in sealed plastic bags (Harborne 1984). However, where possible, freshly collected plant material is carefully sun-dried and sent for extraction in this dried state. The extraction process itself generally involves maceration of the material in a solvent or range of solvents to produce crude extracts containing a wide range of chemicals. These crude extracts are subsequently purified through a series of *fractionation* and *bioassay* processes to produce a pure solution of the active principle(s). For example, using *bioactivity directed fractionation* the US National Cancer Institute (NCI) has identified a compound known as *prostratin* as the active principle in crude extracts of the Samoan plant *Homolanthus nutans* (Euphorbiaceae), an antiviral remedy traditionally used against yellow fever (Cox 1994). However, as increasing evidence suggests that in some cases, synergistic relationships exist between two or more components of a crude plant extract, it has been suggested that in future functional remedies should be validated scientifically prior to detailed chemical and pharmacological studies (Lewis & Elvin Lewis 1994; Lozoya 1994).

Plants as Therapeutic Agents

Traditional therapies range from practices such as acupuncture and therapeutic massage, to the ingestion of chemical remedies based on natural resources. Some of these remedies include chemicals derived from minerals, animals and fungi, yet those which are most widely reported in the ethnomedical literature are largely plant-based. Indeed, in traditional societies throughout the world, plants are used in the treatment of many ailments—particularly infectious and parasitic diseases, diarrhoea, fever and colds, as well as in birth control and dental hygiene. In addition, many of the psychoactive substances used by shamans are of plant origin.

Plants Used in Traditional Medicine

Like phytochemicals used in other aspects of traditional culture, the compounds exploited in ethnomedical systems are chemically extremely diverse, and are derived from numerous plant taxa. For example, in 1990 the NAPRALERT database (see Chapter 4) contained 1249 ethnobotanical records with information on plants used as contraceptives while 1521 contained details of plants with bacteriocidal properties (Farnsworth 1990). Even within a given society, species from a wide range of plant families are used in herbal medicine, as demonstrated in Amazonia, where the Waimiri Atroari utilise at least 134 plant species from 48 distinct families. As with other plant chemicals, certain plant families are often more useful than others, and in some cases specific taxa can be associated with particular types of treatment. For example, in the highlands of southern Mexico, traditional Mayan peoples use at least 32 different plant species in the treatment of gastrointestinal conditions, more than one-third of which belong to the Compositae (Berlin & Berlin 1994), while the Jivaro of the Peruvian Amazon employ 25 species in blackening their teeth, 16 of which are members of the Rubiaceae (Lewis & Elvin-Lewis 1984). The chemical and pharmacological characteristics of some of the plants in common usage in traditional pharmacopoeia are outlined in Box 8.5, while the nature and functions of some common psychoactive plants are detailed in Box 8.6.

Box 8.5   Chemical and medicinal characteristics of traditional herbal remedies.
Like the plant chemicals used in other aspects of traditional life, those used in traditional herbal medicine are chemically extremely diverse and are isolated from a wide range of plant taxa. In some cases, the active principles governing pharmacological activity have now been elucidated.

On a global scale, current dependence on traditional medical systems remains high, with the majority of the world's population still dependent on medicinal plants to fulfil most of their health care needs. For example, it is estimated that throughout India, between 75 and 80 per cent of the population depend on *traditional medical practitioners* (TMPs) for their primary health care (Jain 1994), while among the Maasai pastoralists of Africa, about 78 per cent commonly use herbal medicines (Sindiga 1994). In recent years, the plant remedies used in traditional medical practice—both in traditional herbal medicine and in shamanic healing—have received considerable attention from ethnobotanists, and today the chemistry and pharmacology of many of these are well understood.

_ continued _|

___ continued ___

SOURCES OF TRADITIONAL HERBAL REMEDIES

Within any traditional society, the range of plants used in the treatment of common ailments, is generally extremely broad, the species used often representing a large number of plant families:

Group	Total number of useful plants		Plants used in medicine
Bardi (Australia)	Species	150	26
	Families	56	ns
Seri (Mexico)	Species	384	90
	Families	80	41
Waimiri Atroari (Brazil)	Species	319	134
	Families	64	48
Pokomo (Kenya)	Species	97	21
	Families	38	14

ns, not specified.

In some cases, a given remedy may have a range of applications; others may have much more specific uses. For example, among the Haya of Tanzania one member of the Compositae, *Bidens pilosa*, is used in the treatment of many ailments (including wound healing, malaria, constipation, intestinal worms, conjunctivitis, rheumatic pains, inflammation and colic), while *Cyanotis foecunda*, a member of the Commelinaceae, is used specifically as a female contraceptive (Chhabra & Mahunnah 1994). Similarly, while many ailments—such as fungal infections or coughs—may be treated by a number of plants, others may have much more specific remedies.

CHEMICAL AND PHARMACOLOGICAL ASPECTS

In recent years, there has been considerable interest in the chemical and pharmacological analysis of traditional plant remedies, and there now exists a rapidly expanding volume of ethnopharmacological data on traditional therapeutic agents. For example, recent analyses of a number of Brazilian medicinal plants (see Craveiro *et al.* 1994) have revealed the nature of the chemicals behind their perceived therapeutic activity as outlined below:

Species	Uses	Chemical and pharmacological analysis
Vanillosmopsis arborea	Anti-inflammatory	Essential oil rich in a-bisabolol which is reported to have anti-inflammatory activity

___ continued ___

continued

Species	Uses	Chemical and pharmacological analysis
Croton sonderianus	Antimicrobial	Root extracts contain several terpenes with antimicrobial activity
Wilbrandia sp	Anti-rheumatic	Root extracts contain two novel cucurbitacins with anti-inflammatory activity
Sapindus saponaria	Antimicrobial	Ethanolic extracts contain acetylated saponins with activity against *Pseudomonas aeruginosa*, *Bacillus subtilis* and *Cryptococcus neoformans*
Myracroduon urundeuva	Anti-inflammatory	Extracts demonstrate anti-ulcer activity

Box 8.6 Nature and activity of traditional psychoactive plant drugs.
Psychoactive plant chemicals are used in a variety of roles; as hunger suppresssants, pain killers and tranquillisers, and as a means of entering altered states of consciousness during shamanic healing ceremonies. Like many of the other plant chemicals used throughout traditional society, the nature and activity of many of the bioactive compounds have now been elucidated (see Mann 1994).

For the sake of clarity, psychoactive plants may be broadly divided into *stimulants, inebriants, psychomimetics, analgesics* and *tranquillisers*. Psychoactive chemicals act in a variety of ways and continue to fulfil a number of roles within traditional medical systems. For example, stimulants—ranging from caffeine to cocaine—are commonly used as hunger suppressants and to extend physical endurance; sedatives are used for relaxation and for the treatment of certain psychiatric disorders; powerful *plant teachers* are more restricted in their application, their use being confined to specialists such as shamans and sorcerers.

SHAMANISM AND PSYCHOACTIVE DRUGS

Shamanism is a Siberian term used to describe a complex of religious and ethnomedical beliefs and practices. Its spiritual leaders, or *shamans* exhibit specialist skills in divination and in healing—particularly of conditions which are believed to be of spiritual origin. For example, the shamans of South American cultures are both healers and sorcerers, who act as an intermediary between society and the supernatural world. Unlike the priests of more orthodox relig-

continued

_ continued _

ions, whose power is vested in him by his religious order, the shaman develops his (or her) powers through his own experiences, while subject to some *altered state of consciousness* (ASC). ASCs may be induced either by various means of extreme sensory deprivation, or through stimulation—most frequently, through the use of hallucinogenic substances from a range of plants, fungi and animals.

SOURCES OF TRADITIONAL PSYCHOACTIVE DRUGS

A recent survey of the NAPRALERT database (Farnsworth 1990) reveals a large number of reports concerning traditional psychoactive drugs, including plants which are used as sedatives (606 records), to stimulate the central nervous system (321), in epilepsy (299), as hallucinogens (213), to produce narcosis (115), or as tranquillisers (68). In many cases, psychoactive phytochemicals are more restricted in their distribution than other plant chemicals. For example, many of the hallucinogenic snuffs used among Amazonian communities are based on two genera—*Anadenanthera* and *Virola*—whose use is widespread throughout South America. Indeed, the use of *A. peregrina* has been reported recently among the Mataco Indians of Argentina, while three species of the same genus—*A. peregrina*, *A. collubrina* and *A rigida* are used by several traditional communities in Paraguay (Costantini 1975; Torres 1992). Similarly, snuffs prepared from various species of *Virola* are used throughout the Orinoco basin in Venezuela, and by the indigenous peoples of the Columbian Amazon. Sources of additional drugs which affect the central nervous system include the analgesic *Papaver somniferum* (Papaveraceae), the tranquillising *Rauvolfia serpentina* (Apocynaceae), and stimulants such as *Coffea arabica* (Rubiaceae) and *Catha edulis* (Celastraceae).

CHEMICAL AND PHARMACOLOGICAL ASPECTS

Like other plant drugs, many psychoactive plants have also been examined using chemical and pharmacological analysis. Today, the active principles of many of these plants have been isolated and characterised, and in many cases their mode of action has been elucidated, as outlined in the table below:

Activity	Active principles and mode of action
Stimulants	Most are based on alkaloids known as *xanthines* that increase the production of various *secondary messengers*, which in turn cause the general stimulation of many biochemical processes.
Inebriants	The mode of action of alcohol, the most well known of plant-derived inebriants, is not clear, although some degree of alteration in neurotransmission is evident, probably following the metabolism of alcohol itself to acetaldehyde which, in turn, reacts with neurotransmitters.

_ continued _

_ continued _

Activity	Active principles and mode of action
Analgesics	For example morphine from *Papaver somniferum*, binds to specific receptors which occur in high concentrations in areas of the brain and which are associated with the perception of pain.
Tranquillisers	For example, reserpine from *Rauvolfia serpentina* depletes levels of the *catecholamine neurotransmitters* in the brain, leading to reduced activity of the *autonomic nervous system*, and associated reduction in heart rate and blood pressure.
Psychomimetics	For example, the tropane alkaloid *hyoscine* from plants such as the deadly nightshade (*Atropa belladonna*) compete for certain *acetylcholine* receptors to alter normal neurotransmission.

Collection and Processing of Medicinal Plants

Like many plants used by traditional peoples, plant species used in ethnomedical systems are often collected from the wild. For example, the vast majority of plants used in traditional African medicine are obtained from the wild populations by TMPs who hold specialist knowledge about the nature, ecology and availability of medicinal plants (Cunningham 1993a). Likewise in Amazonian Peru, the Bora agriculturalists cultivate only two medicinal plants—coca (*Erythroxylum coca*) and tobacco (*Nicotiana tabacum*)—while the remainder occur spontaneously during swidden regeneration (see Denevan & Padoch 1988). Nevertheless, in many cases, various local practices facilitate the sustainable management of these wild plants, either through indirect methods such as taboos on collection and other social or religious restrictions; or through more direct methods such as plant protection during ecological succession, and careful harvesting after which plants remain viable (Cunningham 1993a).

Once harvested, medicinal plants may be processed in a number of ways. Like other phytochemicals many plant drugs are obtained using various extraction techniques, including *infusion* and *fermentation*, others may be smoked or inhaled while others still may require no processing. For example, in Venezuela the Maypure Indians prepare an hallucinogenic snuff from the seeds of *Mimosia acaciaoides* (Leguminosae) which are powdered and fermented before lime (calcium oxide) is added (Ott 1993); in contrast a treatment used in curing skin infections in Peru is prepared simply by crushing the leaves of *Jacaranda copaia* (Bignoniaceae) and applying these directly (Milliken *et al.* 1992). Again these preparation techniques can often be explained once the active principles are known. For example, in the case of *Mimosia acaciaoides* the active principles are *aromatic amines*—nitrogen-containing compounds which are chemically related to both mescaline, and in the neurotransmitters noradrenaline and serotonin (Harborne 1984). In this example, enzymic activity during fermentation and the subsequent addition of lime both increase the potency of the snuff, through facilitating extraction of the active principles from the vegetable matrix.

The Efficacy and Modern Uses of Traditional Plant Remedies

As we have seen, there exist many differences in belief systems regarding health and disease, which may significantly affect the cross-cultural efficacy of a given type of treatment. For example, spiritual beliefs inherent in shamanic healing form an important aspect of the treatment itself. Yet while problems in clinical anthropological methodology have made research into the *in situ* efficacy of different types of traditional therapies rather difficult (Anderson 1992), the objective pharmacological evaluation of traditional plant remedies has revealed numerous examples where empirical evidence supports ethnomedical use. For example, Cox (1994) presents a list of 50 allopathic drugs in current usage, which have been discovered through the examination of plants used in traditional medical systems, while Holland (1994) presents a persuasive argument to support the concept of prospecting for new drugs in ancient medical texts. In addition, a number of recently discovered therapeutic compounds have been specifically identified using an *ethnodirected approach*, including the antiviral compound *prostratin*, which is currently under investigation at the NCI (Cox 1994). These newly discovered therapeutic compounds are discussed in greater detail in Chapter 11.

SUMMARY

By virtue of their specialised biochemical capabilities, plants are able to synthesise and accumulate a vast array of primary and secondary chemicals which are beneficial to human societies. In the plant itself, these chemicals fulfil a variety of roles, involved in functions such as storing energy, attracting pollinators and other dispersal agents, or else in protecting plants against a range of environmental stress factors. In human societies also, these chemicals fulfil a variety of functions, their characteristic physical and/or chemical properties facilitating their use in many aspects of material culture, personal grooming and primary healthcare. In some cases, a wide range of plants may equally be used for a given purpose; in others, more specific taxa must be used. Yet in either case, a knowledge of the plant sources themselves, and of the necessary collection and processing techniques is often vital to ensuring the effective preparation of a particular chemical type.

Chapter 9

Understanding Traditional Plant Use and Management: Indigenous Perceptions of the Natural World

We cannot pretend to understand man on earth without some knowledge of what is in the mind of man ... Decision makers operate within an environment as they perceive it, not as it is.

Brookfield (1969 in Butzer 1982)

INTRODUCTION

As we have seen in previous chapters, decisions regarding the use and management of plants and other natural resources are based on a combination of biophysical and socio-cultural factors, which together determine an individual's *decision-making environment*. This, in turn, is based on how the natural world is perceived by a given community or individual. For example, where a specific plant is regarded as sacred, or where social controls determine rights of access to certain areas of land, this will influence how community members behave towards those resources (Table 9.1). Similarly, if a disease is believed to be of spiritual rather than natural origin, the remedies used may well possess more symbolic significance than pharmacological activity. Therefore, if ethnobotanical data are to prove meaningful to workers from distinct cultural backgrounds, it is essential that these various influences are recognised and understood. Unfortunately, this aspect of ethnobotanical study is one which has often been overlooked in utilitarian surveys, yet which is of singular importance, particularly where traditional

Table 9.1 Knowledge is doing[1]—environmental perception and human behaviour. In human societies throughout the world, it is the environment as it is perceived by an individual which determines the behaviour of that individual. In turn, the way in which the environment is perceived results from the combined influences of empirical observation and a range of complex socio-cultural constraints. For example, unlike Western societies which tend to regard nature as something to be tamed and exploited, many folk traditions emphasise the mutual dependence of these apparently opposing forces and highlight the fundamental need to maintain a balance between the two. Such beliefs, in combination both with other social influences and with empirical observations, therefore determine the behaviour of individuals within their local environment (see: [1]Silberbauer 1994; [2]Ellen 1994; [3]Karim 1981; [4]Richards 1994)

Group	Environmental perceptions and resource use
Nuaulu Swidden cultivators of Indonesia	While there are no Nuaulu words to express the concepts of 'nature' and 'culture', the symbolic relationships between these two phenomena are frequently expressed in rituals. For example, while forest clearance is fundamental to Nuaulu subsistence, clearance itself is perceived to create problems. This anxiety is expressed in activities such as *sasi*—the ritual restrictions on harvesting certain products at certain times.[2]
Ma'Betisék Swidden cultivators of Carey Island, Malaysia	Inherent in the origin myths of the Ma'Betisék, is the idea that a balance must be achieved between the opposing forces of *tulah* (which expresses the distinction between humans and other organisms, and the fact that the latter were cursed to become food for humans) and *kemaii* (which expresses the fundamental similarities between humans, plants and animals). These concepts reflect the notion that all life is ultimately interconnected, and the Ma'Betisék manage their resources on the basis of their belief that, because of these complex interrelationships, humans are likely to become susceptible to misfortune or illness where natural resources are abused.[3]
Mende-speaking communities Agriculturalists of Sierra Leone	While rice-breeding here is based largely on the essentially scientific methods of empirical observation and experimentation, some aspects of rice cultivation are dominated by more cultural factors. For example, where Asian rice (*Oryza sativa*) becomes heavily contaminated with African rice (*O glaberrima*), this is considered a mild social disgrace. Hence farming practices will normally ensure that segregation between different rices are maintained.[4]

knowledge is to be applied *cross-culturally*—as in the search for novel bioactive chemicals or in the determination of resource management practices which are both effective and ecologically sustainable in the long-term. This chapter outlines both the major factors which can influence an individual's cognition of the natural world and discusses the ways in which those perceptions can be examined. In this respect, particular emphasis is placed on the central role of ethnotaxonomic study in elucidating traditional perceptions of the plant world.

UNDERSTANDING THE DECISION-MAKING ENVIRONMENT

Research aimed at understanding how the natural world is perceived by a given community, or by individuals within that community, can involve two distinct approaches. First, studies may involve the identification of the range of factors that influence the way in which the environment is perceived locally—including the objective nature of the local biological environment and the *socio-cultural constructs* which might modify that 'objective reality' (an *etic* approach); secondly, researchers may choose to attempt to elucidate how the environment is actually perceived by local people (an *emic* approach). As we have seen in Chapter 3, the identification of a community's social controls can involve several aspects of anthropological study, including the *symbolic analysis* of ritual and myth, and the structural and functional analysis of *social organisation*, while the *ethnotaxonomic study* of traditional systems of biological classification can reveal how relationships between different organisms are perceived locally. In addition, methods from the *natural sciences* can prove valuable in assessing factors such as the nature, quantity and distribution of the natural resources available and the ecological implications of a given subsistence decision.

Factors Affecting Environmental Perception

In a broad sense, the factors which are likely to influence an individual's perception of the local environment—and therefore his (or her) behaviour within that environment—can be divided into four main categories: biophysical, spiritual, sociological and personal. While biophysical factors involve external factors such as the relative size or density of a given plant species, the remaining influences are culturally or individually based, relating to sociological variables such as age or gender, and internal influences such as personal ambition or temperament.

Biophysical Characteristics

Biophysical characteristics include both practical properties such as the durability of a species' timber or the biological activity of its secondary products (see Chapters 7 and 8), and those factors which determine its *ecological apparency*. For example, *physiognomic* features—such as the *density* or *frequency* of a species within a particular plant community—as well as morphological and behavioural features such as size and growth characteristics may all contribute to the relative apparency of a particular type of plant. So far, relatively little research activity has attempted to assess the extent to which different biophysical factors may influence how a given plant is perceived, although Phillips' and Gentry's landmark study into the useful plants of Amazonian Peru, suggests that data based on their statistical use-value may prove a useful tool in future studies (see Box 9.1).

Box 9.1 Biophysical factors and environmental perception. In earlier ethnobotanical studies, it has often been observed that those plant species which are more abundant or more accessible are generally more useful, while lianas have been assumed to be especially important medicinally on account of their high content of bioactive compounds. The attempt described here to assess such relationships on a quantitative basis, certainly suggests that ecological apparency may have a significant effect on how plants are perceived and hence on how they are used; however, other factors such as the similarities between *phylogenically* related species also appear to influence the ways in which particular taxa are both perceived and used (see Phillips & Gentry 1993b).

> While many of the factors which influence how the natural world is perceived are culturally determined, the objective biophysical nature of a given environment must also have a profound effect on local environmental perception. For example, it is unlikely that tropical forest-dwelling peoples would possess a general term for specific life forms such as cacti, while tree species found at high density are unlikely to be overlooked (see Martin 1995).
>
> PLANT APPARENCY AND ENVIRONMENTAL PERCEPTION
>
> Although the relative *apparency* of a given plant species is almost certain to affect how it is perceived by local communities, few investigations into the relationships between apparency and cultural significance have been carried out to date. One study which has addressed this issue in some detail, is Phillips' and Gentry's recent investigation into plant use-values as perceived by *mestizo* populations of Amazonian Peru (Phillips & Gentry 1993b). Here, the contribution made by eight different factors to plant usefulness was assessed quantitatively, to reveal a
>
> _ continued _

_____ *continued* _____

number of relationships between a species' use value and its ecological characteristics.

	Density	Freq	Mean d.b.h.	Max d.b.h.	Max GR	Stem habit	Family
All uses	***	***	**	***	**	***	***
Construction	***	**		*		***	***
Technology	*	s		***	*		**
Food	**	**				s	***
Medicine							**

(Density refers to the mean number of individuals per hectare across seven forest types; Freq = frequency and is equal to the number of 1 ha plots (and therefore of different forest types) in which each folk species occurs; d.b.h. = diameter at breast height for each individual with a stem ≥ 10 cm at breast height; GR = growth rate as measured by increase in diameter per month. Level of significance: *** = $P < 0.001$; ** = $P < 0.01$; * = $P < 0.05$; s = $P < 0.10$).

Using the use-value data obtained from their initial studies (Phillips & Gentry 1993a), these workers have used statistical methods to identify any correlations between a folk species' taxonomic or growth characteristics and its relative usefulness. Although their results should be interpreted with care (see Phillips & Gentry 1993b) these quantitative tests have demonstrated significant correlations between plant use-values and factors such as stem habit and plant size as shown in the table above. Significantly, while factors such as density and size were strongly correlated with uses in construction and technology, taxonomic relationships appeared significant for all types of plant use. Clearly then, while greater *ecological apparency* may influence some plant uses, factors such as the relationships between different species also play an important part in how different plant species are perceived.

Socio-cultural and Personal Factors

Whatever its biophysical characteristics, human perceptions of a given plant species will frequently vary according to a range of socio-cultural influences. For example, in Aboriginal religious ideology, the identification of an individual with their spiritual *totem* will significantly influence their attitude towards that totem. Indeed, where an individual's totem is a potential foodstuff—be it a plant or an animal— then they will eat that species only sparingly if at all (Cowan 1992). Further examples of how magical and religious beliefs can affect environmental perception are outlined in Box 9.2. However, in addition to social controls based on spiritual belief, secular aspects of social organisation can also modify significantly, the way in which the natural world is perceived. For example, social status has been found to exert a significant influence on the ways in which particular crop plants are perceived (Box 9.3).

Box 9.2 Magical and religious beliefs and environmental perception. In many traditional societies, plants and other natural phenomena are often imbued with spiritual or supernatural powers. This can have a significant effect on the ways in which specific taxa are used and managed (see: [1]Fieldhouse 1986; [2]Felger & Moser 1985; [3]Balick 1984; [4]Matose & Mukamuri 1995; [5]Elisabetsky & Posey 1994).

The cultural domains of both *religion* and *magic* are intimately associated with the spiritual and supernatural aspects of a community's cosmology. Religion, which includes concepts such as *animism, ancestor worship* and *polyheism,* is often associated with attributing natural objects with supernatural characteristics; magic, on the other hand, is largely concerned with the deliberate manipulation of supernatural forces (see Seymour-Smith 1986). Throughout human society, where a resource has been invested with either religious or magical significance, this can have a profound effect on its local use or management.

RELIGIOUS OR MAGICAL SIGNIFICANCE AND RITUAL AVOIDANCE

To date, the ritual avoidance of particular people, places or objects has been recorded in many traditional societies, and in several cases, examples involving specific plant *taboos* have been well-documented. In some cases, these avoidances may involve a particular plant species; in others, a plant may be used only at certain times, or must be prepared in a specific, ritual fashion:

Group	Belief and avoidance behaviour
Tamil Nadu (India)	Papaya (*Carica papaya*) is thought to symbolise the female breast on account of its shape. Consequently, this fruit is avoided by pregnant women for fear of interfering with the reproductive process. Such behaviour can be regarded as a facet of *sympathetic magic.*[1]
Seri (Mexico)	The boojum tree (*Fourquieria columnaris*) is associated with strong spirit power—*icor*—and causes strong winds. The Seri therefore avoid cutting its branches.[2]
Amazonian (South America)	A number of palm species—including the corozo palm, *Scheelea zonensis*—are believed to be the home of certain evil

_ continued _

continued

Group	Belief and avoidance behaviour
	spirits. Hence these species are avoided by many forest dwellers.[3]
Zimbabweian (Africa)	It is believed that the conservation of large trees is crucial to ensuring adequate rainfall for two reasons: first, that the cuckoo bird (_hwaya_) which sings for rain, likes to rest in such trees, and secondly, that the high god _Zame_ may punish those who cut down such trees by preventing subsequent rains. Hence, large trees are protected among traditional communities.[4]

POWERFUL PLANTS AND PLANT USE

While certain plants may be _avoided_ according to local custom, others are _used_ for specific purposes on the basis of their spiritual associations. In particular, in many cultures, shamanic healing involves the use of plants which are regarded as having powerful spirits, often based on concepts similar to those encompassed within the _Doctrine of Signatures_—a theory, popular in sixteenth century Europe, which proclaimed that plants indicated by some physical sign, the medical use to which they might be put ([6]Evans 1984).

Group	Belief and plant use
Seri (Mexico)	Plants such as the elephant tree (_Bursera microphylla_) and desert lavender (_Hyptis emoryi_) are used during shamanic healing practices on the basis of their powerful _icor_; equally plants with strong spiritual associations have also been used for a range of other purposes, ranging from shamanic _vision quests_ to controlling the weather.[2]
Kayapó (Brazil)	The inate _energies_ associated with both plants and animals are regarded as integral to human health, and play an important part in shamanic healing.[5]
Medieval Europe	Red beet juice was regarded as useful for treating anaemia and yellow celandine for treating jaundice, while mandrake root—on the basis of its similarity to the human body—has long been heralded as an aphrodisiac.[6]

Box 9.3   Social organisation and environmental perception. In addition to social controls based on religious or magical belief, many secular rules also contribute to the overall perception of the environment, which in turn has a fundamental influence on the decisions made by individuals and on the way in which they behave within that environment.

Within any society, *social organisation*—like religious belief—can have a profound effect on an individual's behaviour. This includes the social institutions which are responsible for ensuring the continuity of both that *society* and for protecting its means of production; it also includes any competitive interactions between individual *members* of that society (see Seymour-Smith 1986).

ENVIRONMENTAL PERCEPTION AND MEANS OF PRODUCTION

Both a community's *mode of production*, and the social controls which protect that society's *means of production* have a profound influence on environmental perception. For example, the perception of an area as *optimal* or *marginal* for production is fundamentally different for hunter-gatherers and peasant farmers, while factors such as land ownership can play an important part in regulating the availability of natural resources.

Mode of production and environmental perception	
Aleut (hunter-gatherers)	The Aleut carry out an annual seal harvest which represents less than 2 per cent of the 1 million seals which inhabit their islands, during which they are careful to take only non-breeding males.
Animal rights campaigners (industrialised)	In 1976, the global animal rights coalition noted a decline in local seal populations which they attributed to the Aleuts' annual harvest of 16 000 seals; following a well-funded campaign, the protesters succeeded in stopping the local seal harvest, thus destroying the economic base on which the Pribilof Aleut had depended for more than 200 years.
	Unfortunately, the externally located animal rights protesters failed to notice that certain species of birds and sea lions had also begun to decline at about the same time—a decline which only the Aleut noticed and which they believed to be ultimately connected with the declining numbers of seals.

See Merculieff (1994).

_ *continued* __

— continued —

SOCIO-ECONOMIC STATUS AND ENVIRONMENTAL PERCEPTION

The significance of socio-economic status on environmental perception has been illustrated clearly in a recent report from the Phillippines, where a number of crops were evaluated by a range of informants. The report shows that while some responses were common to all informants—so-called *prototypic evaluations*—others were peculiar only to certain social groups. The latter, *marginal comments*, are neither completely shared by all informants nor totally idiosyncratic, but are apparently characteristic of informants who share a common sociological niche (Nazarea-Sandoval 1995).

Marginal comments from various social groups			
High status		Low status	
♂	♀	♂	♀
Bell pepper (Capsicum annuum)			
Destroyed by floods	No comments	Difficult to sell; unsuitable for children	Difficult to find seed
Sweet potato (Ipomoea batatas)			
Good food source	Rich in vitamin A	Causes colic in young children	Roots and leaves edible; sensitive to over harvesting
Rice variety IR 36 (Oryza sativa)			
Good quality grain if a little hard	Rapid maturation; needs fertilisation for good grain-fill	Seeds easy to find; low price but easy to sell; good in dry season	Poor cooking quality; too soft; unattractive

Finally, while most of the influences discussed here are essentially external, an individual's perception of the world is further modified by internal factors. For complex relationships between an individual's psychology and life experience dictate that different members of a given community will often have different beliefs and feelings towards both the sacred and secular external controls, as well as different levels of personal ambition or greed. Clearly, these internal modifiers will contribute to the individual's overall perception of the world, and can also have a considerable influence on their behaviour within it.

Investigating the Perceived Environment

While elucidating the nature of an individual's decision-making environment is inherently extremely difficult, there exists a range of distinct approaches which can prove valuable in this important area of ethnobotanical research. These methods may be broadly divided into two major areas: *behavioural research*, which analyses the events and activities resulting from any decisions made, and *cognitive research*, that examines the psychological factors which influence those decisions (Guillet *et al.* 1985).

Behavioural Research: Land-use Activity and Resource Perception

One significant behavioural manifestation of decision-making lies in local patterns of land use, the details of which can reflect not only present concerns, but also past failures and future successes (Nazarea-Sandoval 1995). However, as we have seen in Chapter 3, it is not sufficient simply to map details of land use at any given time, as decisions made in traditional resource management are made in response to ongoing change throughout a given growing season (Alcorn 1989; Richards 1989). To assist, therefore, in the investigation of complex land-use strategies, some researchers have recently begun to supplement conventional data-gathering techniques—such as open-ended interviews—with the construction of resource maps and other diagrams based on both spatial and temporal information provided by local collaborators (see Conway 1989; Gupta & IDS Workshop 1989).

In a useful paper aimed largely at workers involved in agricultural development, Gordon Conway (Conway 1989) outlines five different types of diagram: *maps, transects, calendars, venn diagrams* and *flow charts*. Each has the potential to contribute enormously to the understanding of how vital resources are perceived by local communities. For while the construction of venn diagrams and flow charts can elucidate sociological influences such as the availability of labour or the co-operative relationships between neighbouring groups; maps, transects and calendars can provide details of the spatial and temporal distribution of both wild and domesticated biological resources. If the construction of such diagrams is supplemented by *contrastive elicitation*—where informants are asked why each land-use option is used in favour of any other (see Nazarea-Sandoval 1995)—this can provide considerable insight into the cognitive factors on which particular land-use decisions are made. For example, one study has demonstrated that Fulani herders will specifically avoid certain sources of water, despite the fact that water is often an important limiting factor in the lives of pastoralist societies. In response to contrastive elicitation, however, it was found that these specific ponds or streams were regarded as responsible for inducing outbreaks of the cattle disease known as *sammore*; and indeed these mapped water sources were later shown to be associated with populations of the sleeping sickness vector tsetse (*Glossinia* spp) (Smith 1992).

Ritual Behaviour as a Reflection of the Perceived Environment

While it is now widely acknowledged that intra-cultural diversity in the attitudes and beliefs of individuals exist, *ritual behaviour* is nevertheless regarded as indicative of the general *collective sentiment* of a given culture. Indeed, as the practice of ritual behaviour largely represents a means for expressing or reinforcing cultural beliefs, its analysis can often prove helpful in elucidating the nature of the culturally constructed world of a given community. This relationship between ritual behaviour and environmental perception can be illustrated by the Pitjantjatjara Aboriginals of central Australia. Like other indigenous Australians, the Pitjantjatjara regard humans as an integral part of the environment and have an essential and reciprocal role in maintaining the natural order of the world (Silberbauer 1994). These views are recorded, not only in the rich mythology of *Tjukurpa* (The Dreaming), but are also manifest in ritual conduct and the performance of ceremonies which are regarded as essential both to the survival of flora and fauna, and to the well-being of people. For example, the analysis of Pitjantjatjara ceremonies and totemic relationships with mythical heroes, reveal the importance of *reciprocity* between humans and the land, and hence the essential role of humans in the maintenance of natural processes.

Cognitive Behaviour: Ethnotaxonomy and Human Perception

In his early theories of *structural anthropology* Lévi-Strauss proposes that human culture represents a surface phenomenon which, on analysis reveals the universal human tendency to organise and classify perceived phenomena or experience (Seymour-Smith 1986). He argues that just as language is understood on the basis of a series of contrasting sounds, human cognition generally is based on a series of contrasts—such as culture/nature, right/left, day/night—each concept proving meaningless in isolation from its contrasting element. From the analysis of the social institutions of kinship and marriage systems in his early works, Lévi-Strauss later turned his attention to the analysis of symbolic representations in myth and ritual (Lévi-Strauss 1963–1966); meanwhile this approach has been adopted also by several workers in the field of *folk systematics*, which examines primarily the ways in which individuals categorise and name the organisms around them. In relation to environmental perception, ethnotaxonomic studies can therefore yield valuable information regarding local perceptions both of the relationships which are perceived between different plant or animal taxa, and the nature and significance of individual taxa themselves.

ETHNOTAXONOMY AND THE PERCEIVED ENVIRONMENT

In an attempt to avoid the ethnocentric bias of early anthropological works, cognitive anthropologists have, over the second half of the twentieth century, worked towards the *ethnographic presentation* of a distinctive *world view* from *inside* a given culture. From the 1960s onwards, a considerable proportion of cognitive anthropological research has focused on how different cultures classify and name—and therefore conceptualise—the natural world. Like other areas of cultural anthropology, students of *ethnotaxonomy* have borrowed heavily from linguistic analysis, both in the emphasis placed on recording and studying those categories which are linguistically defined, and in the focus on identifying sets of *contrasts* as outlined earlier. For example, while *Monterey pine* occurs at the same level as other types of *pine* it is recognised as distinct from any other type of pine, and is contrasted from them by virtue of its specific label (Keesing 1981). Nevertheless, it is now widely recognised that not all conceptual categories receive linguistic recognition, and today, folk taxonomists are beginning to look, not only at the *structural* details of linguistically defined classification systems, but also at their *substantive* nature (Ellen 1994a).

The Nature of Folk Taxonomic Systems

The term *taxonomy* was coined originally in 1813 by the Swiss botanist Augustin de Candolle, who used it to describe the placing of biological organisms into particular groups or *taxa* on the basis of shared relationships (Hellemans & Bunch 1988; Ingrouille 1992). However, the *systematic* study of biological diversity dates back at least to the time of Aristotle (381–323 BC) who, under the patronage of Alexander the Great, undertook the large-scale *classification* of biological organisms, arranging them into a hierarchical pattern of categories ranging from plants to humans. In botany, this work was later continued by Aristotle's pupil Theophrastus, who classified numerous plants and coined many botanical terms (Hellemans & Bunch 1988). Throughout the following centuries, students of *systematic botany* have focused their attention not only on the identification, nomenclature and classification of diverse types of plant, but also on their putative *phylogenetic relationships*, their overall goal being to identify natural groupings of plant species which reflect actual relationships according to their descent from common ancestors (Bold *et al.* 1987).

Biological organisms and the relationships between them have therefore occupied scientific thinking for millennia, and taxonomy itself is regarded as one of the oldest fields of biological science. However, it is clear, that this tendency to organise the biological world is not peculiar to Western cultures, but is common to all human cultures which have been studied so far. It is these alternative classification systems which have attracted the attention of *ethnotaxonomists* over the last 50 years, and there now exists a considerable literature which describes these *folk taxonomies*, and in many cases discusses how such studies can contribute to the elucidation of local environmental perception.

Ethnobotanical Systematics

Today, ethnotaxonomic studies may focus on a range of factors— including systems used in classifying different types of disease or in the categorisation of soil types (Box 9.4). Nevertheless, considerable research attention remains focused on traditional systems of biological classification. Indeed, since the 1970s and 1980s, studies in ethnobiological classification have flourished and many diverse publications now discuss the biological taxonomic systems of several traditional communities. As far as ethnobotanical studies are concerned, research into the folk classification of plant species has highlighted a number of significant features which appear to be universal. For example, in each case, a proportion of the botanical species that occur locally are recognised and regarded as distinct folk taxa, each of which has normally been designated a characteristic label. In each case also, certain taxa are perceived as more closely related than others, in which case this larger group of related taxa may, or may not receive linguistic recognition. Indeed, among the Tzeltal Maya, two botanical species of mint, *Mentha spicata* and *M. citrata* are recognised as distinct but related taxa and this is reflected in their linguistic labels—*skelemal tul pimil* and *yach'ixal tul pimil*, respectively. Hence, while these botanical species are regarded as separate folk taxa, both are regarded as types of the higher-order category *tul pimil*. In contrast, while the Peruvian Aguaruna apparently regard two folk taxa *yaís* and *chuáchua* as related, there exists no linguistically labelled group to which both belong (Berlin 1992). Based on studies such as these, it has been possible to draw a number of *general principles* about the ways in which human beings perceive biological reality, as well as the ways in which they organise detailed information for later use.

Box 9.4   Studies in folk taxonomy. Through the analysis of the structural characteristics of human systems of classification, it is now widely believed that the human tendency to categorise, and organise their knowledge and experiences is universal, and that classification systems throughout the world show certain structural and substantive similarities. However, while such comparative work can play an important part in examining human cognition *per se*, individual ethnotaxonomic studies can also reveal much about an individual culture's perception of the natural world.

In recent decades, interest in folk systematics has expanded far beyond the ways in which traditional societies classify biological organisms, and now a number of studies have been carried out in areas ranging from disease classification to the categorisation of local soil types.

DISEASE CLASSIFICATION IN THE MEXICAN HIGHLANDS

As interest in traditional medicinal plants has grown among phytochemists and pharmaceutical scientists, many anthropologists are increasingly emphasising the need to understand local perceptions of disease in order to understand specific treatments in Western terms. In a recent paper Berlin & Berlin (1994) describe how more than 250 health conditions are recognised and categorised among the Maya-speaking Tzotzil and Tzeltal peoples of southern Mexico. Their findings indicate that any health condition is categorised firstly according to whether it is *naturalistic*—in which case it is likely to be treated using medicinal plants, or whether it is *personalistic*—in which case it is more likely to require ceremonial or shamanic healing. More than 1600 distinct plant species—many of which are condition-specific—are purported to have medicinal value for the treatment of naturalistic conditions.

Disease category	Number of reports	Disease category	Number of reports
Gastrointestinal	1840	'Air' (*aire*)	230
Respiratory	1030	Skin infections	130
Fever	700	Personalistic	120

continued

— continued —

Disease category	Number of reports	Disease category	Number of reports
Headache	430	Eye disorders	110
Musculoskeletal	250	Cuts and wounds	100

These data, which are taken from *ethnoepidemiological* reports of health problems of the Highland Maya, are grouped according to major *ethnomedical category*, illustrating clearly not only which health conditions are regarded as serious problems by the Maya, but also the perceived frequency of a given type of illness (Berlin & Berlin 1994).

SOIL CLASSIFICATION AMONG THE MOSSI OF BURKINA FASO

The Mossi farmers of Burkina Faso, West Africa, identify at least 17 soil types based on characteristics such as texture, colour and permeability (Dialla 1993). These soil types are further classified in terms of cropping potential, leading to four major classes of soil which allow the farmers to make optimum use of their land.

Major soil type and cropping potential	Soil types included
Zĭ-kugri (stony soil) good for millet	*Zĕka* (lateritic soil; *Rasempuiiga* (gravelly soil); *Tānga* (mountainous soil); *Zĭ-miuugu* (red soil)
Bīsri (sandy soil) good for peanuts	*Bīs-miuugu* (red sandy soil); *Bīs-sabille* (black, sandy soil); *Zĭ-peele* (white soil)
Bolle (clay soil) good for sorghum	*Dagre* (hard clay soil); *Zĭ-sabille* (black soil); *Zĭ-kotĕka* (lowland clay where water stagnates)
Bāoogo (Loamy soil good for sorghum)	*Zĭ-bugri* (very soft soil); *Zĭ-naare* (wet loamy clay soil); *Kāongo* (black soil with dense growth of bushes)

General Principles of Systematic Ethnobiology

Following the publication of a number of ethnotaxonomic studies from around the world, several workers—most notably anthropologist Brent Berlin of the American University of Georgia—have proposed

Table 9.2. Berlin, Breedlove and Raven's general principles, 1973. The overall proposal emcompassed within these general principles is that groups of related plants (or animals) not only receive linguistic recognition, but that these groupings are semantically related to each other in some structured, analysable fashion. Moreover, these labelled taxa are perceived, and arranged hierarchically in terms of increasing inclusiveness—for example, *plant>tree>oak*. While there has been some criticism of some of these ideas (see Berlin 1992), Berlin's updated proposals (see Boxes 9.5 and 9.6) are gaining widespread acceptance among many researchers (Martin 1995)

General principles of ethnobiological classification*

- In all languages it is possible to identify groups of organisms (taxa) which are recognised linguistically, and which are based on varying degrees of inclusiveness—for example, the linguistic labels *oak*, *tree* and *plant* represent plant taxa of increasing inclusiveness.
- Biological taxa are grouped into a number of *ethnobiological categories* similar to Western *taxonomic ranks*. In 1973, these were designated *unique beginner, life form, intermediate, generic, specific* and *varietal* in order of decreasing inclusiveness.
- Mutually exclusive ethnobiological categories are arranged hierarchically, each encompassed by the single *unique beginner* taxon—which is roughly equivalent to the Western kingdom.
- Taxa of the same *ethnobiological rank* commonly occur at the same taxonomic level in any given system of classification
- The *unique beginner* taxon (in this case *plant*) is not normally named with a single, habitual label. There normally exist between five and 10 taxa at the next level of organisation—the life form—and these are usually labelled, for example, *tree, vine, grass.*
- Most folk taxonomies appear to contain around 500 taxa at the *generic* level—these taxa represent the basic building blocks of any folk taxonomy and are usually the most salient psychologically. While most generic taxa are included within a given life-form category, a few—often those which are either morphologically unique or economically important— are *aberrant* and are not conceptually regarded as *affiliated* to any life-form category.
- *Specific* and *varietal* taxa are less numerous than generic taxa, members of a given *contrast set* often differing from other members in terms of a few, often verbalisable, characters. Unlike taxa at higher ranking levels, these are commonly labelled using *secondary lexemes.*

Intermediate taxa generally encompass a number of related generic taxa. These are rare—or at least difficult to detect—in folk taxonomies, and are seldom named; hence they are referred to as *covert* or unnamed taxa.

*Based on Berlin *et al.* (1973)

a series of structural principles of both classification and nomenclature which appear to be universally shared. Indeed, more than two decades ago, Berlin *et al.* (1973) originally put forward a total of nine major features of ethnobiological classification and nomenclature, which were regarded as general *principles* of folk taxonomy (Table 9.2). Since then, Berlin (1992) has revised and expanded these principles and has recently proposed a total of seven principles of *ethnobiological categorisation*, and five of *ethnobiological nomenclature*, providing a scheme of folk classification which is now beginning to receive increasing acceptance among systematic ethnobiologists (see Boxes 9.5 and 9.6). These principles have been summarised recently by Martin (1995) who highlights succinctly, the major features of traditional systems of *general purpose classification* (Table 9.3).

Box 9.5   Principles of ethnobiological categorisation, 1992. In his recent discussion of the structural features of the ways in which human societies tend to categorise plants and animals, Berlin (1992) proposes a revised set of universal principles which characterise the way groups of organisms are perceived and grouped on the basis of gross morphological affinities or differences.

In his recent revision of his general principles of ethnobiological classification, Berlin (1992) presents a total of 12 features which appear to characterise folk taxonomic systems from diverse parts of the world. Of these, seven relate to the *psychological conceptualisation* of biological organisms by humans, while the remaining five (outlined in Box 9.6) concern the *linguistic reflections* of this underlying conceptual structure.

PSYCHOLOGICAL CONCEPTUALISATION OF BIOLOGICAL ORGANISMS

According to Berlin (1992), traditional societies normally name and categorise only a specific subset of the total number of species found locally, each of which is apparently cognised according to a number of general principles.

- The subset of species which are recognised comprises those species which are most *salient*—which in this case refers to those which are most *biologically distinct*.
- The categorisation of individual plants into distinct groups or taxa is based primarily on observed affinities or differences in their morphological and behavioural traits, with groupings based on pharmacological properties, nutritional value or symbolic significance apparently playing a secondary part.
- Recognised taxa are grouped into ever more inclusive groups to form a hierarchical structure comprised of a small number of taxonomic ranks, which range from the *folk varietal* level to the level of *kingdom*.

continued

_____ *continued* _____

- These six mutually exclusive *ethnobiological ranks* (which are comparable to the ranks of Western systematics) are: the *kingdom* (previously unique beginner), *life form, intermediate, folk generic, folk specific* and *folk varietal*.
- In all folk taxonomic systems, significant similarities are found in both the relative numbers of taxa at each rank and their biological content:

Ethnobiological rank	Relative numbers and biological content
Kingdom (e.g. *plant*)	The rank of kingdom is unique including only one member—in this case the taxon *plant*—and it incorporates all taxa of lesser rank.
Life form (e.g. *tree*)	Life-form taxa are based on gross morphology and are almost invariably *polytypic*.
Intermediate (unnamed)	Intermediate taxa are groups of small numbers of folk generics which show marked perceptual similarities.
Generic	Folk generics constitute the most numerous taxa, with an upper limit of about 500 to 600 distinct taxa.
	About 80 per cent of folk genera are *monotypic*, the remaining 20 per cent including taxa at the folk species or folk varietal levels.
	While most folk generics are included in taxa of life-form rank, a small number remain *unaffiliated*.
	Folk genera are among the first taxa learned by children.
Subgeneric	Subgeneric taxa—folk species and folk varietals—are less numerous than folk genera in all systems examined to date.
	There is some evidence to suggest that recognition of subgeneric taxa is motivated in part by cultural significance in that many of them are domesticated species.

- Where taxa of generic or subgeneric rank are polytypic, there are generally some members which are regarded as more representative of a given taxon than the others—these are referred to as *prototypic* taxa. The perception of prototypicality may be due to a number of factors including frequency of occurrence and cultural significance.
- In any system of ethnobiological classification, many recognised taxa correspond closely in content to those recognised in Western biological systematics. Compared with other ranks, the taxa included within a given life-form category show the lowest degree of corrrespondence to Western biology, while those of generic rank show the highest.

Box 9.6   Principles of ethnobiological nomenclature, 1992. Ethnobotantical nomenclature clearly reveals much information, not only about the relationships which are perceived to exist between distinct plant taxa, but also about plant properties or their cultural significance. In addition, where secondary names are found, their occurrence in contrast sets can lead to the discovery of previously unknown folk specific or folk varietal categories (see: Berlin 1992; Martin 1995).

While the principles of *ethnobiological categorisation* deal with the conceptual organisation of biological organisms into distinct, but related taxa, the principles of *ethnobiological nomenclature* focus on linguistic patterns which reflect not only these conceptual categories, but also certain characteristics of plants such as their bioactive properties or cultural significance. Berlin (1992) proposes five general principles of folk nomenclature which have also been summarised by Martin (1995).

LINGUISTIC REFLECTIONS OF PSYCHOLOGICAL CONCEPTS

Just as humans apparently cognise biological organisms according to a number of general principles (see Box 9.5), they seem also to name organisms according to a number of general rules.

- Categories at the intermediate, kingdom and, occasionally, life-form levels, receive no linguistic recognition, even where such taxa are clearly perceived. For example, in Amazonian Peru, the Aguaruna regard three distinct folk generics—*tséke* (*Cecropia engleriana*), *satík* (*C. tessmanii*) and *súu* (*C. sciadophylla*) as 'companions' on the basis of their similar appearance. However, as there appears to be no linguistic label to describe a higher order taxa encompassing these three related generics, this intermediate taxon is regarded as unnamed and therefore *covert*.
- The names, or *labels* for plant and animal taxa are of two basic structural types—*primary* and *secondary*—which are further characterised according to a number of general principles:

Label	Structural characteristics
Primary name	Primary names consist of a *semantically unitary*, or single expression, which is often composed of a single constituent—for example, *oak, rose, foxglove*. They may be either *simple* or *complex*.
Simple primary	A simple primary name consists of a single word—for example, *oak*.
Complex primary	A complex primary name may be broken down into substituent parts—for example, *foxglove*. These names may be either *productive* or *unproductive*.

_____ continued ___

_ *continued* _

Label	Structural characteristics
Complex productive	Here, a complex name includes specific reference to the relevant life-form category—for example, the labels *crabgrass* and *relief tree* include reference to the life form to which these plants belong.
Complex unproductive	Here, a complex name does not refer to a relevant higher order category—for example *redwood* does not include the name *tree*, the life-form category to which this plant belongs.
Secondary name	Secondary names are invariably complex, their constituents indicating the existence of a higher level category. Each is composed of a semantically binary expression—including a *generic appellation* plus a *specific epithet*—which names the higher category to which it belongs, and which is found within a *contrast set* of taxa with similar names—for example, *red oak, white oak* and *cork oak*.

- Label types are generally distributed in a consistent fashion:
    primary names—generic taxa, and any life-form and intermediate taxa which are labelled,
    secondary names—subgeneric taxa.
- Exceptions to these patterns of labelling, occur only in certain circumstances—for example, *prototypicality* and cultural significance can lead to the abbreviation of secondary names (see Box 9.9).
- The labels applied to plants and other organisms commonly allude to some morphological, behavioural, ecological or other qualitative characteristic of their referents. In addition, many complex primary names are formed on the basis of analogy with the name of another, conceptually related taxon. For example, the Peruvian Aguaruna name the yam *Dioscorea trifida* with the label *kéngke*, while the conceptually related wild species of *Dioscorea* are known as *kengkengkéng*. This process is known as *generic name extension*.

Scientific Classification and Folk Taxonomy

To date current data suggest that all human societies use essentially the same groupings in conceptual recognition—particularly for the generic and subgeneric taxa of folk systems (Berlin 1992). This in turn suggests that objective biological reality allows for little variation in its perception. Nevertheless, important differences do occur between distinct taxonomic systems, and in particular, intrinsic differences exist between

Table 9.3   Summary of the major features of ethnotaxonomic systems.
Brent Berlin's revised scheme of *general purpose* folk taxonomy can be summarised as shown, into a hierarchical system of ethnobiological ranks which
show decreasing levels of inclusiveness. At each rank or level in the hierarchy,
patterns which appear universal in folk taxonomies may be discerned both in
the number of taxa occurring at each level, and the ways in which these taxa are
named

Rank	Label	No. of taxa	Examples
Kingdom	Often covert	1	Mixe—unnamed Seri—*hehe* (polysemous) Plant, wood, log, tree, pole, harpoon handle
Life form	Primary	5–10	Mixe—*aa'ts* vine Seri—*xpanáams* seaweeds
Intermediate	Often covert	NK	Mostly unnamed
Generic	Primary	300–600	Mixe—*maj xajk* large bean Seri—*haamxö* agave
Specific	Usually secondary	typically, *c.* 20% of folk generics contain specifics	Mixe—*yak maj xajk* black large bean Seri—*haamxö caacöi* large agave

*Data taken from Martin (1995) and Felger and Moser (1985); NK, not known.

the taxonomic levels as represented in folk, *versus* scientific classification systems. Indeed, in many cases, folk generics often show a *one-to-
one correspondence*, not with scientific genera, but with scientific
species. For example, the Mixe generic *tsaa'k* (dogwood) corresponds
with the single botanical species *Cornus disciflora*. The fact that this
folk genus is further divided into two folk species—*poop'p tsaa'k* (white
dogwood) and *yak tsaa'k* (black dogwood) reflects the occurrence of
different morphotypes of this single species which are differentiated linguistically by local communities (see Martin 1995). However, while the
greater percentage of folk generics do exhibit a one-to-one corrrespondence with scientific species, there are many examples where this direct
relationship does not occur. For example, scientific species of particular
cultural significance are often *over-differentiated*, or split into many distinct folk generic taxa, as in the case of the Aguaruna classification of
manioc, where a single species, *Manihot esculenta*, is divided locally
into more than 100 distinct folk categories. In contrast, there are also
examples where local taxonomic systems *under-differentiate* botanical
species, which is to say that several similar species may be grouped
together into the same folk generic taxon. In such cases, the folk taxon
may contain either two or more species of the same scientific genus

Table 9.4 Correspondence between ethnotaxonomic and scientific systems of classification. In those ethnotaxonomic systems which have so far been analysed, it appears that most folk generic taxa include only one scientific species, whereas a much smaller proportion are either over- or under-differentiated. For example, among the Tzeltal Maya, 61% of folk generics exhibited a *one-to-one correspondence* with botanical species, of which 21% were under-differentiated (type 1), 14% were under-differentiated (type 2), and only 4% were over-differentiated. Significantly, in some cases where a folk genus is already over-differentiated, further distinction into folk species (and in rare cases folk varieties) has been observed (data taken from Mixe communities, and Tzeltal Maya, Mexico; see Martin 1995)

Botanical species	Folk generic	Folk specifics
One-to-one correspondence		
Cornus disciflora	*tsaa'k* (dogwood)	{ *poop'p tsaa'k* (white dogwood) *yak tsaa'k* (black dogwood)
Over-differentiation		
Persea americana	*kooydum* (avocado)	{ *kooydum* (typical avocado) *pa'ajk kooydum* (sugary avocado)
	xijts (avocado)	{ *poo'p xijts* (white avocado) *tsapts xijts* (red avocado)
Under-differentiation, type 1		
Inga latibracteata *I. oerstediana* *I. schiedeana* *Inga* sp	*ii'k* (small-fruited *Inga*)	{ *poo'p ii'k* (white *Inga*) *tsapts ii'k* (red *Inga*)
Under-differentiation, type 2		
Pinus spp *Abies* sp	*tah*	

(*under-differentiation type 1*), or else it may contain representatives of more than one scientific genera (*under-differentiation type 2*). Examples of each of these situations are outlined in Table 9.4.

Special Purpose Folk Taxonomic Systems

Despite the apparent universality of Berlin's principles of folk classification, confusion may arise in many studies where plant categories used by local people do not fit into this general system. For example, the Ka'apor of Brazil have a category, *kanei*, that includes members of several scientific genera including *Protium* (Burseraceae) and *Hymenaea* (Leguminosae), which are apparently unrelated. However, closer examination has revealed that the taxon *kanei* seems to include those species of plant which produce a combustible resin, latex or sap and which are used as a source for fuels used in illuminating their homes (Box 9.7). Similarly, the Mexican Chinantec have terms for fruit (*'o^{L}huii^L*) and for firewood (*kuii^{LH}*), while the Mixe have words for edible greens (*tsu̱'up̱*) and medicinal plants (*tsoojy*). Unlike the general purpose systems of classification where plant taxa are categorised largely according to perceived morphological similarities, these systems of *special purpose classification* are delimited primarily according to criteria such as their use or *humoral properties*, and do not correspond to life-forms, generics or categories of any other folk rank (Martin 1995).

Box 9.7   Resin classification among the Ka'apor (Amazonian Brazil). On the basis of linguistic and mythological evidence, Balée and Daly (1990) have proposed a hierarchical system of resin classification, which they regard as both parallel to, and similar in structure to, the Ka'apor general classification of non-cultivated plants. They suggest that Ka'apor resin classification differs from the more general plant classification, in that the plants are categorised according to their use rather than on the basis of stem habit, or other morphological features, and therefore constitutes an example of a *special purpose* system of plant classification.

According to Balée and Daly (1990), some aspects of plant classification among the Brazilian Ka'apor Indians reflect strictly utilitarian themes. For example, these workers have proposed the existence of a *covert complex*, which although labelled *kanei*, contains several taxa which are not actually labelled with this name.

THE COVERT COMPLEX, *KANEI*

Conceptually, the suprageneric complex *kanei* (glossed as 'any combustible resin, latex or sap'), appears to include plant species from several unrelated botanical genera, all of which produce a combustible resin which may be used in illuminating homes:

Folk taxa and botanical range of the covert complex *kanei*

irati-hik—*Symphonia* spp (Guttiferae)	*kirihu-hik*—*Trattinnickia berserifolia* (Burseraceae)
yeta-hik—*Hymenaea* sp (Fabaceae)	*kanei-hik*—*Protium* spp (Burseraceae)
trapai-hik—Hymenaea sp (Fabaceae)	*irikiwa-hik*—*Manilkara huberi* (Sapotaceae)

continued

___ *continued* ___

Although only those species belonging to the scientific genus *Protium* are labelled *kanei*, there is some non-linguistic evidence to suggest that all of these folk taxa are regarded locally as part of a single suprageneric special-purpose taxon. For example, one Ka'apor origin myth relates how, long ago, one man named *Kanei-yar* (lit. *kanei*-owner), was killed by another—*Sarakur*—who needed a source of light. Significantly, pieces of bone removed from *Kanei-yar* proved to be pieces of *irikiwa-hik* (latex from *Manilkara huberi*), which were lighted to provide illumination. These ethnobotanical references, clearly suggest that *irikiwa-hik* (*Manilkara huberi*) is regarded as a source of *kanei*, despite the fact that it is not actually labelled with the term *kanei*.

FUNCTIONAL CHARACTERISTICS SUPERSEDE MORPHOLOGICAL SIMILARITIES

Within the functionally determined taxon *kanei*, it appears that the properties of the resins produced supersede morphological characters in this system of classification. For in several cases, Ka'apor groups reflect functional rather than morphological similarities. This is illustrated by three morphologically distinct species—*Protium giganteum, P. pallidum* and *P. spruceanum*—which are grouped into the folk taxon *kanei'i-tuwir* on the basis of their comparable *activity signatures*. Similarly, although very similar to *P. giganteum* in appearance, a fourth species *P. decandrum* is grouped into the distinct taxon *kanei'i-pitag* on the basis of its distinct functions.

PROPOSED HIERARCHY OF RESIN CLASSIFICATION

On the basis of these types of evidence, Balée and Daly have proposed a hierarchical tree of Ka'apor resin classification, within which six ranks can be distinguished (*designates covert taxa):

Rank	Constituent categories
1: plant parts*	leaves; roots; plant fluids*
2: plant fluids*	watery plant fluids; resins, saps and latexes (*hik*)
3: *hik*	*kanei**; non-*kanei**
4: *kanei**	6 folk taxa, each suffixed by '*hik*', including *kanei-hik*
5: *kanei-hik*	5 folk taxa: *ara-kanei-hik; kanei-aka-hik; kanei ape-hik; kanei-pitaq-hik; kanei-tuwir-hik*
6: e.g.: *ara-kanei-hik*	*Protium* sp

* designates covert taxon

In some cases, these special purpose or *cross-cutting* systems of classification are extensive, and may provide an alternative system for categorising the majority of plants known by certain communities. In such cases, these classificatory systems are known as *parallel classifications* and unless ethnobotanists are aware of these alternative taxonomic systems, confusion regarding the identity or characteristics of a given scientific species may arise. For example, the Shipibo-Conibo of the Peruvian Amazon classify many plants in two ways: both as members of general purpose folk generic categories, *and* as *rao*—a group of plants, animals and minerals which are classified according to their effect on human health. Hence, depending on the context in which a particular plant species is discussed, the Shipibo-Conibo use either its generic name, and/or its *rao* name. One example of this is illustrated by *Hura crepitans* (Euphorbiaceae) which may be referred to either by its generic name *anà* or as *peque rao* in the context of its use as a treatment against leishmaniasis (Martin 1995).

Ethnotaxonomy and Cultural Significance

Throughout the preceding discussion, it is apparent that certain patterns of folk systematic organisation are closely related to the cultural significance or *salience* of a given plant taxon. In particular, there is considerable evidence to suggest that the subdivision of folk generic taxa into subgeneric categories occurs largely among those plants which are cultivated; similarly, the rare occurrence of folk varietals is normally associated with important crops such as manioc or corn. In addition, where certain exceptions to the general rules of folk taxonomy occur, this variation in many cases, appears to be related to the cultural significance of a given plant taxon. For example, where a particular folk genus is not affiliated to a labelled life-form category, these taxa are often regarded as distinct from other groups on account of their particular economic importance, including 'corn' among the Mexican Tzeltal and 'bamboos' among both the Tzeltal, and the Philippino Hanunóo. As much of the current research into folk classification systems is aimed at elucidating how traditional people perceive and value the world around them, it is clearly of interest to identify any taxonomic structures which consistently reflect high levels of cultural salience. Hence, while current theories relating to this concept are relatively new and are yet to be tested thoroughly, it is worth noting some of the more common observations which have been discussed in the anthropological literature to date.

Table 9.5 Mode of production and ethnobotanical vocabulary. Following Cecil Brown's contentious report (see text), Brent Berlin has presented data from 17 relatively complete systems of ethnobotanical classification for a number of agricultural and non-agricultural peoples (Berlin 1992), which reveals that traditional non-cultivators labelled a mean of 197 folk generic taxa, compared with a mean of 520 recognised by traditional cultivators. Yet while these data appear to support Brown's assertion that total numbers of ethnobotanical taxa are closely correlated with mode of production, Berlin highlights the discrepancies in the biological diversity of these groups' habitats, pointing out that in Brown's paper no hunter-gatherer system is compared with an agricultural group living *in the same environment*. Equally, the problem of comparing data collected by different authors—for a range of purposes—is also emphasised

Cultivators	No. generic taxa	Non-cultivators	No. generic taxa
Quechua, Peru	238	Lillooet, Canada	137
Mixe, Mexico	383	Bella Coola, Canada	152
Ndumba, New Guinea	385	Haida, Canada	167
Chinantec, Mexico	396	Anindilyakwa, Australia	199
Tzeltal, Mexico	471	Navajo, USA	201
Wayampi, French Guiana	516	Sahaptin, USA	213
Aguaruna, Peru	566	Seri, Mexico	310
Taubuid, Philippines	598		
Tobelo, Indonesia	689		
Hanunóo, Philippines	956		
Mean	519.8	Mean	197.0
Standard error	63.4	Standard error	21.5

A simple Student's t-test carried out on these data presented here reveals that the differences between the means are highly significant ($P < 0.002$).

Mode of Production and Folk Taxonomy

In 1976, Brian Morris, an anthropologist based at Goldsmiths College of the University of London, published a report in which he proposed that communities with significantly different modes of production might exhibit patterned differences in their systems of folk classification (Morris 1976). Since then, Cecil Brown has put forward a rather controversial argument (Brown 1985), which suggests that small-scale cultivators exhibit an expanded inventory of biological terms. Comparing data collected by a number of disparate workers (Table 9.5), Brown suggests that the larger botanical vocabulary of horticultural societies reflects the cultural importance of a greater number of plants, since any cultivated species must have added to an existing list of wild plant terms. In addition, several workers have suggested that small-scale agricultural-ists also recognise comparatively greater numbers of wild taxa, thus implying a greater overall knowledge of both cultivated and wild plant species (see Brown 1985). Brown, and two other authors (Hunn & French 1984) have proposed independently, that the increased inventory of wild plant names can be explained on a strictly *utilitarian* basis in that as agricultural societies are more vulnerable to periodic catastrophe than hunter-gatherers, knowledge of a range of wild 'famine foods' would prove a selective advantage. However, others argue that the increased inventories of agriculturalists reflects a more *intellectual* need (see Berlin 1992); others still, maintain that a lack of lexical recognition for a given taxon does not necessarily imply a lack of its perceptual recogni-tion (Ellen 1994).

In addition to its discussion of the observed relationship between the size of ethnobiological inventory and mode of production, Brown's study also explores the apparent correlation between subsistence methods and the proportion of *binomial plant labels*. Again he suggests that agricultural societies show greater ethnotaxonomic differentiation than hunter-gatherers, in that the former tend to include a much greater proportion of binomial labels, suggesting that cultivators recognise a significantly greater number of *polytypic* genera. For example, while horticulturalists such as the Philippino Hanunóo, and the New Guinea Ndumba use binomial labels for about half of their perceived plant taxa, the non-agricultural Tasaday—also from the Philippines—apply bino-mial labels to only 11.6 per cent of their folk categories (see Berlin 1992). Despite inherent problems with the assumption that all subgeneric taxa have a binomial label and *vice versa* (see Box 9.6), recent surveys do suggest a strong relationship between levels of management and the incidence of polytypy, which in turn suggests that those folk taxa of greater cultural significance are more likely to receive sub-generic recognition (Box 9.8).

Cultural Significance and Plant Labels

In the previous section, we have seen how certain structural character-istics of folk taxonomic systems—for example their size and the inci-dence of polytypy—are apparently associated with differences in the cultural significance of different plant species. Equally, several aspects of folk biological nomenclature also appear to be strongly cor-related with levels of cultural salience, and might therefore act as indi-cators of particularly significant plant taxa. Aspects of nomenclature

Box 9.8   Folk specific taxa and cultural significance. It seems clear from these data that the polytypy observed in folk generic taxa is driven primarily by cultural, rather than biological concerns. Moreover, more detailed examination of ethnobotanical data has revealed a positive correlation between the degree of management and the number of specific taxa recognised within a single folk genus. For example, in a review of the polytypic genera recognised by various cultural groups, Berlin (1992) demonstrates that 89 per cent of generic taxa which contain more than 20 folk species, are cultivated. While there exist well-reasoned concerns about the validity of using these rather arbitrary groups (cultivated, protected, significant and not-utilised) as a measure of *cultural significance*, these data do demonstrate a consistent pattern which merits further examination.

As the vast majority (about 80 per cent) of folk generic taxa are *monotypic*—i.e. they are not divided into subgeneric taxa—a number of researchers have attempted to identify the factors motivating the cognitive subdivision of those folk genera which are *polytypic*. Significantly, in this respect, the examination of the monotypic generics has revealed that many are biologically diverse (*under-differentiated*) whereas a significant number of polytypic folk generics are *over-differentiated*—both of which suggest that the recognition of folk specifics is not dependent simply on inherent biological variability (Berlin 1992). Several workers now suggest that while ethnobiological *genera* are recognised primarily on the basis of morphological distinctiveness, the recognition of ethnobiological *species* may depend more on their relative cultural significance.

MANAGEMENT AND POLYTYPY: CULTURAL SIGNIFICANCE AND THE RECOGNITION OF FOLK SPECIFICS

In a recent publication Berlin (1992) has used the degree of management received by a given plant species as a rough index of its cultural significance. On this basis, he has demonstrated that the proportion of polytypic genera is consistently higher for those taxa which are cultivated, compared with those which receive lower levels of management; equally, in most cases, those genera which have no reported use are least frequently divided into subgeneric taxa:

Group	Total genera	Cultivated	Protected	Significant	Not utilised
		Level of management/utility			
Tzeital					
no. genera	471	63	41	195	172
% monotypic	85	51	78	86	97
% polytypic	15	49	22	14	3
Aguaruna					
no. genera	563	61	45	268	189
% monotypic	82	61	69	80	94
% polytypic	18	39	31	20	6
Quechua					
no. genera	238	102	20	65	51
% monotypic	86	78	90	94	90
% polytypic	14	22	10	6	10

which appear to be strongly influenced by cultural factors include characteristics both of the labels themselves, and of the consistency with which the label is applied. For example, labels which are either abbreviated or *semantically opaque* are often associated with particularly salient taxa (Boxes 9.9 and 9.10) while taxa whose labelling exhibits considerable lexical variation often tend to be less important to local people (Box 9.11).

Box 9.9   Lexical reflections of cultural significance—mononomial labelling of folk specific taxa. While most subgeneric taxa are labelled with binomial names, exceptions to this general rule occur quite commonly within ethnotaxonomic systems. In many cases, abbreviated labelling of subgeneric taxa is accounted for either by the prototypic nature of a given folk specific, or else by its high level of cultural significance. Hence, it might be assumed that most plants which are recognised at the level of folk species, yet which bear a primary name, are in some way significant to local people. However, recent work suggests that while mononomial labelling of folk specifics often corresponds to culturally important taxa—particularly to domesticated species—a small number of exceptions to this pattern have been reported (see Berlin 1992).

As we have seen earlier, most taxa at subgeneric levels of ethnotaxonomic organisation are labelled with a *binomial* secondary name. However, this general rule may not apply where a specific taxon is considered *prototypic* of its genus, or where a folk species is imbued with high cultural importance.

PROTOTYPICALITY AND MONONOMIAL LABELS

Where a subgeneric taxon *x* is considered the *prototype* of the folk genus, then *x* may in some cases be labelled with a primary name. In addition, the primary name used to designate the prototype is generally *polysemous* with that of its superordinate generic. For example, the Mexican Seri, recognise several folk generic taxa which contain prototype subgenerics, each of which bears the same name as the generic category to which it belongs (see Felger & Moser 1985):

Folk generic label	Folk-specific label	
xtoozp	*xtoozp*	*Physalis crassifolia* (prototype)
	xtoozp hapéc	*Lycopersicon esculentum* (lit. cultivated *xtoozp*)

continued

— continued —

Folk generic label	Folk-specific label	
hamíp	hamíp	Boerhavia coulteri (prototype)
	hamíp caacöl	B. erecta (lit. large hamíp)
	hamíp cmaam	Allionia incarnata (lit. female hamíp)

Where: folk genus *xtoozp* = 'native tomatoes'; folk genus *hamíp* comprises a group of closely related genera belonging to the family Nyctaginacea.

CULTURAL SALIENCE AND MONONOMIAL LABELS

Where a subgeneric taxon *y* is of major cultural importance it may be labelled with a primary name, which is linguistically distinct from the label of its superordinate generic. For example, the Peruvian Aguaruna recognise several subdivisions of the generic taxon *máma* (manioc) which are labelled with a primary name only (see Berlin 1992).

Folk-specific manioc taxa with secondary names	Folk specific manioc taxa with primary names
yakía máma—high manioc	puyám—thin (one)
shímpi máma—Shimpi's manioc	suhíknum—stingy stick
yusanía máma—Yusania's manioc	kanús—Santiago River
piampía máma—sandpiper manioc	suhítak—stingy (one)

In this example, folk specifics with primary names are not only cultivated but constitute the staple crop of the Aguaruna, and are clearly of considerable local importance.

Box 9.10   Lexical reflections of cultural significance—the concept of semantic opacity. In general, the more descriptive a plant name, the less cognitive effort is required to associate the name with its appropriate referent; hence, plants with *opaque* labels are likely to be those of particular local significance. The data presented here support this notion, implying a positive correlation between *semantic opacity* and cultural significance—a principle which might be useful in the identification of folk taxa of particular importance to local people.

While many of the labels applied to distinct folk taxa are essentially descriptive or *semantically transparent* others are essentially *semantically opaque*. Closely related to the phenomenon of applying primary labels to specific taxa, this aspect of folk nomenclature also appears to vary largely according to levels of cultural significance.

— continued —

_____ *continued* _____

MANAGEMENT AND SEMANTIC OPACITY

In a recent review, Brent Berlin has presented data which strongly support the notion that the degree of semantic transparency of a given label is largely inversely related to the cultural significance of the taxon to which the name refers (Berlin 1992). For example, among the Tzeltal Maya of Mexico, up to 78 per cent of cultivated genera have opaque labels, compared with only 9 per cent of taxa which are unmanaged:

		Level of management			
	Genera	Cultivated	Protected	Significant	Not managed
Total number	381	27	35	177	142
% Opaque	27	78	51	28	9
% Semi-transparent	33	19	29	38	31
% Transparent	40	4	20	34	60

These data clearly suggest that the greater the level of management received by a particular taxon, the more likely it is to be labelled with a name which is semantically opaque.

USEFULNESS AND SEMANTIC OPACITY

Among non-horticultural societies—where distinct levels of management are less clear—opaque labels appear to be largely associated with those taxa which are particularly useful and/or which have strong spiritual associations. For example, of more than 100 plant taxa to which the Seri Indians have attributed *unanalysable names* (Felger & Moser 1985) only 3 per cent have no reported use, while 68 per cent have two or more uses:

No. of uses per plant	% of taxa per with opaque labels	No. of uses per plant	% of taxa with opaque labels
0	3	6	7
1	29	7	1
2	21	8	0
3	21	9	1
4	7	domesticated	8
5	2		

Significantly, of the opaquely labelled taxa, 72 per cent are vital to subsistence, providing food, fodder or medicine, and include several particularly important species such as the organ pipe cactus (*Stenocereus thurberi*), mesquite (*Prosopis glandulosa*) and ironwood (*Olneya tesota*). The remaining 15 per cent have uses in technology and/or construction, or else have spiritual or mythological associations. Again these taxa include particularly important species such as *Guaiacum coulteri* which is used to make the important Seri Blue pigment.

Box 9.11   Lexical variation and cultural significance—phonological variation and synonymy. As with other patterned variation observed in systems of ethnobotanical classification, variability in the labelling of plant taxa appears to be closely correlated with their cultural significance. However, while these observations provide a useful point for further investigations, it is worth bearing in mind that in most of the examples presented here, cultural significance has to date been estimated solely on the basis of the *level of management* received by a given taxa.

One further aspect of ethnobiological nomenclature which appears to be closely correlated with cultural significance is that of the consistency with which a particular taxon is labelled. Again, a number of examples, suggest that the greater a taxon's level of local salience, the less likely it is to vary in its labelling.

PATTERNS IN LEXICAL VARIATION

Variability in plant labels may be due either to simple *phonological variation* or to the occurrence of *synonyms* or distinct *lexical variants*. For example, in one Tzeltal dialect, the cardinal flower (*Lobelia laxiflora*) is known by at least 10 phonological variants, and three lexical variants (see Berlin *et al.* 1974):

Variation in labelling of *Lobelia laxiflora*		
Phonological variants		Lexical variants
príma najk	*prinajk*	*príma najk*
príma najk'	*pírma najk*	*we'el t'ul*
prímo najk	*piríma najk*	*tulesnail wamal*
príwa najk	*ririm najk*	
prímajk'	*pirinajk*	

While the underlying factors leading to these forms of lexical variation have not been fully explored, it appears likely that taxa enjoying high levels of cultural significance are associated with lower levels of linguistic variation (Berlin 1992).

COGNATES AND CULTURAL SIGNIFICANCE

That lexical variation is lower for more significant taxa is further reflected in the occurrence of *cognates*, where similar labels in different communities refer to the same object. For example, among several Mexican groups the botanical genus *Erythrina* is known as *tsejst* (Totontepec), *tsejcht* (Northern Highland Mixe) and *tsejchk* (Lowland Mixe). Similarly, cognate terms are commonly found between the Tzeltal and Tzoltzil Maya, many of which exhibit regular *phonemic correspondence*:

_____ continued __|

_ continued _____

Language	One	Two	Three
		Examples	
Tzeltal	*tah*	*mahtas*	*may*
Tzoltzil	*toh*	*matas*	*moy*
English gloss	pine	'beggar ticks' herb	*Nicotiana tabacum*

In a comparison of more than 100 cognates between the Tzeltal and Tzoltzil, Berlin and Laughlin (cited in Berlin 1992) were able to demonstrate a strong correlation between the occurrence of cognates and the level of management. Indeed of those taxa which were either cultivated or protected, 87 and 80 per cent respectively were cognate between the two groups. In contrast, of those taxa which were wild but useful, 45 per cent were cognates, while of those which were wild and had no use, only 17 per cent were cognate between the two groups. Similar patterns have also been reported among different Tupian groups in South America (see Berlin 1992).

SUMMARY

As we have seen in previous chapters, a given community's approach to the use and management of their botanical resources is determined both by local empirical knowledge of the floral resources available, and by socio-cultural constructs which influence the ways in which the world is perceived. Therefore, if ethnobotanical data are to be understood cross-culturally, it is important to consider the specific nature of local decision-making environments. Approaches to studying different perceptions of the natural world can involve either the study of the ecological, spiritual, psychological and sociological influences which could affect a community's world view, or else can aim to elucidate the nature of local cosmology itself. Either approach can involve a range of methodologies, including the symbolic analysis of ritual and myth, the structural and functional analysis of social organisation, and the ethnotaxonomic study of traditional systems of biological classification. Ethnotaxonomy, in particular, has been widely used in the study of the ways in which human beings actually perceive their environment, and recent studies have demonstrated a number of important points: that the ways in which human societies categorise and name biological organisms are often very similar in both structure and substance; that folk systems of classification often parallel the formal scientific system; and that certain taxonomic features may consistently reflect high levels of cultural salience.

Chapter 10

The History of Plant–Human Interaction: Palaeoethnobotanical Evidence

... archaeobotany is more than the study of palaeoenvironmental indicators or even economic residues that reflect dietary intake and the seasonality of subsistence activities. It also provides a critical record of the reciprocal relationships between people and plants in dynamic ecosystems remarkably sensitive to most forms of human activity.

Karl Butzer 1982 in *Archaeology as Human Ecology*

INTRODUCTION

It is clear from the preceding chapters that the systematic study of traditional plant knowledge can provide an immense volume of information concerning the uses and management of thousands of plant species. However, while this discussion has centred so far on the botanical knowledge of those societies which exist today, much may also be learned from examining the historical and prehistorical interactions between people and their floral environment. For example, through elucidating the environmental impact of human activities in the past, future environmental degradation might be avoided. Such studies may be loosely encompassed within the term *palaeoethnobotany* (Helbaek 1960), which although used most frequently to refer to *archaeobotanical* studies, is used here to refer to the study of any aspect of the past relationships between plants and people. Data relevant to palaeoethnobotanical studies can be obtained from various sources—ranging from the examination of fossilised plant remains and historical documents, to the interpretation of traditional folk tales and prehistoric art. This chapter outlines the major types of evidence which are currently available, introduces some contemporary ideas on the evolution of plant–human relationships, and describes some of the more recent developments in this complex field of research.

RECONSTRUCTING THE PAST: THE NATURE AND INTERPRETATION OF THE PALAEOETHNOBOTANICAL RECORD

Within the context of this book, the *palaeoethnobotanical record* comprises any evidence which can be used to infer details of the historical or prehistorical relationships between plants and people. This includes not only the physical or chemical evidence provided by the study of preserved plants and artefacts, but also *phytogeographical*, genetic and linguistic data, as well as the evidence found in both written accounts and oral histories. On the basis of such wide-ranging sources of data, a vast body of information regarding many aspects of the nature, evolution and consequences of plant–human interactions has now accumulated.

Physical and Chemical Evidence: Archaeological Records and Modern Plant Populations

Throughout the twentieth century, research into the prehistoric and historic relationships between people and plants, has produced a wealth of information regarding the evolution and consequences of plant use and management. Much of this information has come from two important sources of evidence—the archaeological record and current plant populations. Archaeological evidence includes both preserved plant and animal remains and residual cultural artefacts, the analysis of which can reveal details of activities such as forest clearance and plant processing. Equally, additional factors such as the nature of settlement patterns, types of social organisation and changes in human health can also be determined through the examination of archaeological sites. The nature and distribution of contemporary plant communities can also provide considerable insight into factors including the history of plant dispersal and the origins of early domesticates (Harlan 1992). Since the beginnings of systematic palaeoethnobotanical research towards the end of the nineteenth century, advances in methodology have played a crucial part in these studies—for example the development of radiocarbon dating by *accelerator mass spectrometry* has revolutionised palaeoethnobotanical research by allowing the dating of samples as small as individual seeds, while recent developments in techniques such as *phytolith research* and *coprolite analysis*, also look set to provide powerful tools within this expanding area of research (Table 10.1).

Evidence from Plant Fossils

Over recent decades, plant fossils have been recovered from many archaeological sites where they occur within various *archaeological contexts*. For example, preserved pollen, seeds and tissue fragments have been found in association with specific contexts such as storage vessels, grinding tools or human faecal matter. Preserved plant materials have also been recovered from a range of other sites—such as peat deposits and lake sediments—many of which have yielded considerable evidence regarding prehistoric climate change and human interference.

Plant fossils may be preserved in various different ways (see Ingrouille 1992). Those preserved over several periods of geological time—may be *permineralised*, or may occur as either *compressions* or *impressions*. In permineralisation the plant tissue is infiltrated by soluble minerals which later precipitate out to form a rock matrix thus embedding the organic tissues. Such fossils are particularly useful in discovering the internal structure of early plant species, as they can be used to prepare thin sections which can be examined microscopically. In contrast, coalified compressions result from the collapse of the plant material and the chemical alteration of the residues. In some cases compressions contain pollen or spores, whose sporopollenin coat increases their chances of preservation; others may bear a thin film of waxy cuticle which has avoided degradation. In either case, chemical treatment and maceration of fossiliferous rock can yield spores, pollen or cuticular fragments which can then be examined microscopically. Finally, fossils known as impressions occur where a *mould* forms around the soft tissues of a plant; as these soft tissues later rot away the space becomes filled with sediment to form a *cast*, which provides a detailed representation of the external morphology of the original plant.

While mineralised plant materials from various geological periods have provided a wealth of information regarding plant evolution, more recent plant remains—those recovered from archaeological sites—generally occur in either a charred, desiccated or waterlogged state. These preserved plant materials vary from *microfossils* such as pollen, spores and phytoliths, to *macrofossils* which include fragments of wood

Table 10.1   Major developments in the history of archaeobotanical research. Archaeological methodology has advanced enormously over the last 100 years facilitating the study of many aspects of prehistoric life, including the functional uses of specific tools, the major components of the palaeolithic diet, and any major changes in modes of subsistence. Today, much of this information is interpreted with the help of botanical data from modern plants—including factors such as their geographical distribution, and their ecological or phytochemical characteristics

---

Technical Developments in Palaeoethnobotanical Research

Field archaeology and dating methods

Sixteenth century antiquarians documented prehistoric sites with no chronological framework

From the mid-eighteenth century, subterranean excavation became increasingly common

By the nineteenth century, three technological ages—stone, bronze and iron—had been identified, providing a rudimentary chronology within which to work

During the twentieth century, the development of stratigraphic methodology and of later isotope and spectroscopic dating has revolutionised archaeological research (see Darvill 1987)

Analysis of tool surfaces

Since the 1970s, microscope analysis of tool surfaces has been used to determine specific tool uses based on the analysis of distinctive *microwear patterns* (Keeley & Toth 1981)

Recent experimental work suggests that it may be possible to characterise use-wear patterns arising through the harvesting of plants grown on tilled *versus* untilled soils (see Harris 1989)

In recent years, archaeochemical analysis of stone and ceramic surfaces has become increasingly common (Hill & Evans 1989; and see Rudgely 1993)

Analysis of animal remains

The chemical and structural analysis of skeletal tissues and teeth have been used for decades in the study of prehistoric diet and health

In recent years, the analysis of microwear polishes and phytolith residues on tooth surfaces have proved useful in investigating the nature of prehistoric diet

Since the 1960s, the botanical—and later chemical—analysis of human coprolites has been used in the investigation of human nutrition and disease (see Bryant 1993)

Plant microfossil analysis .

During the seventeenth century, the invention of the microscope saw the beginning of an important phase in archaeobotanical research, with the first observations of pollen grains

By 1940 the potential of pollen analysis in archaeological research was widely recognised

Table 10.1   (*continued*)

Technical Developments in Palaeoethnobotanical Research

Phytoliths were first observed during the 1830s—and since the 1980s the archaeological potential of these microfossils too, has been recognised (see Bryant 1993)

Plant macrofossil analysis

Since the mid-nineteenth century, the identification of plant remains from archaeological contexts has formed a significant feature of archaeological investigation

Since the 1950s, interest in macrofossil analysis has increased considerably, particularly since the development of *flotation* methods which facilitate the recovery of preserved plant materials (see Pearsall 1989)

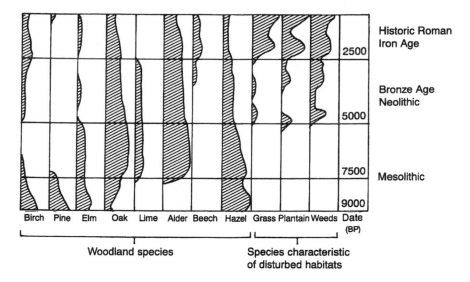

Figure 10.1   Pollen analysis and the reconstruction of past vegetation. Using quantitative methods of pollen analysis, it is possible to construct a total *fossil pollen assemblage* from samples of preserved macrofossils. Using sequential *stratigraphic* sampling, it is therefore possible to identify sequences of pollen assemblages which provide a broad picture of changes in vegetation—whether due to climatic effects, or to human interference. This example, shows a generalised pollen diagram for England from the end of the Palaeolithic epoch, and clearly demonstrates how clearance of woodland during the later stages of the Mesolithic period, is indicated by a reduction in the proportion of tree pollen present, and a concomitant increase in the proportion of grasses and weedy species—both of which are indicative of habitat disturbance. The *pollen curves* are for selected pollen types only, and are expressed as percentage of total arboreal pollen. (Modified from Moore & Webb 1978)

or entire seeds, while softer tissues such as leaves, roots and flowers are found only very rarely. The range of information provided by the recovery and examination of preserved plant materials can be considerable, allowing, for example, the reconstruction of prehistoric changes in vegetation (see Figure 10.1), the examination of the process of plant domestication, and the elucidation of the history of plant use. However, while the discovery of charred seeds found at an archaeological site could indicate the deliberate roasting of that species as part of some plant processing method, the presence of carbonised remains could result equally either from natural or artificial fires, or from the burning of animal dung as fuel. Hence, it is clear that if information revealed in the fossil record is to be of value, then methods facilitating the accurate *interpretation* of the fossil record are essential. For example, the nature of the specific archaeological context in which preserved tissue is found may provide valuable information concerning plant use, while the ecological behaviour of modern plant populations can provide a useful model for the interpretation of fossil evidence. Some of the main points which are considered in current interpretative methodologies are outlined in Box 10.1.

Box 10.1   Interpreting the archaeobotanical record. There are many factors to consider when attempting to interpret the significance of preserved plant materials, whether they are recovered from archaeological sites, or from lake sediments and peat deposits. For the physical, chemical and behavioural characteristics of different plant species will have a significant effect on their representation in the fossil record. In order to alleviate problems of interpretation, ecological knowledge of modern representatives has been applied in the analysis of fossil evidence, and in recent years, various experimental approaches have also been adopted.

> The interpretation of the archaeobotanical record can be facilitated using a number of distinct approaches. For example, a plant's use may be clarified by consideration of the specific archaeological context within which its remains are found, while knowledge of a species' current reproductive biology may prove of enormous value in the interpretation of fossil pollen assemblages. In addition, contemporary experiments into factors such as rates of pollen degradation, or the types of fragmentation patterns produced by specific types of plant processing, are also proving helpful in archaeobotanical investigation.
>
> THE SIGNIFICANCE OF ARCHAEOLOGICAL CONTEXT
>
> In most cases, knowledge of the specific archaeological context in which preserved remains are found is crucial. For example, where charred seeds are

_ continued _

continued

discovered inside ceramic cooking vessels, there is an increased likelihood that those seeds were genuinely processed prior to eating; in contrast, charred seeds occurring in less specific contexts could be indicative of natural fire, deliberate burning or even the use of animal dung as fuel.

BIOLOGICAL AND ECOLOGICAL CONSIDERATIONS

Details of the reproductive behaviour of modern plant species can prove vital in reconstructing palaeoenvironments from fossil floral assemblages. For example, factors such as the rate of pollen production for different species and the dispersal properties of that pollen will clearly affect a particular species' proportional representation in the fossil assemblage (see Bryant & Holloway 1983):

Mode of pollination	Relevant features
Anemophilous (wind-pollinated)	10 000–70 000 pollen grains produced per anther, and show adaptations to long distance travel
Zoophilous (animal-pollinated)	1000 or less pollen grains produced per anther, often 'glued' in place
Hydrophilous (water-pollinated)	as the pollen of these species are less likely to become dehydrated, their pollen grains often lack a durable outer wall of sporopollenin pollen grains
Cleistogamous (self-pollinating)	pollen grains show little dispersal

EXPERIMENTAL EVIDENCE

In addition to characteristics such as pollen dispersal properties, different species also vary in factors such as their pollen's susceptibility to degradation, or the ways in which their seeds fracture during processing. In recent years, a number of controlled experiments have yielded valuable insight into such characteristics in a range of plant species.

- Characteristic rates of preservation have been demonstrated for different pollen types under a range of physiochemical conditions. For example, while the conditions experienced in clay river soils lead to the rapid degradation of species such as alder, ash and elm, only minor degradation of oak and beech is observed (see Bryant & Holloway 1983)
- Recent monitoring experiments have elucidated characteristic pollen spectra which occur currently, in response to specific management practices such as mowing, grazing or burning (Gaillard _et al._ 1992)
- Experimental work—for example, investigating characteristic fragmentation patterns produced by distinct methods of processing—may allow the identification of the residues of processing among archaeobotanical materials (see Stahl 1989)

Evidence from Human Remains

Owing to the relative rarity with which plant materials are actually preserved, the archaeological evidence for plant exploitation and manipulation relies not only on direct archaeobotanical evidence, but also on indirect evidence including that derived from human remains. Human remains provide three major sources of evidence regarding past plant–human relationships—namely, skeletal remains, teeth, and both preserved stomach contents and faecal matter. While the physical and chemical analysis of skeletal remains can reveal much about the general nature and quality of the human diet, the examination of microwear patterns on tooth surfaces can provide information regarding dietary composition and methods of plant food processing (Table 10.2). Finally, *coprolite analysis*—the analysis of preserved faecal matter—is also beginning to yield insight into factors such as dietary composition, food processing techniques and the use of medicinal plants.

Originally applied to the study of dinosaur diet as early as 1829, the study of coprolites has become an increasingly valuable tool in the study of human evolution (Reinhard & Bryant 1992). Indeed, following the early identification of a number of dietary components based on plant macrofossils, more systematic analyses of human coprolites began in the late 1940s, with Janius Bird's examination of preserved faecal matter and intestinal contents excavated from *Huaca Prieta de Chicama* in coastal Peru (Bryant 1994). Between 1960 and 1970, coprolite research was advanced considerably by the pioneering work of the late Eric Callen, a plant pathologist who spent his professional career at McGill University, Montreal. Callen developed many of the methods now used in the preparation and analysis of coprolites, and carried out the first pollen analysis in preserved faecal matter. These early studies were subsequently extended by experimental work which examined, for example, the rate at which different types of pollen move through the gut. Such studies have yielded data allowing not only the identification of the basic elements of prehistoric diet, but providing also, some insight into which foods were eaten together (see Bryant 1994). Finally, chemical analysis of coprolites began in the 1970s with the application of gas chromatography, and it now seems likely that current research into the potential of immunological analysis may ultimately allow the identification of diagnostic plant and animal proteins where no visible residues exist (Reinhard & Bryant 1992; Sutton 1993).

Table 10.2   Evidence from human remains. Several methods used by students of *palaeopathology*—who investigate the nature and occurrence of prehistoric disease—can yield valuable information to the palaeoethnobotanist. For example, the incidence of annual stress may explain, or reflect, certain modes of subsistence, while the relative importance of plant foods may be inferred from the chemical makeup of fossil bones. In addition, methods such as microwear analysis can provide an insight into other aspects of prehistoric subsistence, including the development of plant food processing, or the occurrence of significant changes in diet

---

Palaeoethnobotanical data from human skeletal remains

Chemical analysis of fossil bones

Levels of micronutrients such as iron, calcium, magnesium, lead, zinc, copper and strontium can all be determined, giving some insight into dietary intake.

In particular, the ratio of strontium:calcium can be used as an indicator of the relative proportion of animal and vegetable foods consumed, as higher proportions of strontium reflect a high level of plant foods (Eaton *et al.* 1988).

More recently, it has been suggested that stable carbon isotope analysis might be used to establish the presence of certain cultigens in the diet (Goodman *et al.* 1984).

Skeletal signatures of stress

General features such as mortality, dental crowding and random asymmetry may all be interpreted as indicators of stress (Goodman *et al.* 1984).

General, chronic dietary stress can apparently be inferred from the occurrence of *enamel hypoplasias* and *enamel microdefects*—both of which are manifest as deficiencies in enamel thickness due to some disruption in development.

More specific evidence includes the appearance of *Harris lines* or lines of growth arrest observed in longbones. These are thought to be related to periods of acute episodic stress such as annual periods of famine (Yesner 1994).

Skeletal pathologies restricted to *cribra orbitalia* reflect the occurrence of iron-deficiency anaemias (Yesner 1994).

Tooth surface analysis

Extensive toothwear is suggestive of general use of teeth as grinding surfaces—often suggesting little use of food processing techniques.

Detailed analysis of microwear patterns using scanning electron microscopy can reveal the nature of plant foods and level of processing used (Rose *et al.* 1984).

Recovery and identification of phytoliths from tooth surfaces can also reveal the nature of plant foods used (Bryant 1993).

Evidence from Artefacts

In addition to the analysis of bioarchaeological remains, information obtained from the examination of preserved artefacts can also prove useful in palaeoethnobotanical research. The significance of the specific archaeological context from which plant remains are recovered has already been mentioned, and various other lines of evidence are also proving valuable. Indeed, as in the case of dental examination, the careful analysis of use-wear patterns of tool surfaces has provided considerable insight into the development of plant food processing. These studies are based on evidence from modern experimental work with stone tools which has allowed the identification of microscopically distinct *wear-polishes* or *use-wear patterns*, which can be used to characterise the past uses of specific tool types. The functional interpretation of these microwear patterns is based on various features, including the reflectivity and texture of the polish, the size and direction of any striations, the nature and extent of any edge damage, and the identification of any residual phytoliths. For example, using these techniques, stone tools dating from 1.5 million years ago at the *Koobi Fora* site, Kenya, have been identified as used in cutting soft plant tissues, while others appear to have been used in the scraping and sawing of wood (Keeley & Toth 1981).

Like use-wear analysis, the *archaeochemical* analysis of organic residues found on tool surfaces also represents a palaeoethnobotanical tool with considerable future potential. For example, using *pyrolysis mass spectrometry*, it has been demonstrated that the working surfaces of grinding stones and mortars found at the *Wadi Kubbaniya* site, Egypt, were used in processing tissues which were rich in cellulose, yet poor in protein; this strongly suggests a dependence on root rather than seed plants (see Smith 1989). More recently, UK researchers Edward Hill and John Evans have analysed heptane extracts of organic residues from pottery shards excavated from the site of *Rungruw* in Yap Island, Micronesia, using a range of *spectroscopic* techniques (Hill & Evans 1989). Through comparison with extracts of modern plant samples, these workers conclude that organic residues showing little obvious denaturation can be extracted from preserved artefacts dating back at least 2000 years. In addition they demonstrate that infrared (IR) analysis of these extracts produce characteristic traces of sufficient clarity to allow comparison between modern plant extracts and archaeochemical samples, and that the presence of certain diagnostic peaks can permit the tentative identification of preserved residues (Box 10.2).

Box 10.2   Archaeochemical analysis of preserved organic residues. Preliminary data from the analysis of residues from ancient tool surfaces suggest both that heptane extracts of specific cultigens may be characterised on the basis of their IR spectra, and that archaeological residues suffer little obvious denaturation following their deposition. Hence, it is likely that the application of chemical analysis to the identification of plant residues on ancient tool surfaces offers a promising method in palaeoethnobotanical research.

That the chemical analysis of organic residues found on preserved tool surfaces could yield valuable information on tool usage was first demonstrated more than a decade ago, when infrared (IR) spectroscopy was used to demonstrate the presence of organic compounds in crude preparations made from surface layers (see Hill & Evans 1989). Since that time, this approach has been developed further to produce preliminary diagnostic chemical spectra for a range of traditional food plants.

IDENTIFICATION OF PACIFIC FOOD PLANTS FROM CHARACTERISTIC CHEMICAL-SPECTRA

Following several years of research, workers Edward Hill and John Evans (1989) have now produced 'standard' IR spectra for relatively crude heptane extracts of six major Pacific cultigens: banana, rice, sago, sweet potato, taro and yam. On comparison with similar extracts from archaeological samples, these workers have been able to identify—if at this stage rather tentatively—the presence of specific cultigens in the surface residues of preserved potsherds dating from about 2000 BP.

IR frequency (wavenumber cm⁻¹)	1750	1700	1660	1600	1510	1280	1260	1235	1110	1100	1020	970
Modern plant foods												
Banana						/				/		
Sweet potato					/							
Rice	/	/					/					
Sago								/				
Taro				/				/	/			
Yam			/									/
Archaeological samples												
YP83-2262						/				/		
YP83-2540						/				/		
YP83-2618				/	/			/	/			
YP83-2673	/	/					/					

Data taken from Hill & Evans (1989).

continued

— *continued* —

From these data it appears evident that extracts of modern cultigens may be characterised by the presence of certain diagnostic peaks—for example, banana is recognised by peaks occurring at 1260 and 1020 in IR spectra. By comparing modern extracts with those of archaeological residues, it therefore seems possible to identify the presence of known cultigens in the preserved samples. Indeed, it seems evident that extracts YP83-2262 and YP83-2540 both contain preserved material from banana, while extract YP83-2673 contains organic residues characteristic of modern rice. Sample YP83-2618, on the other hand, appears to contain both taro and sweet potato.

Evidence from Living Plants

Despite the range of archaeobotanical materials available, such evidence is lacking in many important regions of the world—most notably throughout the African continent (Chikwendu & Okezie 1989; Harlan 1989). In addition, certain plant staples, including root crops and cucurbits, are consistently underrepresented in the fossil record on account of their poor preservation. In such cases, researchers have relied largely on the analysis of existing plant populations to suggest probable sites of domestication and the likely *progenitors* or wild ancestors of modern domesticates. For example, Harlan (1989, 1992) has examined botanical and geographical data from wild races of sorghum (*Sorgum bicolor*) to suggest a likely explanation of the origin and subsequent spread of this important cultigen (Box 10.3). Integral to the success of this type of approach, is the application of *cytogenetic analysis* or chromosome studies which allow researchers to trace the phylogenetic relationships between wild and cultivated plants. For example, two *tetraploid* wheat species *Triticum dicoccoides* and *T. timopheevii* have *genomic constitutions* of AABB and AAGG, respectively, indicating that as both have the set of chromosomes designated 'A' they must have one ancestor in common (see Box 10.4).

While providing crucial information where little archaeological evidence is found, the study of modern plant populations also plays an integral part in the interpretation of the archaeological record itself, and as we have already seen (Box 10.1), knowledge of a given species' ecological characteristics is essential to the interpretation of fossil pollen assemblages. In addition, several research teams are currently involved in experiments aimed at elucidating floral indicators of specific human activities. For example, a group of Swedish workers have conducted a recent study into the ways in which activities such as grazing or tilling are reflected in contemporary pollen assemblages (Gaillard *et al.* 1992),

Box 10.3 **Phytogeographical evidence for the domestication of sorghum (*Sorgum bicolor*).** In the work presented below, the details of the origins and development of domesticated sorghum are proposed largely on the basis of the nature and distribution of modern sorghum populations. Significantly, for crops where both archaeological and modern botanical evidence are now available, there is often considerable corroboration between the two. For example, both current plant distribution and evidence from the archaeological record indicate that wild emmer wheat (*Triticum dicoccoides*) could have been domesticated only in the Near Eastern arc; similarly, current populations of the wild progenitor of chickpea (*Cicer arietinum*) and the distribution of fossilised remains, together pin down the domestication of this crop to southeastern Turkey (Zohary 1989).

Sorghum (*Sorghum bicolor*) represents one of the most important cereal crops known today, yet until recently, relatively little was known about the history of its domestication or its subsequent development. However, following years of extensive field work, researchers at the Crop Evolution Laboratory in Illinois have discovered much about the origins and development of this staple species (Harlan 1989, 1992).

THE ORIGINS OF DOMESTICATED SORGHUM

On the basis of its distribution and ecological characteristics, a probable progenitor for domesticated sorghum has been identified from several wild candidates:

Name and characteristics of wild races of sorghum	
arundinaceum	A West African race which is adapted to the tropical forest zone. Any great contribution from this race can be largely discounted on the basis that cultivated sorghum is an essentially dryland crop which does not grow well under the high rainfall conditions of the forest zone.
virgatum	A very small sorghum which grows along the Nile flood plain, this race seems an unlikely progenitor for domesticated forms.
aethipicum	This race may be a secondary derivative produced by hybridisation between cultivated sorghum and its wild relatives, and, like *aethipicum,* its seems an unlikely ancestor of cultivated sorghum.
verticilliflorum	This extremely productive race is fully fertile with cultivated sorghums and seems a likely candidate as the progenitor of cultivated sorghums. Extremely abundant in eastern Africa, particularly in Sudan and parts of Chad it seems likely that this is the area where domesticated forms first appeared.

continued

_ continued _

THE DEVELOPMENT OF DISTINCT SORGHUM CULTIVARS

Based on both experimental evidence and the close examination of herbarium collections, domesticated sorghums have been classified into five major races, and the distribution patterns of each has been analysed. Together, these analyses have facilitated the proposal of a history of domesticated sorghum in the African continent:

Name and characteristics of the major domesticated races of *Sorghum bicolor*	
bicolor	The most primitive of the five races of sorghum, this race is found almost everywhere that sorghum is grown. Morphologically it is very similar to wild sorghum, yet its grains are larger and its inflorescences do not shatter so much as in the wild form. On the basis of its primitive nature and wide distribution, it is assumed that this represents the first race to be domesticated.
guinea	Primarily a West African race, this form has characteristic modifications which allow it to grow under conditions of relatively high rainfall—presumably derived through gene flow between early domesticates and the wild *arundinaceum* race.
caudatum	This race which yields a distinctive grain is found largely from Lake Chad to the Ethiopian border.
kafir	This is a Bantu sorghum and belongs to southern Africa.
durra	Another distinctive race, this is highly derived with very dense, compact inflorescences. It is grown primarily by Islamic peoples along the edge of the Sahara, while its main centre of cultivation is in India.

while experimental research in northern Norway has facilitated the floral characterisation of specific sites, such as field margins, footpaths or superficially manured land (Vorren 1986). Finally, modern plant populations are often crucial to investigations into plant domestication, many of which are dependent on the morphometric comparison of modern cultivars with their wild relatives as outlined in Box 10.5.

Documentary Evidence: Plants in Art and Literature

During the later stages of human evolution, from around the end of the Pleistocene epoch, human societies began to demonstrate increasing levels of cultural complexity, as illustrated by increasingly

Box 10.4   Cytogenetic studies and crop evolution—the chromosomal make-up of modern wheats. Over recent decades, chromosomal analysis has been used to investigate the development of several important crop species, including maize (*Zea mays*) and cotton (*Gossypium* spp). In one review of a number of important New World domesticates, this approach—often in combination with additional techniques such as *isozyme analysis*—has demonstrated several significant features. For example, there is strong evidence to suggest that multiple domestications of *Capsicum* occurred over different parts of its geographical range, resulting in the *annuum-chinense-frutescens* complex observed today (Pickersgill 1989).

Many current ideas concerning the wild progenitors of cultivated plant species are based on *cytogenetic* studies in which the chromosomal composition of related species are visually compared. For within a given *genome* the constituent chromosomes each have a characteristic size and shape which can be observed after staining, using light microscopy. In recent years this ability to observe chromosome characteristics has provided a valuable tool in tracing the phylogenetic relationships between domesticated plants and their wild relatives.

THE CHROMOSOME COMPOSITION OF WHEAT

While most plants have two sets of chromosomes *per* cell and are said to be *diploid*, certain cultivated species possess four, six or even eight sets of chromosomes and are known as *polyploid*. In the case of modern wheats, the ploidy levels range from diploid in the primitive domesticate *Triticum monococcum* to hexaploid in the widely cultivated *T. aestivum*. However, in these polyploid derivatives, the sets of chromosomes present are not necessarily identical, but may be derived from separate ancestral species, thus producing hybrids which contain genetic material from a number of distinct, and characteristic genomes. Within a group of related organisms, these distinct genomes are labelled specifically using alphabetical symbols.

Species	Ploidy level	Genomes present
T. monococcum (einkorn)	Diploid	AA
T. speltoides	Diploid	BB
T. tauschii	Diploid	DD
T. turgidum (emmer)	Tetraploid	AABB
T. aestivum (bread wheat)	Hexaploid	AABBDD

Data taken from May *et al.* (1991).

Archaeological and genetic evidence suggest that by the eighth century BC, a cultivated einkorn and another species—possibly *T. speltoides*—hybridised to produce the tetraploid emmer *T. turgidum*. One variety of emmer, *T. turgidum* var. *dicoccum*, subsequently underwent mutation which caused the bases of the *glumes* to collapse at maturity—a change which made the separation of the seed

continued

continued

from the surrounding chaff relatively easy, and thus gave rise to the free-threshing durum wheat, _T. turgidum_ var. _durum._ Shortly after the appearance of the tetraploid, a hexaploid species arose, which bore the AABB chromosomes of the tetraploid emmer, and the DD chromosomes of a diploid species, believed to be _T. tauschii._ These free-threshing hexaploids—_T. aestivum_—have surpassed all other wheats, with more than 20 000 cultivars of this species in existence today (Simpson & Conner-Ogorzaly 1986).

Box 10.5   Comparative analysis of living plants—investigating domestication in the fossil record. It is evident from the data presented here, that the management, cultivation and human selection of useful plants can lead to detectable changes in morphology which are characteristic of the domestication process. By extrapolation, it therefore seems reasonable to assume that the detection of diachronic morphological changes in the archaeological record could provide a means of tracing the domestication process. Indeed this method has proved valuable in investigating the domestication of a number of important crop plants including sunflower (_Helianthus annuus_) and sumpweed (_Iva annua_), both of which were apparently domesticated in eastern North America before 1500 BC (see Gremillion 1993).

Just as the domestication of animals has been traced in the fossil record through the detection of quantifiable changes in morphology, so too the effects of artificial selection of plant cultivars can be discerned through morphometric analysis. This method depends largely on determining differences between modern populations of wild and cultivated plants, and may involve the identification of significant changes in quantitative traits such as grain size, or more qualitative attributes such as whether inflorescences remain intact on ripening.

MORPHOLOGICAL CHARACTERS REFLECTING DOMESTICATION

In most cases, the morphological changes which characterise domestication, reflect human preferences and needs. For example larger grains clearly provide a more useful source of nutrition, while maize cobs which retain their kernals facilitate much more efficient grain harvesting.

Morphometric changes associated with crop domestication	
Echinochloa utilis	Barnyard millet; a 15 per cent increase in seed size occurred over one millennium, facilitating both the harvesting and processing of this staple crop.[1]

continued

_____ *continued* _____

Morphometric changes associated with crop domestication	
Triticum spp	Wheat; rachises changed from brittle morphotypes which shattered when ripe, to those which remained intact, while seeds became larger; both characteristics facilitate harvesting and processing.[2]
Helianthus annuus	Sunflower; seeds have increased in size from 4 to 5.5 mm in length 2850 years ago, to 7.8 mm in length today.[3]
Chenopodium belandieri	Seedcoat thickness has reduced from between 40 and 80 μm in wild populations to 20 μm in domesticated forms—a change which is related to reducing dormancy.[4]

See: [1]Crawford (1992); [2]Miller (1992); [3]Smith (1992); [4]Gremillion (1993).

EXPERIMENTAL EVIDENCE FOR MORPHOMETRIC CHANGE: RECONSTRUCTING THE DOMESTICATION OF THE AFRICAN YAM

Evidence that rapid changes in plant morphology can occur as a consequence of *anthroselection* has been presented recently by a group of workers from the University of Nigeria, who have attempted to simulate the process of early yam (*Dioscorea* spp) domestication in Africa. Using wild yams—whose huge tubers and large, thorny roots required several days to uproot a single tuber—tubers were grown under a range of experimental conditions, and any genetic, ecological or morphological changes were monitored over a period of 8 years (Chikwendu & Okezie 1989). In addition to various changes both in the morphology of the plants' aerial parts and in nutritional features such as fibre content and starch levels, these experiments demonstrated a number of significant, measurable changes in root and tuber morphology.

• Significant reduction in the number and length of thorns occurring on tubers
• Reduction in enormous size of tubers and thickness of subterranean stems
• Formation of more uniform, rounded tubers
• Reduction in the branching of tubers
• Reduction in the branching of roots, and roots themselves become thinner and less thorny

sophisticated technology, the emergence of decorative items and works of art, and by the ultimate emergence of writing. Hence, the body of evidence available to students of palaeoethnobotany expands to encompass not only biological and archaeological data, but also artistic and later, written evidence of the relationships between plants and people.

Evidence from Art History

In the archaeological record, evidence of artistic behaviour dates back to the Upper Palaeolithic (beginning *c*. 40–35 000 BP), a period which, throughout much of Europe, represents a time of unusually rapid technological change. For example, from about 35 000 years ago, stone tools exhibit much more explicit symbolic significance, while tools and artistic items fashioned from two novel materials—bone and ivory—appear suddenly in the fossil record at this time. Throughout the history of art, both naturalistic and more abstract representations have commonly occurred (Table 10.3), and both have received considerable attention from students of both art history and archaeology. The interpretive analyses of such art historians have revealed much about the roles and significance of plants in human culture from Palaeolithic times onwards.

Palaeolithic art is often divided into two main forms—*parietal* art, which refers to the paintings and engravings found on the walls of caves, and *mobiliary* art, which describes depictions found on small portable items (Mellars 1994). Among the most impressive manifestations of the parietal art of this period, are the cave paintings of *Lascaux*, France and *Altamira*, Spain (20 000 and 15 000 BP respectively), the basic features of which are now fairly well documented. Elaborately worked using mineral-based pigments such as iron ochre and manganese dioxide, these paintings largely depict representations of animals of economic importance —such as reindeer, bison, wild cattle and mammoth—and with few exceptions, human figures are almost lacking. In addition to these naturalistic figures, geometric designs also, are found frequently at various points over cave walls and ceilings.

Since their discovery in the 1870s the interpretation of paintings has already preoccupied several generations of specialists, and hypotheses explaining the motivation behind the paintings range from the notion of 'art for art's sake', through their use as a form of sympathetic magic, to more recent interpretations which emphasise the role of art as a means of carrying information (see Mellars 1994). Some of these explanations have suggested that the geometric designs found alongside naturalistic forms are representations of the tracks or hoofprints of the respective animals; other authors argue that many of the signs

Table 10.3 Artistic developments reflected in the archaeological record.
In the Old World, the Upper Palaeolithic (*c.* 35 000–10 000 BP) marked a period of
rapid and repeated change in technological expertise, including a sudden burst
of bone and ivory technology. Concomitant with this expansion of technological skills, this period also saw a remarkable proliferation of artistic behaviour,
some of the most impressive manifestations of which are seen in the cave paintings of *Lascaux* and *Altamira*. Both naturalistic and other artistic representations which have appeared in the archaeological record have received
considerable attention from art historians and archaeologists, who have sur-
mised much about Palaeolithic life from their interpretations. Equally, the interpretation of artistic items from both the New World, and from later in the
archaeological record, has provided some insight into the possible uses of
plants in both prehistoric and historic times (see [1]Cunliffe 1994; [2]Dumond 1987;
[3]Miller & Taube 1993; [4]Taube 1993)

Old World developments[1]	New World developments
35 000–20 000 BP	**35 000–20 000 BP**
Stone tools became more ornate	Some evidence suggests possible
Appearance of items of personal	human occupation of
decoration	northernmost parts of America[2]
Appearance of ivory figurines	
20 000–10 000 BP	
Appearance of cave paintings bearing	
naturalistic figures and geometric	
designs	
Proliferation of mobiliary art objects	
and expansion of rock art	
10 000–6500 BP	
Ceramics appear in the archaeological	**10 000–1000 BP**
record—initially plain and later	Tools and bone carvings appear in
decorated	northernmost America
Appearance of tablets inscribed with	Olmec peoples (*c.* 3200 BP) begin to
enigmatic symbols	make permanent records using
Beginnings of metallurgy leading to	complex standardised iconography[3]
gold work and copper engravings	Glyphs used in Maya art, portray
	plants of economic and cultural
	significance[4]

defy such explanation, and propose that these symbols depict *entoptic
phenomena*—light-independent visual percepts which can be observed
during certain altered states of consciousness (see Rudgely 1993).
Within the context of palaeoethnobotany, this latter hypothesis is of
particular significance, as a role for psychoactive plant use during
Palaeolithic times could be implied.

While the interpretation of any work of art will inevitably remain controversial, the analysis of plant motifs found within various contexts has been used as a source of palaeoethnobotanical evidence. For example, at the 35th Annual Meeting for the Society of Economic Botany, held in Mexico City in June 1994, several papers presented data on plants in art history, including a discussion of species of the psychoactive genus *Datura* which are depicted in the Aztec herbal (Delfeld 1994), and a survey of domesticated plants and fungi in contemporary Latin American art (McMeekin 1994). Elsewhere, evidence of plant use has been determined from the plant motifs identified from various sources including a 2000-year-old Peruvian textile (Cordy-Collins 1982), Mesoamerican pottery *spindle whorls* carved with fruits and flowers (McMeekin 1992), and Egyptian ceramics which are shaped in the form of specific plant parts (see Rudgely 1993). In each case complementary data from chemical, ethnographic or other sources have been used to test hypotheses, revealing considerable insight into past relationships between plants and humans (Box 10.6).

Box 10.6   Art in ethnobotanical research—evidence for the prehistoric use of psychoactive plants. The examples of the interpretation of prehistoric and ancient art presented below, illustrate how plant motifs and other artistic representations can prove of value in palaeoethnobotanical enquiry. Significantly, in both cases, researchers have used contemporary studies to support their hypotheses: Lewis-Williams and Dowson used a combination of laboratory research and ethnographic study; Cordy-Collins also uses ethnographic evidence to support her claims.

In addition to more conventional sources of palaeoethnobotanical evidence, the symbolic analysis of floral motifs in both parietal and mobiliary art can also prove a useful source of information. One area of plant use which has so far been investigated through the interpretation of Palaeolithic and more recent art, is the use of psychoactive plants—a practice which may reflect the prehistoric development of shamanic practices.

ENTOPTIC PHENOMENA IN PALAEOLITHIC ART

To date, much of the research into Palaeolithic rock art has focused on the caves of the Franco-Cantabrian area of southern France and northern Spain, where such sites are both abundant and relatively accessible. The paintings themselves generally occur deep inside the caves, often in relatively inaccessible and narrow passages, and consist largely of animal figures—such as bison and wild cattle— which are frequently found in association with various apparently abstract designs (Mellars 1994). While some researchers have interpreted these geometric patterns as tangible objects such as nets for catching animals, or hoofprints of

continued

_____ *continued* _____

particular animals, a recent thesis has proposed that these designs depict *entoptic phenomena* or *phosphenes*—a series of luminous percepts which derive from the human nervous system, which are commonly perceived by people in altered states of consciousness (Lewis-Williams & Dowson 1988). On this basis, these workers suggest that geometric designs in Palaeolithic art may provide evidence (outlined below) for the use of psychoactive plants:

* Entoptic phenomena may be generated by various conditions, including the use of psychoactive drugs, extreme fatigue and sensory deprivation, and anyone—regardless of their cultural background—is liable to perceive them
* Lewis-Williams and Dowson describe six basic entoptic forms—grids, parallel lines, dots, zigzag lines, nested curves and meandering lines, each of which has been recognised during the initial stages of hallucinogenic drug use, during self-experiments by the authors
* These six basic entoptic forms have also been described in an ethnographic account of the use of intoxicants among the Tukano Indians of Amazonia
* Contemporary rock art of several extant shamanic cultures also exhibits geometric patterns strongly resembling entoptic forms (see Rudgely 1993).

PSYCHOACTIVE PLANTS—THE SHAMANISM TEXTILE FROM ANCIENT PERU

In 1969 an ancient Peruvian textile was discovered among a cache of 200 textiles found on the south coast of Peru. Apparently, originating from a Chavín site located in the eastern Andes, the iconographic analysis of the motifs painted on this textile, has revealed the shamanic use of several psychoactive plants dating back to at least 2000 years ago (Cordy-Collins 1982). Again the evidence supporting these claims is outlined below:

* Based on earlier studies of Chavín shamanism, Cordy-Collins has identified that all the basic elements of shamanism are represented on the textile—the shaman's animal familiar, his means for entering the spirit world, and a deity from that world
* Of a total of 65 painted motifs, 15 (almost one-quarter) represent hallucinogenic plants; these are identified as the flowers of the San Pedro cactus (*Trichocereus pachanoi*) and, more tentatively, the seed pods of *Anadenanthera peregrina*
* The columnar cactus *T. pachanoi* contains the psychoactive alkaloid mescaline and, according to modern ethnographic accounts, this plant is still used by coastal Peruvians today to achieve a state of trance; equally, *A. peregrina* has a documented use in South America where the ground seeds are used to prepare an hallucinogenic snuff

Historical Anthropology: Evidence from Written Records

A final source of information available to palaeoethnobotanists lies in the historical documents of ancient and historical civilisations which

Table 10.4 Literary developments reflected in the archaeological record. Dating from about 3400 in Mesopotamia, the earliest writings known were inscribed on clay tokens; by 1400 BC, pictograms gave rise to an alphabet in Ugarit—present day Ras Shamra—in Syria. Later still, permanent records were increasingly made using various plant materials such as the manuscripts made from *Cyperus papyrus* in Ancient Egypt (see Cotterell 1980; Wenke 1990; Taube 1993)

Old World developments	New World developments
5400–4000 BP	
Pre-Sumerian development of pictographic forms on clay tokens	
Appearance of Sumerian *cuniform* writing on clay tablets	
Development of Egyptian *hieroglyphics* first on stone and later on papyrus	
Appearance of Akkadian cuniform, based on Sumerian style	
4000–2850 BP	**2600–557 BP**
Development of the Minoan hieroglyphic/syllabic 'Linear A' and 'Linear B'	Appearance of Zapotec pictographic forms on stone
First alphabet appeared—in Ugarit culture	Appearance of Maya hieroglyphs on stone, plaster and bark paper; also development of complex *calendrics*
Appearance of Chinese characters Sanskrit alphabet of Aryan culture appeared	Development of Mixtec and Aztec pictographs
Appearance of the Phoenician alphabet	
Emergence of the Greek alphabet	

are examined largely by students of *historical anthropology*—the historiographic study of cultures. From at least 5 000 years ago, writing appeared throughout the ancient Near East, with a general increase in literacy following the development of the alphabet in the 2nd millennium BC (Douglas 1988). In the New World, writing developed much later—about 600 BC—among the Zapotec peoples of Oaxaca, and by between AD 300 and 900, writing in Mesoamerica had reached a high level of complexity (Table 10.4). Of these earliest writings, those of interest to ethnobotanists are often medical documents, which commonly contain frequent references to the use of remedies based on medicinal plants. For example, the Chinese *Pen Ts'ao* allegedly written by legendary herbalist Shen Nung about 4 800 BP, describes the use of 366 distinct plant drugs, while a series of carved stone tablets produced under King Hammurabi of Babylon in the eighteenth century BC also refers to the use of several healing plants (Griggs 1981).

In the Old World, one of the most seminal of these ancient texts, is Dioscorides' treatment on medicinal plants *Peri hulas iatrikes* (or *De materia medica*) which dates back to the first century AD. Born in Asia Minor, Pedanius Dioscorides became one of the most famous physicians in Ancient Greece, while his *materia medica*—which described the medicinal properties of about 600 plants—formed the prototype herbal or *pharmacopoeia*, which later became widespread throughout Europe. In the New World, much of the ancient literature is associated with the Maya, Aztec and other great civilisations of Mesoamerica. Specific to the study of Mayan history, there remain four pre-Hispanic *codices* or *screenfold books* which date back to the Late post-Classic period (*c*. AD 1250–1521); information on Aztec cultures, on the other hand, is found largely in the writings of Spanish colonists who wrote prolifically about native customs and beliefs. The most renowned of these sixteenth century chroniclers of Aztec society, was Fray Bernardino de Sahagún, whose most important piece of work—the encyclopaedic *Historia General de las Cosas de Nueva España*—represents the most comprehensive and detailed treatise on any pre-Hispanic culture (Taube 1993). However, the interpretation of these early documents is not always easy. For example, from ancient Egypt, medical knowledge from about 3600 years ago has been determined from the examination of documents such as the *Ebers papyrus* and *Smith papyrus*. Yet, despite extensive studies since their discovery more than 100 years ago, many words for specific plants cannot be translated, and to date many of the plants referred to have remained unidentified (Reeves 1992). Clearly, the interpretation of historic texts represents a very difficult area of study, yet their careful examination can provide direct evidence of the historical uses and cultural significance of many plant species; hence the value of collaboration between students of textual history and of palaeoethnobotany is potentially enormous.

Interpreting the Palaeoethnobotanical Record: Lessons from Living Primates

Until fairly recently, ideas on the origins and development of plant–human interactions have been based largely on rather biased interpretations, which have tended to be both anthropocentric and ethnocentric. For example, recent observations of stone-tool use by chimpanzees both in captivity and in the wild, suggest that the interpretation of some Palaeolithic sites as human may prove inaccurate (McGrew 1992). Equally, modern ethnographic studies clearly demonstrate that the conventional classification of hunter-gatherer societies as 'primitive' compared with their 'advanced' agricultural counterparts, is far too simplistic (Harris 1989). In recognition of these methodological flaws, recent developments in palaeoethnobotanical research have incorporated the study of extant primates—both human and non-human—as a key to allowing more accurate interpretation of the existing archaeological record.

The Ethnographic Record: Evidence from Extant Human Societies

Earlier in this chapter, we saw how modern ethnographic evidence can help in the interpretation of archaeological evidence. For example, in her analysis of the Peruvian *Shamanism textile* (outlined in Box 10.6) Cordy-Collins (1982) identified two recurring plant motifs on the basis of current ethnographic evidence. Of these, the identification of the San Pedro cactus (*Trichocereus pachanoi*) is perhaps the least controversial, its recognition on the basis of its four ribbed stem and its open blooms being supported by two pieces of evidence from extant Peruvian societies. First, this plant, which is used during shamanic healing ceremonies, is used traditionally at night when the flower blooms, and significantly, in most cases, this plant motif is portrayed as blooming. Secondly, while known specimens of *T. pachanoi* possess between six and eight ribs on its columnar stem, modern Peruvian shamans believe that four-ribbed cacti do exist, and that these individuals are particularly potent. Hence, it seems significant that the painted San Pedros represent mythical four-ribbed forms which have retained magical connotations for modern shamans.

In addition to this type of cultural comparison, modern ethnographic studies can provide valuable linguistic evidence to palaeoethnotaxonomists. For example, in an ambitious research project based in South America, the distribution of cognate forms of plant names is being used as a means of tracing the history of population dispersal, and of determining the historical significance of certain plant taxa. This innovative work forms part of the South American Indian Languages Documentation Project, in which Brent Berlin and his collaborators have been developing ways of exploring the spatial distribution of modern ethnotaxonomic data through the use of a novel graphics database, SAPIR (South American Prehistory Inference Resource). Using data from about 350 South American languages, it is hoped that the recognition and mapping of cognate plant labels will allow the identification of spatial patterns which can provide some insight into the cultural history of the region. For example, evidence from eight

Table 10.5 Spatial distribution of cognate terms—linguistic evidence for past events. A survey of eight language groups distributed throughout northwestern South America, reveals eight distinct terms which describe the ethnobiological taxon of 'collared peccary' (*Tayassu pecari*). These terms are linked into groups (*linked polygon sets*) based on their relative similarity, to reveal patterns of relatedness between different parts of the region (see Berlin 1992). For example, here we see the distinction of three regions which are characterised by long-term interactions between their inhabitants. It is anticipated by the creators of the SAPIR database, that the recognition of such patterns will allow linguists and historians to explore, in detail, the cultural history of a given region

Native term	Linked polygon set	Location
pák	A	Midwest
páki	A	Midwest
yungkipák	A	Southwest
máki	A	Southwest
piráchi	B	North central
pirátsi	B	North central
mríti	C	South central
miríshi	C	South central

language groups in northwestern South America reveals eight distinct terms for 'collared peccary' (*Tayassu pecari*), some of which are clearly more closely related than others (Table 10.5). Based on their distribution and similarity, these terms have been grouped to form several distinct units, each of which links those language groups sharing a high proportion of cognates, and which presumably, therefore, have been in close contact during their development (see Berlin 1992).

Finally, two further approaches to palaeoethnobotanical research involve the study of *ethnohistory* and *ethnoarchaeology*—two related fields of study which are concerned with documenting a people's own representation of its past. For many extant groups, this type of evidence is found in their oral history—the oral equivalents of the archives of those cultures which are intrinsically literate. As we have seen (in Chapter 3), the durability of oral traditions can be inferred from myths or legends which appear to be correlated with known prehistoric events; hence it seems reasonable to assume that ethnohistoric data from essentially oral cultures can prove invaluable to archaeologists. As in any other area of ethnobotanical study, ethnohistorical research must first consider local views—in this case their perception of the past. For example, whether the passage of past events is regarded as essentially linear and/or cyclic, or where the distinction between the recent and distant past is drawn (Layton 1989). Where such factors have been considered, collaborative research between archaeologists and local peoples has already proved successful, as illustrated in a recent project in the Mitongoa region of Madagascar. Here, local Betsileo peoples have a considerable knowledge of local sites of archaeological interest—many of which were previously unknown to Western archaeologists; in addition, many local names for these sites provided useful insight into the functional and symbolic significance of both the sites themselves and their preserved artefacts (Raharijaona 1989).

Ethological Data: Evidence from Animal Behaviour

The notion that an increased understanding of the social behaviour of extant great apes—the closest living relatives of human beings—could complement data found in the fossil record, has provided the impetus for considerable research over the last few decades. Two of the most influential early studies are those of Jane Goodall who worked on chimpanzees (*Pan troglodytes*) in Tanzania, and Diane Fossey's studies of gorillas (*Gorilla gorilla*) in Rwanda. Together with other recent primate studies, these works have already provided a means of discussing the possible social structures of human ancestors, particularly in relation to factors such as body size, sexual dimorphism and food sharing. By extrapolation, therefore, evidence for relationships between non-human primates and the natural world, could theoretically provide considerable insight into areas such as the role of plants in human evolution (McGrew 1992). In relation to the interactions between extant primate populations and plants, two areas of study in particular have received increasing attention in recent years—the observation of non-human methods of food collection and processing, and *zoopharmacognosy*, an emerging discipline which examines the concept of self-medication by animals (McRae 1994).

While early research into prehistoric food processing has focused largely on the analysis of stone tools, recent studies into non-human primates have demonstrated that chimps and other modern primates use many plant-based tools in obtaining and processing foods. For example, chimpanzees in Assirik, Senegal have been observed using tree roots in order to open hard-shelled palm fruits, while chimps in Gombe, Tanzania are known to use modified branches to dip for termites (McGrew 1992). On the basis of such observations, zoologist William McGrew has suggested that comparative work between the behaviour of modern primates and the enthnographic literature might ultimately lead to new ideas on the evolution of technology and subsistence change during human evolution.

In addition to contributing to ideas on the development of human technology, it now seems likely that the observation of the activities of living primates may also yield some insight into the development of plant detoxification processes. For example, it appears that several primates—including gorillas (*Gorilla gorilla*) and colobus monkeys (*Colobus guereza*)—are able to eat plants containing significant levels of toxic alkaloids, due to their use of a behavioural process known as *geophagy*. Geophagy—the eating of clay—has so far been observed in at least eight species of modern primate, as well as among several extant human societies. For example, among traditional Andean cultures, certain species of potato are eaten with clay, which is said to eliminate bitterness and to prevent the occurrence of stomach pains; equally, many African famine foods, including the wild yam *Dioscorea dumentosum*, are commonly eaten with clay. In recent years, the consequences of this behaviour have been examined experimentally, revealing that a range of phytotoxins—including various akaloids, tannins, quinones and saponins—are efficiently bound to clays, thus rendering these toxic compounds inactive. Together, these chemical and behavioural data have led to the recent proposal of a chemical ecological model as a driving force in the evolution of the human diet (Johns 1989).

Finally, one of the most recent developments in the study of animal behaviour is the phenomenon of zoopharmacognosy. Initiated as a field of academic study in 1978 by biologist Daniel Janzen, the international research group CHIMPP (the group for Chemo-ethology of Hominoid Interactions with Medicinal Plants and Parasites) was later set up in 1989 with the aim of combining local cultural knowledge with data from primate behaviour, in order to explore how early *Homo* might have learned to use medicinal plants (McRae 1994). For example, the ingestion of the bitter, toxic sap of *Veronia* sp by chimps showing signs of parasitic infestations has been connected with the medicinal use of the same plant by local Tongwe peoples; and significantly, chemical and pharmacological analysis of the plant extracts has demonstrated their potent antiparasitic activity. On the basis of a number of similar studies, zoopharmacologists are becoming convinced that the study of primate behaviour may have much to offer to workers interested in the evolution of traditional medical systems.

Nature and Evolution of Plant–Human Relationships in History and Prehistory

Throughout this century, palaeoethnobotanists have collected a vast body of diverse evidence relating to the past relationships between plants and humans. The processes of plant domestication have been determined through the study of plant fossils; the effects of human impact on the natural environment have been determined using pollen analysis; the development of plant use has been examined using a range of biological, sociological and linguistic techniques. On the basis of this evidence, it is becoming possible to produce a general picture both of the role of plants in the emergence and development of human beings, and of the role of humans in shaping the current world flora.

Plant Evolution and the Development of Modern Humans

Despite the difficulties inherent in interpreting the fossil record (see Chapter 4), evidence from biological and geochemical sources can be used to construct a generalised scheme of world history, the main points of which are presented in Figure 10.2. In particular, the examination of preserved biological materials has made it possible to chart not only the evolution of the major groups of terrestrial plants, but also to suggest ways in which the changing floral environment has influenced—and been influenced by—the evolution of the human species. Indeed, from the first appearance of the early primates about 50 million years ago, to the emergence of *Homo erectus* well over 40 million years later, it seems likely that plants played a fundamental part in the diet of ancestral primates, most of which appear to have been vegetarian.

Much later, following the emergence of modern humans (*Homo sapiens*), pollen evidence suggests that climatic changes associated with the end of the great Pleistocene glaciation saw the spread of open grassland and patches of woodland over much of the world, with ancient rainforest flora surviving only in small refuges. These changes had a profound effect on the subsistence practices of early human populations, leading both to changes in diet, and to an extensive migration into new habitats. For example, the formation of extensive grasslands in southern Europe and the Middle East, led initially to the large-scale cultivation of wild cereals, and ultimately to the domestication of important modern grain crops including wheat (*Triticum* spp) and barley (*Hordeum* spp) (Hillman *et al.* 1989). Significantly, in terms of human evolution, the domestication of plant (and animal) species was often closely associated with profound changes in social organisation—including increased sedentism and concomitant increases in social complexity. Some of the major roles of plants in the evolution and development of modern humans are outlined in Table 10.6.

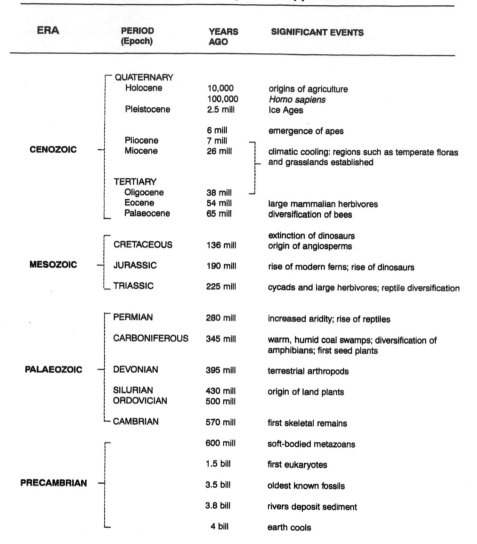

ERA	PERIOD (Epoch)	YEARS AGO	SIGNIFICANT EVENTS
CENOZOIC	QUATERNARY		
	Holocene	10,000	origins of agriculture
		100,000	*Homo sapiens*
	Pleistocene	2.5 mill	Ice Ages
		6 mill	emergence of apes
	Pliocene	7 mill	
	Miocene	26 mill	climatic cooling: regions such as temperate floras and grasslands established
	TERTIARY		
	Oligocene	38 mill	
	Eocene	54 mill	large mammalian herbivores
	Palaeocene	65 mill	diversification of bees
MESOZOIC	CRETACEOUS	136 mill	extinction of dinosaurs origin of angiosperms
	JURASSIC	190 mill	rise of modern ferns; rise of dinosaurs
	TRIASSIC	225 mill	cycads and large herbivores; reptile diversification
PALAEOZOIC	PERMIAN	280 mill	increased aridity; rise of reptiles
	CARBONIFEROUS	345 mill	warm, humid coal swamps; diversification of amphibians; first seed plants
	DEVONIAN	395 mill	terrestrial arthropods
	SILURIAN	430 mill	origin of land plants
	ORDOVICIAN	500 mill	
	CAMBRIAN	570 mill	first skeletal remains
PRECAMBRIAN		600 mill	soft-bodied metazoans
		1.5 bill	first eukaryotes
		3.5 bill	oldest known fossils
		3.8 bill	rivers deposit sediment
		4 bill	earth cools

Figure 10.2 Generalised scheme of world history. Using biological and geochemical evidence, a general history of the world can be reconstructed, illustrating the approximate times at which major events—such as the origin of angiosperms and the emergence of modern humans—apparently took place. Much of this evidence is based on the collection and identification of fossilised plant materials, which provide information relating to factors such as the evolution of particular groups of plants, significant changes in world climate, and the impact of human activity on natural vegetation (modified from Bold *et al.* 1987; Goonatilake 1991; Ingrouille 1992; Lewin 1993)

Table 10.6 The role of plants in human development. From the origins of large mammals early on in the Cenozoic, the evolution of ancestral primates—and ultimately the emergence of modern humans—has been largely dependent on sustenance provided by the plant world. However, even as the development of lithic technologies facilitated increased hunting activity and a greater dependence on animal protein, the relationships between humans and their floral environment remained closely intertwined. For example, it seems likely that the combination of nutritional richness and stability in supply provided by the new omnivorous diet, facilitated human migration into new habitats and hence the exploitation of new vegetation types. Equally, climatic changes during the great Pleistocene glaciation led to profound changes in flora and fauna, which in turn had dramatic effects on human subsistence activities, and ultimately their social development (see: Flood 1983; Lewin 1993; Mithen 1994)

Plants in human evolution and development

Plants in human evolution: from mammals to *Homo sapiens*
  Primates of modern aspect appeared about 50 mya, the *omomyids*—ancestral to the anthropoids—were noctural and *frugivorous*

  Australopithecus afarensis, the earliest known hominid appeared about 5.6 mya; both its bipedal locomotion, and its frugivorous diet are thought to represent adaptations to habitat, which, as a result of global cooling, was changing to open bushland and savanna with food scattered in patches.

  Based largely on pollen analysis and the examination of dentition, it seems likely that ancestral hominids remained essentially vegetarian until the emergence of *Homo erectus* about 400 000 years ago.

Plants in human culture: from the emergence of *Homo sapiens*
  While the analysis of early lithic technology—from about 1.7 mya—suggests a predominant role in hunting game, the analysis of later, more sophisticated technologies, implies an increasing use in the harvesting and processing of plants.

  Pollen analysis indicates fluctuating climates during the Pleistocene epoch which would have a profound effect on global vegetation; in turn these changes greatly influenced the subsistence practices of early humans, leading to changes in diet and extensive migration into new habitats.

  Tool analysis reveals that the increasing importance of wild grasses in the human diet in Africa (18 500 BP) and Australia (15 000 BP)

  By about 10 000 years ago, in many parts of the Old World, archaeological evidence reveals increasing plant domestication is often associated with important socio-economic change—including greater sedentism and social complexity

  Although plant materials appear only sporadically in the fossil record, direct and indirect evidence of plant use reveals the importance of plants from early on in the material culture of modern humans:

  • Preserved artefacts reveal the existence of woven textiles from about 9000 BP; grasses used as roofing materials from 7000BP

Table 10.6 *(continued)*

Plants in human evolution and development

- The discovery of the *Iceman* in the Ötzaler Alps (Austrian–Italian border) has revealed the extensive use of plants in clothing and technology from 5000 BP
- Symbolic analysis of plant motifs and symbolic representations, imply the shamanic use of psychoactive plants from Palaeolithic times
- From around the 4th millenium BC, written evidence provides much information about the use of plants in diet and medicine

mya = million years ago.

Anthropogenic Effects: Human Influence on the Natural World

While plants have played a significant part in the evolution of modern human societies, humans have equally played a vital part in the development of modern plant populations. For human activities such as burning, and land clearance have led to pronounced changes in vegetation types, while artificial selection has determined the physical, chemical and behavioural characteristics of domesticated plant species. For example, a huge increase in the presence of charcoal and the expansion of fire-tolerant *Eucalyptus* species has been interpreted as the result of human activity from about 120 000 BP (see Flood 1983); meanwhile, on Easter Island, pollen analysis demonstrates a strong reduction in the proportion of tree and shrub species, and a concomitant increase in grass pollen from about 1200 to 800 BP—both of which are strongly suggestive of deliberate forest clearance (Flenley *et al.* 1991). Equally, significant changes in the seed morphology of *Chenopodium* sp have been related to the domestication of this New World crop (see Gremillion 1993), while marked changes in plant distribution reflect the development of trade in economic plants.

In most cases, both extensive land clearance, and plant domestication have been closely associated with the development of agricultural communities—the origins of which have attracted particular interest from palaeoethnobotanists. Indeed, philosophical speculation regarding the nature and causes of agricultural development dates back to Classical times (Harris 1989). More substantive research into the history of domesticated plants and their cultivation has been carried out, however, for little more than a century, during which time this topic has received considerable attention, not only from biologists and anthropologists, but also from historians, economists and demographers, all of whom have speculated about the causes and consequences of this apparently dramatic change in human subsistence. In recent years, increasingly sophisticated analyses of archaeobotanical and other evidence have revealed much about the origins and development of domesticated species (Box 10.7); yet despite decades of intense research and lively speculation, theories explaining the causes behind the origins of domestication and the development of agricultural modes of production remain largely unsubstantiated. Hence more recent interest has begun to focus more on documenting the actual processes which occurred, rather than on attempting to explain these dramatic events in the history of plant–human evolution (Box 10.8).

Box 10.7   Tracing the origins of domesticated plants. Influenced originally by the ideas of the Swiss botanist Alphonse de Candolle (1806–93) who published his *Origine des plantes cultivées* in 1833, and later by the Russian plant geneticist Nikolai Vavilov (1887–1943), this century has seen an enormous research effort focused on the origins of domesticated plants. Today, a vast literature documenting an impressive array of bioarchaeological, phytogeographical, molecular and linguistic evidence has made it possible to trace the domestication of many important crop species, demonstrating clearly, that plant domestication occurred independently in many parts of the world, throughout the early Holocene epoch.

In a biological sense, agricultural evolution has involved three main factors: the *cultivation* and *domestication* of useful plant species and the establishment of *agricultural economies*. Throughout the academic literature, *cultivation* is used to refer to various aspects of plant management—ranging from incidental weeding to systematic tillage and the deliberate propagation of domesticated crops. *Domestication*, on the other hand, refers to the artificial manipulation of plant growth and reproduction, which ultimately leads to permanent modification of a species' natural life cycle. For example, unlike its wild predecessors, domesticated maize (*Zea mays*) has no effective means of seed dispersal, as long-term selection by human populations has resulted in the production of a cob which retains its seeds—a morphological feature which facilitates efficient harvesting. Finally, *agriculture* is used to describe the systematic modification of the environment—by activities such as weeding, tilling, and pest control—with the specific intent of increasing crop productivity.

EARLY OCCURRENCES OF DOMESTICATED SPECIES

Much of the research into agricultural origins has focused on pinpointing the emergence of domesticated plants in the archaeological record. As we have seen (Box 10.5), the domestication process can be monitored through the detection of patterned changes in significant morphological features, which are suggestive of traits which have been *fixed* as a consequence of *artificial selection*. On the basis of such evidence, the domestication of several important species has been traced worldwide:

continued

_____ *continued* _____

Crop species	Archaeological site	approx date (BC)
Wheat (*Triticum*)	Tell Mureybit, Syria	8000
Barley (*Hordeum*)	Jericho, Jordon	7000
Lentil (*Lens*)	Hacilar, Turkey	7000
Bean (*Phaseolus*)	Guitarrero Cave, Peru	5600
Maize (*Zea*)	Tehuacán, Mexico	5000
Rice (*Oryza*)	Non Nok Tha, Thailand	4000
Millet (*Pennisetum*)	Yang-shao, China	3500
Potato (*Solanum*)	Chiripia, Bolivia	400

Data taken from Wenke (1990).

Box 10.8   The origins of agriculture—theories and evidence. As explanations of the causes of agricultural activity remain unresolved, a number of key workers have begun to put forward new ideas on the actual processes involved. One of the most notable of these (see Harris 1989) envisages a continuum of people–plant interaction from the burning of vegetation to the propagation of domesticated crops, evidence for several stages of which can be inferred from the archaeological record (e.g. Hillman 1989)

> While the emergence of many domesticated plants is becoming increasingly well documented, it is becoming clear that the occurrence of plant domesticates is not always indicative of a strong dependence on agricultural subsistence. For example, in Mesoamerica, it seems that at least five plant species were in the process of domestication well before the appearance of the first sedentary communities. The identification of such unconventional patterns of agricultural development have added greatly to the complex debate on the causes of the so-called *agricultural revolution*.

MODERN THEORIES ON THE ORIGINS OF AGRICULTURE

As intensive research into the origins of agriculture has proceeded throughout the course of this century, many workers have speculated on the causes of the significant changes in the *palaeoeconomies* of many human societies since the end of the last ice age. Most of the theories which have been proposed have centred around the themes of climate change and/or population pressure (see Wenke 1990 for review).

- In 1904 the *Oasis hypothesis* proposed late Pleistocene climatic change as the major factor in plant domestication
- In 1926, proponents of the *natural habitat hypothesis* sought evidence for crop domestication in the natural habitats of crop species' wild progenitors

_____ *continued* _____

continued

- Childe put forward the notion of an abrupt _neolithic revolution_ (1936)
- Boserup's _population pressure theory_ (1965), Bindford's _edge-zone hypothesis_ (1968) and the ecological models of Harris, Cohen and Hayden (1977–1981) incorporate three common concepts: climate change, population pressure and technological advancement, to explain agricultural development
- In recent years, archaeological and palaeopathological evidence suggest that in many cases, population pressure on resources may have occurred subsequent to the emergence of agricultural modes of production, and most workers now agree, that the 'origins of agriculture' represent not so much an abrupt 'neolithic revolution', but more a long-term process in the evolution of plant exploitation (see Harris 1989)

AN EVOLUTIONARY CONTINUUM OF PEOPLE–PLANT INTERACTION

In the light of changing ideas on agricultural origins, a number of new proposals have been put forward. Notably, at the World Archaeological Congress, held in Southhampton, UK, David Harris of University College London proposed a new model of people–plant interaction which is based on ecological and evolutionary assumptions yet which is neither unidirectional nor deterministic (Harris 1989):

Interaction	Mode of production	Energy input (unit area^{-1})
Burning vegetation Gathering wild plants Protecting wild plants	Wild plant-food procurement (Foraging)	Relatively low
Replacement planting Transplanting Weeding Intensive harvesting Storage Drainage/irrigation	Wild plant-food production with minimal tillage	Increasing
Land clearance Systematic soil tillage Domestication	Cultivation with systematic tillage	Relatively high
Cultivation of domesticates Evolutionary differentiation of specialised agriculture	Agriculture (farming)	High—particularly in large-scale industrial systems

SUMMARY

Within the context of this book, palaeoethnobotany is used to describe any study into the historic, or prehistoric relationships between plants and people, and as such, can involve evidence from a wide range of disciplines. Archaeobotanical and other archaeological data provide information on the past roles of plants in diet and material culture; evidence from modern plant populations can provide an insight into the history of the domestication and dispersal of useful plant species; the symbolic analysis of art and myth can reveal much about the past uses and symbolic significance of plant species; evidence from oral or literary records can yield direct evidence relating to the history of plant–human interactions. In addition, the interpretation of the palaeoethnobotanical record can be facilitated using a variety of methods, including ecological, ethnographic and ethological techniques, which can assist through providing some understanding of current plant, human and primate behaviour. Using these diverse techniques, archaeobotanists, art historians and linguistic experts have begun to develop a detailed picture, both of the roles of plants in human evolution, and the effects of humans on the development of the world's present-day flora.

Chapter 11

Applied Ethnobotany: Commercialisation and Conservation

The greatest service which can be rendered any country is to add a useful plant to its culture.

Thomas Jefferson 1821 (see Plotkin 1988 in *Biodiversity*)

INTRODUCTION

Like the approaches which have been adopted in the study of traditional botanical knowledge (TBK), the motivations behind these studies have been extremely diverse. Indeed, while the voyages of sixteenth century explorers were motivated by essentially economic concerns, and the observations of the early North American settlers were largely made in the interests of survival, many of the studies carried out since the beginning of the twentieth century have been of a much more academic nature. For example, while palaeoethnobotanists have attempted to reconstruct the evolution of plant–human relationships since pre-historic times, ethnotaxonomists have explored the nature and development of human cognition. Yet despite this long-term trend towards a theoretical rather than a practical approach, interest in the potential applications of ethnobotanical research has, in recent years, shown a marked resurgence. This renewed interest in *applied ethnobotany* is primarily due to the increased recognition of the value of TBK, as both plant uses and management methods have been validated empirically. For the sake of clarity, the potential applications of ethnobotanical data can be categorised into two major areas—economic development and resource conservation. Within these broad groups, however, the

Table 11.1 Potential benefits of applied ethnobotanical research. To date, traditional knowledge of plant use and management has been applied in a number of practical contexts. For example in relation to economic development some pharmaceutical companies—such as the American *Shaman Pharmaceuticals Inc.*—have focused on traditional medical systems in drug discovery programmes. Meanwhile in several countries, the harvesting and processing of a range of non-timber plant products has begun to provide an alternative source of income where forest clearance due to logging and agricultural expansion has proved environmentally destructive. Equally, in relation to resource conservation, traditional management practices are now being employed in the protection of endangered species and fragile habitats. Finally, and perhaps most significantly, the documentation and application of traditional botanical knowledge has the additional effect of reaffirming cultural identity—a factor which may play a key part in the very survival of the world's remaining traditional societies

National and global benefits	Local benefits
Economic development	
Ethno-directed sampling for biodiversity prospecting	Generating income from renewable plant sources
	Maintaining and improving methods of production which are suited to local environmental conditions
Resource conservation	
Habitat conservation for biodiversity and environmental services	Conservation and recognition of local knowledge which is crucial to cultural survival
Conservation of germplasm diversity for commercial plant breeding programmes	Conservation of species and habitats which are crucial to cultural survival

benefits may be relevant at local, national or global levels (Table 11.1). This chapter examines the potential role of ethnobotanical research in each of these areas, and explores the practical links between them. Where possible, relevant examples of current projects are presented as brief case studies.

PRACTICAL APPLICATIONS OF ETHNOBOTANICAL DATA

From the data presented in the preceding chapters it is clear that ethnobotanical research can provide a wealth of information regarding both past and present relationships between plants and traditional societies. Investigations into the traditional use and management of local flora have demonstrated extensive local knowledge of not only the physical and chemical properties of many plant species, but also their phenological, ecological and—in the case of domesticated species— their agronomic behaviour; palaeoethnobotanical studies, together with modern field experiments, are beginning to reveal much about the environmental impact of particular types of human activity; while cognitive studies have provided insight into the ways in which human societies perceive their environment, and have revealed widespread similarities in the structure and function of their ethnotaxonomic systems. In recent years, much of this information has been put to practical use, yielding a range of benefits both locally and on a wider scale.

External Benefits: National and Global Interests in Ethnobotany

Since well before the voyages of Columbus, Europeans have benefited greatly from the observations of how plants were used by indigenous peoples throughout the Old World. For example, by the fifth century BC, the ancient Greeks were importing Oriental spices such as cassia (*Cinnamomum cassia*) and cinnamon (*Cinnamomun zeylandica*), which, along with many other Eastern spices, became increasingly important with the establishment of the trans-Asian *Silk Road* in the first century AD. Today the volatile oils of these species are worth more than US $5.4 million *per annum* to the countries which produce them—and considerably more to food manufacturers such as the Swiss giant Nestlé who use these oils to flavour many commodities including soft drinks and confectionery (Verlet 1993).

It is not surprising then, that ethnobotanical research today is commonly aimed at identifying novel plants with economic potential. In recent years, however, interest in traditional plant knowledge has broadened considerably, and the concept of applying traditional methods of plant management in conservation programmes has also become important. As a result of these new developments, there are currently three main areas which have attracted international interest in ethnobotanical data: the identification of new plant products of potential commercial value; the application of traditional techniques in conserving vulnerable species and fragile habitats; the conservation of traditional crop germplasm for future plant breeding programmes.

Ethno-directed Sampling in Biodiversity Prospecting

In February 1990, at a Ciba Foundation symposium held in Bangkok, economic botanist Mike Balick presented a paper on the role of ethnobotany in the identification of therapeutic agents from the

Table 11.2 Plant-derived drugs used in orthodox medical practice. The 27 phytochemicals listed here represent the most important plant-derived drugs used in the medical industry today (Farnsworth 1990). It should be pointed out that, in addition, many biologically active phytochemicals which have not been used as drugs *per se*, have provided models for synthesis or structural modification to produce active synthetic derivatives. For example, while the natural product *khellin* was used for many years to treat several conditions including angina, this natural plant chemical displayed many adverse side-effects including nausea and vomiting. However, using khellin as a template, the less toxic *cromolyn* was synthesised. Bearing in mind the importance of the pharmaceuticals which are already derived from, or based on natural phytochemicals, it is evident that future investigations are likely to reveal many more plant species which yield useful biologically active compounds

Important plant-derived drugs		
Atropine	Emetine	Quinidine
Bromelain	Ephedrine	Quinine
Caffeine	Ianatosides A, B, & C	Reserpine
Chymopapain	Morphine	Scopolamine
Codeine	Norpseudoephedrine	Sennosides A & B
Colchicine	Papaverine	Theophylline
Deslanoside	Physostigmine	Tubocurarine
Digitoxin	Pilocarpine	Vinblastine
Digoxin	Pseudoephedrine	Vincristine

rainforest (Balick 1990). To begin, Balick points out that of the 250 000 higher plant species which are believed to exist today, only relatively few have been studied thoroughly for potential therapeutic value. Yet in the USA alone, about 25 per cent of all prescriptions contain active principles obtained from higher plants (Farnsworth 1990), while many synthetic drugs are derivatives based on natural phytochemicals— figures which clearly suggest that the plant world merits further exploration in the search for novel therapeutic compounds (Table 11.2). The focus of Balick's paper is aimed at the unstudied portion of the plant kingdom, and in particular, on the strategies which are currently used in the search for bioactive chemicals. There are essentially three methods used in collecting plants for pharmacological screening: the *random method* involves the collection of all plants found in a given area of study; *phylogenetic targeting* entails the collection of all members of those plant families—such as the Solanaceae or Euphorbiaceae—which are known to be rich in bioactive compounds; *ethno-directed sampling* on the other hand, is based on the traditional knowledge of medicinal plant use.

On the basis of plant collections screened using a preliminary *in*

Table 11.3 The efficiency of random *versus* ethno-directed research. In his comparison of the efficiency of different methods of plant collection, Balick (1990) produced data which suggest that plant collection based on ethnobotanical leads gives rise to the collection of proportionally greater numbers of plants containing bioactive chemicals. In this comparison, plants collected at random were taken from various sites in Honduras and Belize, while the ethnobotanical collection was composed of species identified by Mayan healer, Don Eligio Panti as 'powerful plants'. These were plants which Don Panti used frequently in his practice to treat a range of conditions. All plant samples collected were dried and screened against HIV as part of a programme sponsored by the US NCI (National Cancer Institute), revealing that a comparatively greater proportion of the ethno-directed sample exhibited anti-HIV activity

	Random collection	Ethno-directed sampling
Total species tested	18	20
Proportion active (%)	6	25

vitro anti-HIV screen, Balick provided evidence to indicate which ethno-directed plant collections may well prove more efficient than random sampling methods. This would therefore suggest that ethnobotanists might be able to identify significantly higher numbers of *lead compounds*—compounds which exhibit a required bioactivity—compared with plants selected at random (Table 11.3). However, following more recent work, which involved the *dereplication* of the plant extracts—that is the removal from the crude extracts of certain widespread plant compounds—the percentage of plants showing anti-HIV activity fell rapidly (Balick 1994). Nevertheless, many ethnobotanists have remained convinced that ethno-directed sampling has a useful part to play in *biodiversity prospecting* and one—botanist Paul Cox of Brigham Young University, Utah—has recently begun to evaluate the relative efficacy of ethno-directed sampling—or the use of *ethnobotanical leads*—in relation to specific drug types. On the basis of his findings (Box 11.1), he predicts that properly designed ethnobotanical surveys could prove particularly successful in identifying drugs used in the treatment of gastrointestinal, inflammatory and dermatological complaints (Cox 1994).

Since Balick's presentation in 1990, there have been numerous articles in both the scientific and popular press regarding the use of TBK as a tool in biodiversity prospecting (e.g. Prance 1990; Bennett 1992; Vidal 1993; Cox & Balick 1994). Several of these describe the use of ethnobotantical leads not only for the identification of novel drug compounds, but also in the identification of many other useful plant species.

For example, ethnobotanist Mark Plotkin has described a range of plants which may provide a useful source of foods, pesticides, oils and fibres—all of which are already used by native peoples of Amazonia (Table 11.4). Nevertheless, while the range of potential plant products which might be developed on the basis of ethno-directed screening is fairly broad, much of the current research remains focused on drug discovery. One of the most recent successes in this area is associated with the San Francisco-based company Shaman Pharmaceuticals. Established as recently as 1990, Shaman quickly identified two novel antiviral compounds on the basis of ethnobotanical leads, and both of these compounds are already in clinical trials (Box 11.2). On the strength of such convincing evidence, it does seem reasonable to conclude that careful ethnobotanical research has a valuable contribution to make in the search for novel plant products with considerable economic potential.

Box 11.1   Western *versus* indigenous drugs, and the ethnobotanical approach to drug discovery. Following Balick's initial work on ethno-directed drug discovery, Cox (1994) has compared the importance of certain disease categories in Western medicine with those found in a range of traditional medical systems. He concludes from this study, that the ethnobotanical approach is more likely to prove successful in the identification of drugs used for gastrointestinal, inflammatory and dermatological disorders, than for those used in the treatment of, for example, cancer, as the former disease categories receive considerably greater attention within traditional medical systems.

Following Balick's preliminary presentation on the relative efficiency of *ethno-directed sampling* in the search for new drugs (Balick 1990), Paul Cox has presented both historical and more recent evidence to support the concept of ethno-directed drug discovery programmes (Cox 1994). For example, in a paper which discusses the strengths and limitations of the ethnobotanical approach, he lists 50 plant-derived drugs which have been discovered on the basis of ethnobotanical leads, including many such as aspirin (from *Filipendula ulmaria*), digoxin (from *Digitalis purpurea*), morphine (from *Papaver somniferum*) and quinine (*Cinchona pubescens*). More recent support for the efficacy of ethno-directed research has resulted from Cox's own research into the pharmacological activity of Samoan medicinal plants. Based on evidence from a number of healers, Cox found that extracts from woody stems of *Homalanthus nutans* (Euphorbiaceae)—which is used in Samoan medicine to treat the viral disease yellow fever—exhibited strong activity against the human immunodeficiency virus, HIV-1. Activity-directed fractionation of the extract resulted in the isolation of *prostratin*, a novel compound which strongly inhibits the killing of human host cells *in vitro* by HIV, and which the US National Cancer Institute currently considers as a candidate for drug development.

_____ *continued* ___

continued

ESTIMATING THE VALUE OF ETHNO-DIRECTED DRUG PROSPECTING

Using published data from 15 widespread geographical regions, Cox and his co-workers classified the uses of traditional medicinal plants into nine treatment categories. When compared with Western drug use patterns (based on information from the US pharmacopoeia), a striking distinction was observed between the disorders receiving greater attention in traditional _versus_ orthodox medicine. For example, while anti-inflammatory drugs constitute only 4 per cent of the US pharmacopoeia, on average 12 per cent of the plants used in traditional medicine are used to treat inflammatory disease. Equally, while antimicrobial drugs account for 33 per cent of US drugs, only 9 per cent of indigenous remedies were used in such treatments.

	US (drugs)[1] (%)	Indigenous (plants)[2] (%)	Western diseases with ethno-derived drugs[3] (%)
Cardiovascular	8	2	24
Cancer	5	1	10
Gastrointestinal	4	15	10
Anti-inflammatory	4	12	11
Dermatological	11	15	4
Nervous system	10	10	20
Antimicrobial	33	9	4
Obstretrics/ gynaecology	3	—	—
Other	17	—	18

[1]Percentage of drugs used in the USA to treat particular types of disease; [2]mean percentage of plants used in traditional medicine to treat particular types of disease; [3]proportion of the 50 ethno-derived drugs listed by Cox (1994) which are used in Western medicine to treat particular types of disease.

Cox suggests that these differences in drug-use patterns may reflect a number of factors, including the perceived severity of particular types of disease, the diagnostic ease of certain ailments, the toxicity of specific treatments, and prevailing economic incentives. For example, during this study, Cox also categorised 50 ethnobotanically derived drugs which are used currently in Western medicine, according to their use, and concluded that, on the whole, the use-pattern observed reflects Western funding categories rather than indigenous use categories. From these data, he suggests that future ethno-directed sampling is more likely to succeed for drug categories which are more pertinent to indigenous medicine.

Traditional Plant Management and Environmental Conservation

In Western society, interest in wildlife conservation is a relatively recent phenomenon, the earliest national laws aimed at the protection of living

Table 11.4 The commercial potential of traditionally useful plants. In his discussion of useful tropical plants, ethnobotanist Mark Plotkin (1988) outlines the traditional uses of a number of plant species which may have commercial potential. He points out that many well known plants such as cocoa and corn were first discovered and used by indigenous peoples, and emphasises that their vast knowledge of the plant world remains crucial to both the future use and the continued protection of plant diversity

The commercialisation of traditional plant use

Food plants

Pourouma cecropiaefolia (Moraceae)—native to western Amazonia, the uvilla tree yields tasty fruits which are eaten raw, or else fermented to produce wine.

Solanum quitoense (Solanaceae)—one of the most highly prized fruits in Columbia and Ecuador, the *lulo* or *naranjilla*, is made into a delicious beverage. The *lulo* has already been introduced into Panama, Costa Rica and Guatemala, where it is marketed as a frozen concentrate.

Bactris gasipaes (Palmae)—native to northwestern Amazonia, the peach palm is cultivated widely throughout South and Central America. Each year it produces high yields of fruit which contain carbohydrate, protein, oil, minerals and vitamins in almost perfect proportions for the human diet.

Mauritia flexuosa (Palmae)—the buriti palm produces a fruit which is rich in vitamin C, while its pulp oil is believed to contain as much vitamin A as domesticated carrots. Many Amazonian Indians also eat its edible palm heart and ferment wine from its fruit, sap and inflorescences.

Industrial products

Caryocar spp (Caryocaraceae)—produces a compound which is toxic to the leaf-cutter ant (*Atta* spp) an insect pest which causes millions of dollars of damage each year.

Jessenia bataua (Palmae)—the *patauá* palm produces an edible oil which is almost identical to olive oil in its physical and chemical properties.

Orbignya spp (Palmae)—the *babassú* palm produces high yields of an oil which may either be refined into an edible oil, or which can be used in the manufacture of plastics, detergents and soap.

Astrocaryum tucuma (Palmae)—the tucum palm produces a fibre which is considered to be among the finest and most durable in the plant kingdom, and which is highly valued by Amazonian Indians.

resources being developed only when over-exploitation of particular species led to fears of species extinction (Birnie & Boyle 1992). Among the first organisms to incite such interest were various species of birds, fish and whales, and in 1885 the *Convention for the Uniform Regulation of Fishing in the Rhine* became one of the first treaties to regulate the exploitation of migratory species of economic interest. By the beginning of the twentieth century, the protection of natural heritage on the basis of its uniqueness alone—regardless of any immediate economic concern—became increasingly important in Europe and the USA, leading to the concept of *natural areas*, or habitats which were to be legally protected. For example, in 1928, the Montseny mountain range near Barcelona in Spain came under partial statutory protection on the basis of its culturally and politically symbolic value (Castelló *et al.* 1993). In many other countries, moves to protect wildlife areas were relatively rare before the latter half of this century. Indeed, in Australia the management of protected areas received a major impetus only in 1967 with the establishment of the New South Wales National Parks and Wildlife Service, while in Alaska the *National Interest Lands Conservation Act* was passed as late as 1980 (Table 11.5).

Box 11.2   From Shaman to human clinical trials—the discovery and development of two novel antiviral compounds. According to the financial literature of San Francisco based Shaman Pharmaceuticals Inc., the principle behind the company's investment thesis is that ethnobotanical screening is more efficient than random drug screening (Shaman Pharmaceuticals Inc. 1994). Significantly, after only 4 years in operation, this company has developed two novel compounds with considerable market potential (Cohen & Tokheim 1994).

Established in 1990, Shaman Pharmaceuticals Inc. is one of about 200 pharmaceutical companies searching for novel bioactive molecules in natural plant products. However, unlike many of these companies, Shaman employs an ethnobotanical approach as a primary screening mechanism (Cohen & Tokheim 1994). Using a multidisciplinary team which includes medics, natural product chemists and ethnobotanists, this approach has so far resulted in two products in clinical development: Provir™, an oral product for the treatment of respiratory viral infections; and Virend™, a topical antiviral product for the treatment of herpes (Shaman 1994). In addition, Shaman has continuing programmes which are targeted at identifying novel compounds for use against fungal infections and diabetes, which have so far resulted in two antifungal candidates at the medicinal chemistry stage, while investigations into native treatments for diabetes have generated several promising leads (Cohen & Tokheim 1994).

SHAMAN'S DRUG DISCOVERY PROCESS

Shaman's drug discovery process is based on data generated by ethnobotanical enquiry into areas which fulfil two important criteria: first, they must be rainforest areas which exhibit characteristically high levels of biodiversity; secondly, their traditional inhabitants must demonstrate an historical reliance on plant-based medicine. To date, the company has devoted considerable effort to developing suitable methodologies for the prioritisation of their research such that teams of

continued

_ continued _

ethnobotanists and medical doctors are now able to work together on highly focused data collection strategies thus maximising the probability of new drug discovery. To this end Shaman requires that four key conditions are met:

- There must be an incidence of phytotreatment of the target disease among local populations
- In vitro and/or in vivo models of the human disease must already exist, thus allowing both the initial screening of crude extracts and the subsequent activity-directed isolation of active compounds
- There must exist clearly defined clinical end-points which allow unambiguous determination of efficacy
- A significant market that justifies the expense of a drug's discovery and development must be clearly identified

Having defined suitable targets, interdisciplinary teams are able to present descriptions of specific diseases to the shamans and traditional healers of forest communities. Using photographs of disease symptoms, and with the help of translators, local healers are asked to describe the botanical treatments for any conditions they recognise, and where several independent and reliable informants describe a similar treatment for a given disease, the plant is collected for laboratory analysis (King & Tempesta 1994).

SHAMAN'S SUCCESS

Using the general approach outlined above, Shaman Pharmaceuticals have worked so far with 70 distinct rainforest cultures in various parts of the world— less than 5 per cent of the 1500 rainforest peoples who remain. Through concentrating largely on plants which are used for their antifungal, anti-infective and antiviral activity, on those which affect the central nervous system, and most recently on species used against diabetes, Shaman have now collected more than 640 plants, of which almost 50 per cent have exhibited activity to justify their further investigation.

Drug discovery from ethnobotanical leads (1994)

No. plants collected	640
No. plants screened	471
No. of plants which justify further investigation	290
No. compounds of interest in target areas	113
No. compounds prioritised for preclinical animal studies	10
No. compounds in clinical trials	2*

*One further compound expected to be in clinical trials in the near future. Data taken from King & Tempesta 1994.

Significantly, the management of many of the world's national parks has been based so far on the American 'Yellowstone model', which regards the ideal national park as one which is not inhabited by human populations. This model, encapsulated in the US *Wilderness Act 1964*, defines a *wilderness* as

Table 11.5 Legislating for natural resource protection. Since the beginning of this century there has been considerable legislation aimed at protecting natural resources, including both endangered species and fragile habitats. Some of the more relevant of these are listed here (Birnie & Boyle 1992)

Legislation towards nature conservation	
1902	Convention for the Protection of Birds Useful to Agriculture (Paris)
1911	Treaty for the Preservation and Protection of Fur Seals (Washington)
1931	Convention for the Regulation of Whaling
1933	Convention Relative to the Preservation of Fauna and Flora in their Natural State (London)
1940	Convention on Nature Protection and Wild-Life Preservation in the Western Hemisphere (Washington)
1964	Agreed Measures for the Conservation of Antarctic Fauna and Flora
1968	African Convention on the Conservation of Nature and Natural Resources (Algiers)
1971	Convention on Wetlands of International Importance (Ramsar)
1971	UNESCO Convention Concerning the Protection of the World Cultural and Natural Heritage
1974	Nordic Convention on the Protection of the Environment (Stockholm)
1976	Convention on the Conservation of Nature in the South Pacific (Apia)
1979	Convention on the Conservation of Migratory Species of Wild Animals (Bonn)
	Convention on the Conservation of European Wildlife and Natural Habitats (Berne)
1982	World Charter for Nature
1985	ASEAN Agreement on the Conservation of Nature and Natural Resources (Kuala Lumpur)
1992	Rio Declaration on Environment and Development (Rio de Janeiro)

an area which exists free of human interference—a notion which is strongly influenced by the Western vision of pristine forests and untouched wilderness. As a consequence, many conservation initiatives have involved the exclusion of local peoples from designated sites—a move which not only denies access to the very resources which are essential to traditional life, but which in some cases appears detrimental to the conservation process itself (Box 11.3). Indeed, while conservationists have argued for years in favour of excluding people from protected areas, recent work has highlighted the dynamic nature of natural forest ecology, and the crucial role of human influence on maintaining biodiversity has been increasingly recognised. Consequently, since most habitats have apparently been influenced significantly by humans during the course of their evolution, most are likely to remain dependent on continued human interference for survival in their current state (McNeely 1994). In the light of such new ideas, many ecologists are now beginning to argue in favour of incorporating traditional knowledge into the management of protected areas (Box 11.4).

Box 11.3 Conservation in the Serengeti rangelands. Based on the traditional American concept of a wildlife reserve as an area which excludes human activities, many national parks throughout the world have excluded the local peoples who have managed these habitats for millennia. As illustrated in this example, however, such methods may prove detrimental to conservation objectives, as unmanaged habitats may begin to degenerate.

The establishment of national parks and other protected nature areas has often involved the exclusion of local peoples, preventing their access to essential resources such as wild foods, fuel and medicinal plants. However, recent observations suggest that the protected areas themselves may also suffer where traditional management is withdrawn, resulting in altered habitats and loss of biodiversity. For example, the East African rangelands in the region of the *Serengeti National Park* (SNP), are home to both traditional Maasai pastoralists and migratory ungulates, such as wildebeest, gazelle and zebra, which form the basis of the SNP's wildlife tourism. For years, the Maasai have been excluded (for 'conservation' reasons) from certain areas of their traditional homelands, yet recent evidence suggests that these changes in management practice may be leading to the loss—rather than the preservation—of important habitats.

THE SERENGETI ECOSYSTEM

The Serengeti ecosystem is composed of three distinct land-use types, all of which form homes for both resident and migratory wildlife:

- The formal conservation area of the SNP includes the Maasai Mara National Reserve and the Maswa Game Reserve; here the interests of wildlife are paramount
- The Ngorongoro Conservation area; a multiple land-use area, where the needs of both wildlife and the pastoral Maasai are of equal importance
- There exist extensive agricultural and rangeland areas under more or less formal land tenure systems, where the interests of the landowners or users are paramount

Of particular significance, is the Maasai Mara National Reserve, a 1368 km² area which is owned by the Kenyan government, and within which land-use activities are restricted to wildlife tourism. For this crucial area provides protection for resident wildlife communities, as well as essential dry season grazing resources for migratory populations of charismatic ungulates. Together these animals constitute the premier wildlife attraction in Kenya, generating up to 8 per cent of national tourist revenues—about US$20 million *per annum* in foreign exchange (Norton-Griffiths 1995). The decision to exclude the Maasai peoples from this area, is based largely on the premise that the Maasai's domesticated livestock would compete with wild ungulates, leading to problems of overgrazing, trampling and soil erosion, and that the Maasai's own collection of timber and fuel, and their use of fire would lead to a significant reduction in forest areas (see Homewood & Rodgers 1991).

Yet while little substantial evidence exists to support these arguments, a

_ *continued* _

— continued —

growing body of data suggests that the cessation of traditional management regimes is actually proving detrimental to wildlife. For example, the prohibition of traditional pasture-burning practices (see Chapter 5) has led to the accumulation of coarse, unpalatable forage which is avoided by wildebeest and gazelle, while the numbers of tick pests has risen dramatically (Estes & Small 1981). As a consequence of these observations, controlled burning has been reintroduced into several areas, suggesting that perhaps the best managers of the Serengeti wildlife are the Maasai themselves (Homewood & Rodgers 1991).

Box 11.4   Aboriginal involvement in parks and protected areas. Following the leasing of the Kakadu National Park to the Australian National Parks and Wildlife Service (ANPWS), a later development saw the proclamation of the Uluru-Kata Tjuta National Park at Ayers Rock. Both parks now receive considerable world acclaim, and, unlike other approaches to conservation, the joint management strategy employed here (see below) recognises the importance of both biological and cultural diversity (de Lacy 1992).

In July 1991, a conference was held at the Johnstone Centre of Parks, Recreation and Heritage and Charles Sturt University in Australia. At this meeting on *Aboriginal Involvement in National Parks and Protected Areas* a number of Aboriginal and non-Aboriginal workers discussed the issues surrounding Aboriginal involvement in the management of land. These discussions were initiated in response to a perceived need to include an awareness of Aboriginal issues and knowledge in the development of land and cultural resource management. For as many of the authors of these papers would argue, *'National parks in Australia, despite cherished Euro-Australian beliefs which conceive of them as pristine wilderness or natural landscapes which must be preserved from all human activity, are in fact cultural landscapes brought about by thousands of years of Aboriginal management'* (Birckhead & Smith 1992).

WESTERN PERSPECTIVES ON TRADITIONAL LAND MANAGEMENT

For thousands of years, Aboriginal cultures throughout Australia have used and cared for their land, using a range of traditional ecological and sociological controls. Yet since the arrival of Europeans in 1788, traditional methods of land management have been increasingly restricted in the interests of 'conservation'. However, recent observations—including a number of near extinctions—have prompted considerable interest in reviving traditional management techniques. Three major points have been identified as justification for this approach:

- Aboriginal knowledge of the land is more detailed, complex and extensive than current knowledge based on scientific and management studies (Baker *et al.* 1992). For example, the effective pyrotechnical manipulation of the natural environment is intimately linked with Aborigines' profound ecological understanding (Lewis 1992)

— continued —

___ continued ___

- Concepts of 'wilderness' are not applicable to Australia, where the Aboriginal peoples have been actively managing the land for possibly 50 000 years (Sullivan 1992)
- Cultural ties to the landscape are maintained only through the practice and implementation of Aboriginal knowledge systems (Birckhead & Smith 1992)

ABORIGINAL PERSPECTIVES ON NATIONAL PARKS

Tony Tjamiwa an Aboriginal representative on the Uluru Board of Management says that without the presence of Aboriginal peoples, a national park is *'like a table with one leg . . . It's not very stable. Shove it and it will fall over. Just one leg is not enough for Aboriginal land. It has to have the other legs there: the leg that Aboriginal Law and ownership provides; that an Aboriginal majority on the board of management provides'* (Tjamiwa 1992). However, since Australia's national parks and marine parks have been established, there has been generally little communication with Aboriginal people (see Wallace *et al.* 1992). Abraham Omeenyo, an Aboriginal ranger has identified a number of key points which he regards as essential to the success of these protected areas:

- Aboriginal involvement in conservation and land management
- Communication between government agencies, community elders and Aboriginal rangers
- Greater consultation regarding the zoning and rezoning of protected areas
- Greater information regarding the occurrence of traditional sites within protected areas
- Greater autonomy and traditional rights for Aboriginal rangers
- Greater communication and collaboration between rangers and park authorities

JOINT MANAGEMENT OF KAKADU NATIONAL PARK

Kakadu National Park is one of the most important national parks in Australia and has considerable international significance, having been designated a World Heritage site (Blyth *et al.* 1992). Kakadu conserves not only a range of the major habitat types of Australia's 'Top End'—including many species of animal and plant which are endemic to this area—but also represents one of the few places which has maintained a strong Aboriginal culture. In addition, Kakadu has one of the highest densities of rock art and archaeological sites in the world. In 1979, the traditional owners of the park leased their land to the Director of the Australian National Parks and Wildlife Service (ANPWS) to be jointly managed as a national park, with its traditional custodians maintaining full involvement in all management issues. Five training programmes have now resulted in Aborigines holding 35 per cent of permanent posts and 40 per cent of temporary contracts, with further moves to involve Aboriginal communities in park management envisaged in the future.

Traditional Germplasm Management: in situ *and* ex situ *Conservation*

In addition to recognising the value of traditional knowledge in habitat conservation, agricultural scientists and conservation biologists have recently begun to acknowledge the crucial part played by traditional farmers in generating and maintaining diversified germplasm of important agricultural crops such as maize (*Zea mays*), wheat (*Triticum aestivum*) and rice (*Oryza* spp). Hence, while *ex situ* strategies have frequently been used for the conservation of crop genetic resources, many scientists are now beginning to highlight the additional need for the *in situ* conservation of traditional cultivars if we are to conserve a wide genetic base for future plant breeding (Brush 1991; Soleri & Smith 1994). For example, as we have seen in Chapter 6, unlike formal plant breeding which selects for genetic uniformity, traditional farming practices often allow for continued gene exchange between wild relatives and folk cultivars, thus facilitating the continued evolution and selection of useful genotypes. This point is illustrated clearly in the work of Daniela Soleri and Steven Smith of the Centre for People, Food and Enviroment in Arizona, whose investigations into Hopi agriculture have demonstrated significant morphological, phenological and genetic differences between *in situ* and *ex situ* populations of Hopi maize varieties (Soleri & Smith 1994, 1995). These authors argue convincingly, that only through the recognition and maintenance of traditional methods of germplasm management, can effective strategies for the conservation of diverse landraces be developed (Box 11.5).

Local Benefits: Cultural Survival and Community Development

While external benefits from ethnobotanical knowledge may lie largely in the commercial potential of natural product development and the conservation of potentially important resources, local benefits include the development of strategies for sustainable economic growth, the conservation of TBK and ultimately the cultural survival of traditional peoples themselves. In recent years, an increasing number of ethnobotanical projects has been aimed towards benefiting local communities, such projects ranging from the compilation of traditional pharmacopoeia to the development of improved methods of habitat management through the collation of local ecological knowledge.

Box 11.5   *In situ* and *ex situ* conservation. In his acclaimed book *The Diversity of Life*, science writer Edward Wilson describes the discovery of a wild relative of maize which is resistant to many important diseases, and whose genes—if transferred in domestic maize—could play a vital part in future crop protection (Wilson 1992). Yet wild relatives of crops are not the only source of useful genes, since traditional small-scale farmers in remote environments have selected specialised crop varieties, or *landraces* for generations. As these landraces are genetically extremely diverse, and often contain genes which confer resistance to pests and disease, many have been collected and conserved *ex situ*; however, many researchers are now beginning to believe that complementary strategies based on both *ex situ* and *in situ* methods offer the most effective solution to the conservation of vital crop germplasm.

In recent decades, many conservation initiatives have been set up in response to recent concerns about biodiversity conservation. Several of these are operated on a global scale and involve international bodies including the International Board for Plant Genetic Resources (IBPGR); others operate on a local level, such as the non-profit organisation Native Seeds/SEARCH, based in Tucson, Arizona (Nabhan 1985). One of several organisations now attempting to conserve local cultivars, Native Seeds—together with many other conservation bodies—see a clear need for combining both *in situ* and *ex situ* methods in order to ensure effective long-term conservation of native germplasm.

EX SITU CONSERVATION

Ex situ conservation essentially involves the collection of plant seeds and other propagative materials for subsequent preservation under controlled conditions. These materials may be grown in botanical gardens, or more frequently, stored in seed banks, yet despite considerable investment in such facilities, these methods exhibit several inherent problems:

- Loss of germplasm through *genetic drift* is inevitable for a number of reasons:
  - current collection strategies rarely include a quantitative sampling procedure which ensures the representation of rare genes
  - the uniform environment of the seed bank selects preferentially for the survival of certain genotypes
  - the whole process of repeatedly collecting, storing and regenerating seeds, has an implicit bias towards species or varieties with a long, or inducible dormancy; this feature is generally absent from tropical plants
- Where genotypes are isolated from normal selective pressures—both natural or human-directed—these plants remain relatively static, compared to normal populations which remain responsive to environmental change
- In many collections, there is often so little information recorded about the source, uses, or other characteristics of a particular cultivar, that commercial farmers have generally been reluctant to include these varieties in their breeding programmes (Peters & Galway 1988).

continued

— *continued* —

IN SITU CONSERVATION

While *ex situ* conservation provides a reservoir of individual plants which can buffer the impact of local catastrophes, *in situ* conservation can provide a complementary strategy which can help to overcome the problems outlined above. In contrast to *ex situ* techniques, *in situ* conservation involves conservation conducted by farmers in their own fields, allowing, for example, the ongoing co-evolution with pests, pathogens and other forms of environmental stress. In the past these methods have largely been rejected on the assumption that traditional agriculture is characterised by the rapid and uncontrolled loss of germplasm. However, more recent research suggests that such a view is erroneous (Brush 1991), and that traditional methods of germplasm management can facilitate both the exploitation of genetic variability and the maintenance of desirable genotypes (Bellon & Brush 1994).

IN SITU AND EX SITU CONSERVATION OF HOPI MAIZE

During her recent investigations into Hopi agriculture, US ethnobotanist Daniela Soleri has demonstrated clearly that the long-term *ex situ* storage of two traditional Native North American varieties of maize has had a significant effect on their morphological, phenotypic and genetic characteristics:

- Comparisons between populations maintained for over 40 years at an *ex situ* facility in Iowa, and current populations maintained by the farmers themselves, have demonstrated clear differences in the characteristics of the *ex situ* and *in situ* populations
- These differences include reduced levels of diversity in *ex situ* collections, which could prove a real problem for future germplasm conservation (Soleri pers comm.; Soleri & Smith 1994, 1995).

Ethnomedicine and Primary Health Care

One area of applied ethnobotanical research which has attracted considerable interest in recent years is that of providing adequate health care to rural populations. As we saw in Chapter 8, more than half of the world population is dependent on traditional therapies, many of which involve a heavy reliance on medicinal plants. Hence, several projects have been aimed at documenting local knowledge of therapeutic plants, while others have combined local plant knowledge with scientific analysis, as a means of standardising traditional plant remedies. These two approaches are illustrated clearly among the current projects within the WWF/UNESCO/RBG, Kew initiative, *People and Plants*— the work by the Caribbean group TRAMIL (which is dedicated to studying Caribbean medicinal herbs), and the development of improved health care in Madagascar.

TRAMIL was originally initiated by a multidisciplinary group of researchers who began investigating health care provision in Haiti and the Dominican Republic during the 1980s. Since this time, the project has expanded to include English, Spanish and French-speaking regions of the Caribbean, and by the time of the 6th TRAMIL meeting in November 1992, the compilation *Towards a Caribbean Pharmacopoeia* had been produced (WWF *et al.* 1993). Participants at the meeting reviewed the plants cited throughout the text, and offered new data on their local uses, chemical constituents, biological activity, toxicity and conservation status; this new information is to be included in the new book *A Popular Caribbean Pharmacopoeia*. In addition to documenting the range and depth of local knowledge concerning the uses of medicinal plants, TRAMIL aims also to compile information from local communities on how to harvest plants in a sustainable fashion, to compose a database on the conservation status of useful plants, and to train participants in methods for measuring rates of resource use and the efficacy of a range of plant management techniques (WWF *et al.* 1993).

A second project also aimed at combining both cultural and biological conservation with the development of local health care is currently underway at the Manongarive Special Reserve in Madagascar. In a region where the nearest clinic is 80 km away and communications are poor, adequate health care forms the primary concern of local people. This project therefore aims to improve the care available, not through replacing the existing medical system, but through the evaluation and authentication of the traditional therapies in order to produce an integrated health care system which combines both local and Western practices. Participants in the project include traditional plant specialists, ethnobotanists, medical doctors and pharmacologists, and together this team is able to document and to test the efficacy and toxicity of medicinal plant extracts, to identify active principles, and to return the results of the research to the community as a whole. As a consequence of this work, local knowledge of medicinal plants can be compiled and conserved, recommendations for usage can be based on an empirical understanding of activity and toxicity, while medicinal plants with proven activity can be targeted for conservation. In addition, the project, which was initiated by project leader Nat Quansah in 1989, has so far been able to train at least eight Malagasy students—two doctorates in chemistry; two Masters in botany; two Masters in pharmacology; and two undergraduate botanists. A further 10 postgraduate students are now in training, alongside 26 medical students who are working with qualified medical doctors in the field (WWF *et al.* 1993).

Renewable Plant Products: A Sustainable Source of Income?

A second area of research which has received considerable research attention lies in the identification of marketable plant products which can be harvested sustainably to provide an additional source of income. These *non-timber plant products* (NTPPs) encompass any useful plant products other than fuel, or wood which is used in construction. While many of these, such as fruits, nuts and medicinal plants, may be harvested for subsistence use, others may be sold in local markets. In addition, the increasing demand for natural products from Western societies has also stimulated demand for some of these products in the world market. One of the earliest NTPP to achieve widespread economic status was the latex of the Brazilian rubber tree (*Hevea brasiliensis*) which now accounts for world sales in excess of US$1.7 billion *per annum* (see Chapter 1), while later examples include nuts from the Brazil nut tree (*Bertholletia excelsa*) and cashew (*Anacardium occidentale*), cupuaçu fruits from *Theobroma grandiflorum* and canauba wax from the palm *Copernicia prunifera* (Balick 1984; Clay 1992).

Ethnobotanical research into the economic potential of NTPPs involves both the identification of potentially marketable plant products, and the estimation of their current and potential economic value. Methods used in estimating economic value may involve simply determining existing market prices, or, where a subsistence product has no current market value, an estimated figure based on criteria such as the *contingency value* may have to be used (Box 11.6). In addition, where the current *versus* potential income generated by the sale of NTPPs are to be compared, factors such as the quantities available and possible rates of extraction must also be considered (Box 11.7). On the basis of such quantitative analyses it is hoped that through the sustainable management of *extractive reserves*, traditional communities will be able to generate income from natural resources, while avoiding the generally destructive practice of forest clearance. However, if the income generated from the sale of plant products is to be increased significantly, it is essential that existing markets for these products are expanded considerably (Box 11.8).

Protecting Local Resources

In addition to its potential roles in health care and economic expansion, ethnobotanical research can have a range of benefits in relation to the conservation of local resources—resources which include not only those essential to the livelihoods of local populations, but also those which are fundamental to maintaining the cultural identity of local peoples. In the preceding chapters, we have already seen how local knowledge can contribute to the development of ecologically sensitive methods of plant resource management, yet research into TBK can also be applied in areas such as identifying the conservation status of particular plant species (Box 11.9), predicting the cultural impact of proposed development projects (Box 11.10), and in conserving local knowledge *per se*. For example, the Marovo Lagoon resource management project (Solomon Islands) was initiated in 1985 by the people of the Marovo community (Baines & Hviding 1992). Initially aimed at producing an inventory of both existing local resources and current development activities, the project has now expanded to include the documentation of the local community's environmental knowledge. For one of the main concerns of many of the Marovo people, is the current erosion of traditional knowledge—knowledge which is fundamental to the Marovo culture—particularly among younger residents who spend much of their time away at school. Hence, it is hoped that this ethnoecological project may help both to conserve local knowledge, and to promote residents' interest in their cultural heritage.

Box 11.6 Estimating the value of non-timber plant products. In investigations to determine the economic potential of NTPPs, one of the first steps involves assigning a market value to such products. In some cases this is relatively straightforward; however, for products with no existing market value, either comparative values, or contingency values must be determined. Yet whichever way a plant product is priced, its net value must consider all of the costs associated with its production. Unfortunately in many early studies into the value of traditional plant products, methodologies have varied enormously, and valuations have often been incomplete. It is felt therefore, that more attention should be paid to the development of accurate methodologies if future studies are to produce more broadly applicable data (see Godoy *et al.* 1993).

While policy makers have often assumed that land has no economic value unless it is logged or farmed (Godoy *et al.* 1993), it is clear from the preceding chapters that traditional communities depend on the plant resources of rural habitats to fulfil a wide range of dietary, material and cultural requirements. In recent years, ethnobotanists have become interested in determining the local economic value of these non-timber plant products—both as a means of assessing the economic costs of habitat destruction, and with a view to facilitating local economic development through increased trade in such products.

continued

_ continued _

PUTTING A PRICE ON FOREST PRODUCTS

In many cases, the most difficult part of evaluating the economic potential of NTPP extraction lies in assigning a monetary value to a given product, the methodologies used varying according to the market status of a given product.

- Where a product is sold in local markets, a price can be determined on the basis of prevailing costs
- Where a product has no market price, but is bartered for another which has, a price can be determined on the basis of the latter
- Alternatively, a price can be determined on the basis of the value of close substitute of a given product
- Where no priced substitutes are available, researchers may resort to _contingent valuation_ where price is estimated on the basis of how much local people say they would be willing to pay (or barter) for a given product

ESTIMATING NET ECONOMIC VALUE

However, allocating a market price for forest products is not sufficient to assign a _net value_ to a given product, as total _marginal costs_ associated with the collection, processing, transport and sale of such products must also be considered.

- _Shadow pricing_—the value of a given product must be adjusted according to any nationally or locally determined taxes or subsidies that affect the product's net value; such adjustments are referred to as _shadow pricing_ (see Turner _et al._ 1994)
- _Labour time involved_—time spent procuring and processing goods may be estimated from the gross time spent in the forest, together with recalled information on _time allocation_; alternatively _instantaneous sampling_ or _focal subject sampling_ may be used (Johnson 1975). Total labour time must then be assigned a monetary value based either on local wages prevailing at the time, or on the country's official wage for similar work (Godoy _et al._ 1993)
- _Environmental damage incurred_—the cost of rectifying environmental damage caused by harvesting, processing or transporting NTPPs should be deducted from the overall price fetched in the market-place
- _Costs of materials used_—any investment in materials used in obtaining a final product must also be considered; this may include the financial costs of materials such as arrowheads, or costs such as the time spent in producing a collecting basket
- _Discounting_—the effects of time should also be considered. For example, the harvesting of a given plant _now_ may have a significant effect on the success of harvests in the _future_; equally, the price fetched for a given product sold now, may be significantly lower than the same product sold in the future. These influences of time are considered through an economic process known as _discounting_ (see Turner _et al._ 1994)

Box 11.7　Total economic value of NTPPs—valuing plant products in tropical forests. In the absence of formal quantitative data, estimating the overall value of NTPPs is extremely complex and requires the use of a range of economic and ecological methodologies. For a more detailed discussion of these diverse techniques, see Godoy et al. 1993; Martin 1995.

Among forest dwellers throughout the tropics, a combination of market- and subsistence-orientated agroforestry activities is common. For example, the Kantu' peoples of West Kalimantan cultivate subsistence food crops alongside cash crops such as pepper (*Piper nigrum*) and rubber (*Hevea brasiliensis*) in fallowed swidden land (Dove 1993), while in Amazonia, many cultures produce hammocks and other plant-based handicrafts for sale in commercial markets (e.g. Padoch 1988). Indeed the production and exchange of non-timber forest products (NTFPs) appears to be a long-standing feature of traditional peoples throughout the humid tropics (Dove 1994), and has played a key part in the economic development of forest dwellers throughout much of their history.

CURRENT ECONOMIC VALUE OF NTFPs

A recent survey of the Bora agroforestry system in Peru, has revealed that to these swidden cultivators, NTFPs provide a considerable proportion of their annual cash income (Padoch et al. 1988):

Product category	Mean percentage of annual income
Cultivated fruits	63
Intensively managed crops	21
Charcoal	3
Fibres and handicrafts	2
Forest fruits	1
Medicinal plants	0.5
Animal products	9

Similarly, a participatory rural appraisal conducted in the Middle East and North Africa, revealed that trade and crafts accounted for between 17 and 54 per cent of participants' livelihoods (Theis & Grady 1991 cited in Martin 1995). However, the total value of NTFPs among traditional communities, includes not only the financial returns from the sale of these products in local and international markets but also their subsistence contribution. Hence, in calculating the quantity of each product produced *per annum*, researchers must include the amounts used both in the market-place and at the household level.

- Actual value of products harvested should be based on the *flow* of stock (the quantity actually harvested) rather than the *inventory* (the stock quantity in the forest)

_____ continued ___

continued

- The most accurate method for measuring flow is to identify, count and weigh each item which is extracted from the forest each day
- Seasonal variations in product availability should be considered
- The nature and extent of the extraction area exploited at any one time should be considered
- Detailed surveys of plants and animals sold in market-places should include attention to the types of economic transaction which take place, the approximate quantity of each product available, and an estimate of the apparent demand for each product

POTENTIAL INCOME FROM MARKETING NTFPs

In contrast to estimates of the actual value of NTFPs, estimates of the potential value have been based largely on stock inventories. For example, an inventory of all the useful plants within a given area has often been used to estimate the economic potential of standing forest. Using such an approach, estimates for the potential value of standing forest have been calculated in several regions, the estimated values obtained varying from about US$1 to $420 ha^{-1} per annum according to biological and economic diversity, differences in methods used, and the nature of the products studied (see Peters _et al._ 1989; Balick & Mendelsohn 1992; Panayatou 1992).

Box 11.8 Developing markets for NTPPs—lessons from Cultural Survival Enterprises. While the potential value of a given unit of vegetation may appear to be high, the realisation of its full economic potential can only occur where appropriate, regular markets have been identified. To this end, the commercial organisation Cultural Survival Enterprises (CSE) was founded in order to establish and expand international markets for NTPPs extracted from tropical forests. During its first year of trading, CSE was able to identify a number of key factors associated with successful market development (Clay 1992).

While the economic potential of extracting NTPPs from standing forests seems to present a promising alternative to logging, there are a number of factors which must be considered—in particular the _sustainability_ of rates of extraction which are economically viable (see Chapter 12) and the _size_ of the potential market for a given commodity. This latter point has received detailed attention in a recent paper from the commercial organisation _Cultural Survival Enterprises_ (CSE), which was established in 1989 to expand and develop markets for NTPPs produced specifically by traditional forest residents (Clay 1992).

SOME GENERAL PRINCIPLES FOR NTPP MARKET DEVELOPMENT

CSE is currently working with rainforest communities throughout the world, who are trying to expand their income through the development of sustainable

continued

— continued —

extractive activities. By demonstrating the possibilities for expanding international markets for NTPPs, CSE therefore hope to set a compelling economic
example not only to local producers, but also to national and international
governments, non-governmental organisations (NGOs), international banking
organisations and development agencies, to show that tropical forests are best
protected by their traditional residents once sustainable sources of income are
developed.

During its first year of operation, CSE imported NTPPs (excluding medicinal
plants*) from several tropical regions and sold them for a total of US$349 000 to
17 companies. As a consequence of their experience, CSE have learned a
number of important lessons regarding the practicalities of marketing NTPPs
from tropical forests (Clay 1992).

- As it takes a great deal of time for forest products to develop new international
  markets, efforts should focus initially on products for which markets already
  exist
- Nevertheless, in the long term, the diversification of products sold is fundamental to the overall viability of extraction
- In order to reduce the overall risk to producer groups, attempts should be
  made to diversify the number and type of end users of NTPPs
- Wherever possible, value should be added locally to maximise the local
  capture of product value (see also Chapter 12)
- The marketing system established should maximise the return of revenues to
  local people
- Projects should aim to be large enough, at least eventually, to help all forests
  and forest peoples rather than those in selected areas only
- As no single forest group can fulfil the demands of even a small company in
  North America or Europe, supplying large companies is out of the question;
  unfortunately this often means that prices to consumers may be relatively
  high, thus reducing demand
- Controlling a large market share of a commodity (such as the Brazil nut market
  in the USA) allows for much greater influence over the overall market
- Despite their existing concern over the conservation of tropical plants and
  animals, there is a clear need to educate consumers in North America and
  Europe regarding the cultural erosion of tropical forest peoples
- As the growing *green market* in the USA and in Europe has risen primarily out
  of conservation concerns, the sale of commodities from tropical rainforests
  must be linked with systems which determine the impact of product harvest on
  local environments (see Chapter 12)

(*Out of concern for the sustainability of harvesting wood from tropical forests, CSE
decided not to work with wood products; in addition, they have determined not to become
involved with the trade of medicinal plants until they can negotiate appropriate returns for
forest dwellers through mechanisms such as patents and royalties. This latter issue is discussed further in Chapter 12.)

Box 11.9   Local knowledge and resource conservation—assessing the conservation status of African medicinal plants. While several ethnobotanical research projects include the assessment of the conservation status of useful plant species, the work outlined here specifically involves consultation with local practitioners in identifying vulnerable plant species. In this example, independent bark damage assessments confirmed most of the observations of the herbalists, the exceptions being those species (such as *Acacia xanthophloea*) which were scarce due to their limited geographical distribution rather than their overexploitation (Cunningham 1993a)

The sustainable management of traditional medicinal plant resources is essential in areas where local populations rely on traditional medicine to fulfil much of their health care needs. However, in Africa, where between 70 and 80 per cent of the population depend on traditional medical practitioners (TMPs), supplies of traditional medicines are vulnerable to extensive losses, due to rapid rates of forest clearance for agriculture and uncontrolled grazing of pastures.

CUSTOMARY CONTROLS OVER MEDICINAL PLANT HARVESTS

Taboos, seasonal and social restrictions on gathering, and the tools and techniques used in plant collection have all served to limit harvests of medicinal plants among traditional communities (Cunningham 1993a). These traditional sociological controls include:

- Taboos against the collection of medicinal plants by menstruating women in South Africa
- Mythical proportions accorded to levels of toxicity in plants such as *Synadenium cupulare*. This plant is reported to be so toxic that birds flying over this tree are killed
- The bark of *Okoubaka aubrevillei* is removed using only traditional wooden tools; under no circumstances may a machete or other metal implement be used
- In Swaziland and South Africa, the collection of several species is restricted to the winter months
- In Zimbabwe, permission must be obtained from ancestral spirits before entering forests where the medicinal plant *Warburgia salutaris* occurs

In each case, the medicinal species concerned are popular, scarce and effective, and where pressure on medicinal plants has remained low, the supplies of medicinal plants remain protected. For example, in remote areas and in countries such as Mozambique and Zambia, where commercial trade in traditional medicines remains relatively undeveloped, local supplies of medicinal plants remain adequate. Elsewhere, however, increased demands for both medicinal plants and economically important timber species have precipitated large decreases in the plant resources available.

THE IDENTIFICATION OF VULNERABLE SPECIES

Today, in areas where commercial gathering is taking place, many species have sustained extensive damage—even in State Forest which, theoretically, has been

continued

___ *continued* ___

set aside for conservation (Cunningham 1993a). For example, in the Malowe State Forest, South Africa, assessments of damage due to bark removal from the medicinal trees *Curtisia dentata* and *Octea bullata*, demonstrate that more than half of the trees encountered have had at least 50 per cent of their bark removed. However, official data on the impact of harvesting medicinal plants remain limited, and in recent years some researchers have begun to consult local TMPs about the conservation status of the medicinal plants they employ. As a result of such work in South Africa, it has been possible to compile a list of species which are regarded as becoming rare locally, yet whose demise in some cases, has not been noticed by commercial herb traders:

Species	Zulu name	No. of TMPs[†] (%)	Species	Zulu name	No. of TMPs[†] (%)
Ocotea bullata*	*unukane*	90	*Cassine tranvaalensis*	*ingwavumua*	55
Warburgia salutaris*	*isibaha*	85	*Alepidia amatymbica*	*ikhathazo*	50
Bowelea volubilis*	*igibisila*	70	*Pimpinella caffra*	*ibheka*	45
Scilia natalensis*	*inguduza*	65	*Acacia xanthophloea*	*umkhanyakude*	45
Helichrysum sp	*imphepho*	60	*Curtisia dentata**	*umalhieni*	45
Eucomis sp*	*umathunga*	55	*Gunnera purpensa*	*ugobho*	45
Hawarthia limifolia*	*umathithiibaia*	55	*Cassine papiliosa*	*usehiuiamanye*	45

[†] Proportion of TMPs who regard the species as becoming rare. * Also identified as becoming rare by herb traders.

Box 11.10   Local knowledge and resource conservation—cultural impact assessments. While local knowledge can prove valuable in efforts aimed at conserving diversity, it is important to remember that both traditional knowledge, and traditional resources are fundamental to cultural identity, and the conservation of cultural diversity itself. In the example outlined below, it is clear that plant resources are elements which permeate virtually all forms of social interaction, and that they are certainly of great symbolic significance to the North American Western Shoshone, Southern Paiute and Owens Valley Paiute peoples of the Yucca Mountain region (see Stoffle *et al.* 1990). As a consequence of this work, traditional experts have recommended that plant species identified as significant to local cultures should be conserved, preferably *in situ*, but otherwise through transplantation, or through the conservation of similar stands occurring elsewhere.

While local knowledge may be usefully applied in identifying those plant species which are becoming depleted due to changes in harvesting activities, TBK may also be used in identifying plant species which merit specific conservation efforts on the grounds of their practical value and/or cultural significance. This concept has been clearly illustrated in a recent report which outlines the involvement of a number of native American peoples in carrying out an *environmental impact assessment* (Stoffle *et al.* 1990).

___ *continued* ___

___ *continued* _____

TBK AND ENVIRONMENTAL IMPACT ASSESSMENT AT YUCCA MOUNTAIN, NEVADA

In 1982, the *US Nuclear Waste Policy Act* proposed a plan to select safe disposal sites for high-level radioactive waste, on the basis of both environmental and cultural impact assessments. Following a national search, three sites were recommended for further consideration, the first of which to be studied was that at Yucca Mountain, Nevada. An area of great cultural significance to three Native American ethnic groups—Owens Valley Paiute, Southern Paiute and Western Shoshone—the environmental impact assessment required an investigation of potentially vulnerable cultural resources, including archaeological and other cultural sites as well as any biological resources.

With regard to ethnobotanical resources, the report contains discussions of both traditional and current uses of 76 species of plant found in the study area. It summarises patterns of their use and management, and presents the recommendations made by Native American plant specialists, regarding the conservation of these resources:

Use category	No. of plant species	Comments
Food	46	These food plants include several seed, fruit, bean and nut-producing species, as well as green vegetables; many remain important to all three groups.
Medicinal	20	The majority of the medicinal plants recognised were brewed as teas; others were used as poultices, compresses, and smoking materials. Many of these remain in current use.
Construction	11	These species were used in the construction of dwellings and other structures including animal pens, fences, sweathouses, and irrigation channels.
Technology	25	These species were used to make various items ranging from weapons, sandals and clothing to ceremonial baskets; willow (*Salix* sp) and squawbush (*Rhus trilobata*) remain important in weaving baskets, including the famous Navajo wedding basket.
Fuel	11	These species were collected and used as needed when camping,

_____ *continued* ___

—— *continued* ——

Use category	No. of plant species	Comments
		and served additional functions when burned in sweathouses and during rituals and ceremonies.
Ritual and ceremonial	8	Like other plants used by these cultures, species used in ceremonial contexts are accorded the same respect as humans.
Mythical significance	5	Only those plants which were mentioned in the field are listed in this report; however, it is likely that there are many more species which have mythical significance.

Of the 76 plants mentioned in the report, 60 were used traditionally, and 43 of these remain in common use today. These plants are not only of practical value to local peoples, but also constitute symbols of cultural identity, and permeate the daily, seasonal and annual life cycles of the community members. These 43 plant species thus continue to make a significant contribution to the persistent cultural systems of the Yucca mountain peoples, and their conservation is integral to the cultural identity of those peoples.

Commercialisation and Conservation: Exploring the Links

For the sake of convenience, the benefits of applying ethnobotanical data have been divided here into discrete sections: external benefits have been distinguished from local benefits, while economic aspects have been discussed separately from conservation issues. However, it is clear from many of the examples outlined above, that there exist important links between each of these factors. For example, the identification of novel plant drugs should, theoretically, benefit global and local populations, both economically and in terms of improved health care; equally in each example presented, it is clear that economic benefits from the identification of marketable plant products can be realised only where sources of those commodities are effectively maintained.

Sustainable Development: Economic Growth and Resource Conservation

For many tropical countries, their national forests represent one of their most valuable assets, generating economic returns both through the sale of specialist timber and through the provision of new land for agricultural or industrial expansion. Unfortunately, this fact has often led policy makers to assume that these forests and other wildlife habitats can realise their economic potential only when they are logged and/or converted to an alternative use—even where such activities are clearly unsustainable (Godoy *et al.* 1993). For example, the systematic occupation of the famous rainforests of Brazilian Amazonia began in the late 1950s as generous tax cuts and loans encouraged private investors to clear significant areas of forest for livestock ventures (Hall 1991). Later, the building of the *Transamazônica* in 1971 facilitated the immigration of 64 000 farmers into the forest area. However, both of these major development programmes were initiated prior to technological and economic feasibility studies, and ecological disaster ensued: roads soon proved impassable in the rainy season; settlers demonstrated poor knowledge of local soil conditions and employed inappropriate management techniques; cattle ranches proved to be non-viable after only a few years under pasture.

Following considerable media coverage of these and other examples of environmental destruction, there has been growing concern over issues such as biodiversity conservation and global warming. In recent years, this concern has had a significant effect on the perceived value of standing forests, in terms of their potential genetic and chemical resources, the environmental services they provide, and the aesthetic value they hold. Today, it is hoped that if the global benefits of standing forests can be realistically evaluated, the harvesting of NTPPs together with other non-destructive activities could provide a practical economic alternative to land clearance, facilitating both local economic development and long-term biological conservation.

The Belize Ethnobotany Project

In a recent paper presented at the Ciba Foundation Symposium *Ethnobotany and the search for new drugs*, Mike Balick has looked specifically at these links between commercialisation, local economic development and biodiversity conservation (Balick 1994). Here, he reports on the *Belize Ethnobotany Project*—an ambitious project which aims to link pharmaceutical prospecting with both biological and cultural conservation. The project was initiated in 1988, as a collaborative effort between *Ix Chel Tropical Research Foundation*—a Belizean NGO, and the *Institute of Economic Botany* based at the New York Botanical Garden. Beginning with an ethnobotanical inventory of this region, coupled with valuation studies of the NTPPs available, workers involved in the project have estimated that the value of the standing forest on the reserve ranges between US$726 ha^{-1} for a plot with a 30-year rotation and of $3327 ha^{-1} for a plot with a 50-year rotation.

Based on these studies, and with a view to establishing a sustainable extractive reserve, an area of lowland tropical forest was given 'forest reserve' status by the Belize government in 1993. It is intended that this diverse region, the *Terra Nova Rain Forest Reserve*, will support three main activities: the work of traditional healers, their apprentices and local students, the continuation of ethnobotanical and ecological research, and the development of ecological tourism. It will, of course, be many years before this first *ethnobiomedical reserve* can be judged as a success or otherwise—and indeed there are many potential economic, ecological and ethical problems associated with this type of approach (see Chapter 12). Nevertheless, in view of its innovative nature, there exists, within Belize, considerable optimism and support among its people.

Applications of Palaeoethnobotanical Data

While much of this text has focused on the knowledge of extant cultures, as we have seen in the previous chapter, there exists also, considerable information concerning the historic and prehistoric relationships between plants and people. And just as the TBK of living peoples can be applied in areas such as natural product development and effective resource management, the careful scrutiny of palaeoethnobotanical data can fulfil a similar role. For example, while the ecological consequences of certain types of human activity may be gleaned from the archaeobotanical record, several modern drugs have already been discovered through the analysis of ancient medical texts, including the antihypertensive agent reserpine (isolated from *Rauvolfia* spp), which was found in the texts of Indian Aryuvedic medicine (see Cox 1990). Since then, several workers have begun the systematic analysis of ancient medical systems as a source of 'novel' bioactive compounds, particularly with respect to the widely known systems of India, China, Ancient Greece and Rome (e.g. Holland 1994; Jain 1994; Xiao 1994).

A Role for Folklore in Modern Drug Discovery?

While ancient medical texts appear a reasonable starting point in the search for bioactive plants, another source of evidence lies in folklore. For example, while the examination of herbals such as those of Gerard and Culpeper may reveal much about the *orthodox* medical practices of the time, it is the folklore of historical Europe which may reveal insight into the traditional *folk* medicine of the same era. This distinction is particularly significant in terms of European studies, for here the development of orthodox medicine has led to the historical blurring of the boundaries between folk healing, sorcery, black magic and witchcraft. Hence, just as the symbolic analysis of ancient Polynesian mythology points to the bioactive properties of the *kava* plant (see Boxes 4.6 and 4.7), the analysis of European folklore may equally reveal the bioactive nature of plant species.

One of the most common characters found within European folklore is the witch—usually a witherered old hag, who uses herbal ointments and potions to carry out her sorcery. Written accounts of the sorcery for which the archetypal witch of popular folklore is famed, can be found at least as far back as the mythology of ancient Greece. For example, the goddesses Circe, Medea, Hecate and Diana were all famous for their magical powers, and are mentioned in later medieval accounts of European witches (Simpson 1994). Common motifs occurring within these tales include *lycanthropy*—the witch's ability to change her form—and her ability to fly; historical evidence on the other hand, commonly indicates the use of so-called 'witches' salves' which were based largely on plants. Of the plants used in these ointments, four related species enjoyed particular notoriety—deadly nightshade (*Atropa belladonna*), mandrake (*Mandragora officinarum*), henbane (*Hyoscyamus niger*) and *Datura stramonium*—all of which belong to the family, Solanaceae. Significantly, the numerous accounts of flying sensations and the occurrence of shape-shifting stimulated a number of early workers to investigate these plants further—work which eventually led to the discovery of the tropane alkaloids and their application in modern orthodox medicine (Box 11.11).

Box 11.11   Folklore and the tropane alkaloids. Their interest stimulated by medieval folk tales—of witches flying and changing into animals—a number of scholars investigated the nature of witches' flying ointments. Even as early as the sixteenth century, it became clear that these unlikely tales stemmed from visions experienced by users of flying ointments, and later analysis has revealed the presence of four plant alkaloids which are largely responsible for their psychotropic effects. Through their action on the central nervous system, several of these chemicals have since found a place in modern orthodox medicine, suggesting a role for folklore analysis in the future identification of bioactive plants.

> While ancient medical texts may provide a valuable source of information regarding the nature and traditional uses of bioactive plants, folklore too can
>
> *continued*

_____ *continued* _____

offer a useful source of information. The example outlined here demonstrates how the pharmacological analysis of European folklore supports the notion that folklore analysis may reveal useful information on the bioactive properties of plants.

THE WITCHCRAFT PHENOMENON IN MEDIEVAL EUROPE

Initially, the orthodox medicine of medieval Europe shared its roots with the peripheral practices of folk healing and sorcery. Yet as Christian doctrine spread throughout the continent, sorcery became increasingly associated with heresy—a crime which, by the eleventh century was already punishable by death. The thirteenth century later saw the increased demonification of witchcraft and the concomitant isolation of traditional healers—most of whom were women. By the middle of that century, hostility towards traditional women healers intensified further, marking the start of 250 years of persecution, as male, university-trained doctors increasingly closed ranks. By 1572, following the *Witchcraft Acts* in England (1542) and Scotland (1563), all forms of witchcraft were punished by death, and the witch hysteria characteristic of the Middle Ages and Renaissance period soon developed.

THE STUDY OF WITCHES' FLYING OINTMENTS

It is clear that throughout the centuries, traditional healers were increasingly persecuted as a result of changing sociological factors—sexual hostility, economic marginality and rapidly segmenting society (Oakley 1992). Nevertheless, even as hysteria towards witchcraft reached its peak, a number of scholars were convinced that many of the witchcraft phenomena—such as flying and shape-shifting—were the result of drug-induced visions, rather than any acts of the Devil:

- Sixteenth century Spanish physician AF De Laguna suggests that the flights of witches are the results of hallucinations
- In 1562, Giovanni Battista Della Porta, becomes one of the first scholars to experiment with flying ointments
- Following the successful isolation of the alkaloid morphine from crude opium (in 1803), similar techniques were used to extract alkaloids from other sources, including atropine from *Atropa belladonna*
- In 1869, Dr Jacques Joseph Moreau—who was at that time experimenting with psychoactive drugs in the treatment of mental illness—began a study of medieval remedies used in alleviating symptoms of insanity; he found *Datura stramonium* to be particularly effective
- During the middle of the twentieth century, Siegbert Ferckel and Will-Erich Peuckert began modern self-experiments with flying ointments made from recipes found in old magical *grimoires*; they, like the other workers before them, reported experiencing distinct sensations of flying.

_____ *continued* _____

— *continued* —

It is now established that the main active principles of these witches' ointments are the tropane alkaloids hyoscyamine, atropine, scopolamine and mandragorine. Like most psychoactive compounds they act on the nervous system once they have been absorbed into the bloodstream, these alkaloids acting in a number of ways. Together they are able to control smooth muscle action and to produce hallucinations (including the sensation of flying, and visions of werewolves). As a consequence of their biological activities, atropine, hyoscyamine and scopolamine have all been used in orthodox medicine as antispasmolytics, and in alternative therapies, as antiasthmatics, while henbane is still used today as a sedative (Griggs 1981).

Palaeoethnobotany and Sustainable Development

As in the case of ethno-directed drug discovery programmes, lessons in conservation may be learned, not only from extant peoples, but from the examination of palaeoethnobotanical evidence. For example, integrated studies based on pollen analysis, the excavation of structural artefacts, the examination of colonial accounts, and the interpretation of Mayan iconography, have together facilitated the experimental reconstruction of *chinampas*—a series of raised agricultural fields separated by canals, which were commonly developed in the swamps or river flood plains of ancient Mexico (Puleston 1978; Turner 1978). Within these complex systems, the canals could collect and store water, while the raised fields provided adequate drainage for the growth of a number of crops. In addition, it seems likely that periodic cleaning of the canals and nearby reservoirs could have provided material for fertilising fields, while the canals themselves could have provided ideal conditions for fish breeding (Thompson 1974; Hammond 1978). Significantly, the modern reconstructions of these ancient agricultural systems have so far proved well suited to the low input cultivation of beans, tomato and cotton (Puleston 1978), and it is possible that the adoption of these historic agricultural practices may prove both economically efficient and ecologically sustainable today.

SUMMARY

While early ethnobotanical observations were made largely in the interests of practical factors, such as survival and economic expansion, the formal academic study of TBK has often adopted a much more theoretical approach. However, in recent years, the empirical validation of traditional plant use and indigenous management strategies has led to renewed interest in the practical applications of local people's knowledge. These applications are essentially twofold: knowledge of plant use may be used in the identification of new marketable products, while ecological knowledge may be usefully applied in conservation programmes. In addition, such applications may have both local and external benefits—ranging from the improvement of local health care to the development of a modern pharmaceutical, and from the protection of resources fundamental to cultural survival to the conservation of the World Heritage Sites of the global community.

Chapter 12

Applying Ethnobotany in Sustainable Development: Practical Considerations

Many herbal medicine men will not like this book since it may deprive them of their profession once their secrets are revealed. The majority of them were reluctant to show me the drug plants as a whole for this reason . . . But after some persuasion, I was shown the plant on condition that I would not reveal it to anyone else.

JO Kokwaro 1976 (see Reid *et al.* 1993 in *Biodiversity Prospecting*)

INTRODUCTION

From the preceding chapter, it is clear that the potential benefits arising from ethnobotanical research appear considerable—particularly in relation to the sustainable economic development of impoverished rural areas. For example, the systematic analysis of traditional medical systems may provide solutions to current health care problems in both local and global terms; the identification of marketable plant products may provide an alternative source of income to logging or farming; the recognition of increasingly vulnerable species may facilitate the effective targeting of future conservation efforts. However, as with any other development activity, there are a number of practical problems which must be overcome if this new, ethnobotanical approach to development is to succeed (Table 12.1). While the problem of Western acceptance of traditional scientific knowledge has been discussed earlier (Chapter 3), additional problems are associated with two closely linked issues: the actual realisation of predicted economic benefits and the protection of species with recognised commercial potential. This chapter looks at the

Table 12.1 Problems associated with conventional and ethno-directed development projects. In the past, development initiatives in wildlife areas such as tropical forests have often failed to consider their long-term consequences. For example, in Amazonia, a range of government funded development activities has led to widespread environmental problems. These activities include road building, directed colonisation by small farmers, dramatic increases in livestock ventures, mining activities and commercial logging, which together have led to problems of soil erosion, siltation of rivers soil compacting and leaching, flooding, pollution, loss of species diversity, and global changes in climate (Hall 1991). It is hoped that by adopting alternative development strategies based on local knowledge of useful plants, economic expansion may coincide with the protection of species, habitats and cultural diversity. Nevertheless, before such projects prove viable, appropriate legislation must be developed in order to address a number of economic and ecological problems, while ethical aspects of research into applied ethnobotany must also be considered

Problems associated with conventional development projects	Problems associated with ethno-directed development projects
Novel practices recommended by external organisations may not be readily adopted by local people.	Western society remains reluctant to accept the value of traditional knowledge.
Novel practices may prove inappropriate to a given ecological or sociological environment.	Many benefits of conservation may not be realised in economic terms, such that the economic incentive in
Increased polarisation of landowners and landless may lead to rural violence.	the immediate term often favours land clearance.
Indigenous people have often been driven out during the course of development schemes.	Where plant species attain significant commercial interest, problems of overharvesting may follow.
Often involve extensive land clearance leading to loss of biodiversity and the breakdown of other environmental services.	Where ethnobotanists reveal details of traditional knowledge in the public domain, this may jeopardise a local community's ability to negotiate terms for equitable sharing of returns from any products which become a commercial success.

nature of these concerns, and outlines some of the current legal and ethical movements towards their solution.

SUSTAINABILITY AND VIABILITY: ECONOMIC AND ECOLOGICAL CONCERNS

Fundamental to the success of projects aimed at sustainable development are their economic viability and ecological sustainability. For example, where forest clearance proves, in the short-term, to be more profitable than more sustainable activities, then forest clearance is likely to proceed (Godoy *et al.* 1993); equally, where non-timber plant products (NTPPs) prove a commercial success, problems of over-harvesting are likely to ensue (Cunningham 1993a). Clearly then, it is imperative to the long-term viability of initiatives such as the *Belize Ethnobotany Project*, both that economic returns from sustainable activities are maximised, and that effective controls over rates of extractions are developed.

Realising the Economic Potential of Bioculture Reserves

One of the main problems associated with the economic success of *bio-cultural reserves* lies in the fact that many of their benefits cannot currently be *captured* in financial terms. For example, while agricultural and pharmaceutical industries accept that maintaining biodiversity may prove crucial in providing future genetic and biochemical resources of considerable commercial value (Reid *et al.* 1993), local people receive no financial compensation for protecting resources which are currently of only *potential* value. Equally, although a number of NTPPs are already for sale in external markets, in many cases the raw materials collected locally achieve very little of the market price of the final product (Clay 1992). To address these practical economic difficulties requires two major factors: the effective economic analysis of the overall costs and benefits of any given project, and the development of suitable commercial and legislative strategies for improving economic returns both nationally and locally.

Environmental Accounting: Cost/Benefit Analysis of Biocultural Reserves

To determine the economic viability of any development project, it is not sufficient simply to determine the potential net value of the plant resources available (see Boxes 11.6 and 11.7). Rather, the *total economic benefits* of the proposed project must be compared with the amount of revenue which could have been generated from alternative land-use strategies such as logging or agriculture. These calculations are often complex requiring the use of methodologies which have been developed by *environmental economists*, and which are designed to consider both the immediate and the long-term effects of different land-use strategies.

The benefits associated with standing forests and other diverse habitats, can be broadly divided into two categories: those which stem directly from the habitat itself (including the extraction of NTPPs), and those less tangible, or indirect benefits, which include the provision of *environmental services.* As we have seen in Chapter 11, the direct uses of intact forests include the provision both of subsistence goods for domestic consumption, and of marketable products for local, national and international trade. Other direct benefits stem from activities such as selective or renewable logging, and wildlife tourism. However, while timber still represents one of the most valuable plant products which can be extracted from tropical forests, it is important to bear in mind that even selective logging can cause considerable habitat destruction. Even so, it is anticipated that within biocultural reserves, certain areas within managed *buffer zones* could sustain prolonged timber extraction if balanced by effective re-establishment programmes (Norton-Griffiths & Southey 1994). Many projects also envisage sustainable tourism as a means of generating income. For example, in Belize's *Terra Nova Rain Forest Reserve*, it is intended that interested visitors will be invited to enjoy nature walks and to participate in seminars and classes with traditional healers, while Africa's game park reserves have attracted many tourists since the advent of mass tourism in the 1970s (Balick 1994; Pleumaron 1994). Revenue from wildlife tourism can be generated from a number of sources, including hotels and entertainments, park fees and travel, as well as from foreign exchange (Box 12.1).

Box 12.1   Direct benefits of standing forests. In addition to the overall value of NTPPs, standing forests—or managed areas of forest—may also be used to generate income through the sustainable production of fast-growing timber, and activities associated with wildlife tourism. While these values can be difficult to estimate, a recent study in Kenya estimates an income of $14 million and $27.2 million respectively *per annum* (Norton-Griffiths & Southey 1994).

We have seen in the previous chapter that tropical forests may contain plant species which are harvested both for domestic use and for sale. However, the direct benefits of standing forests also include activities such as wildlife tourism and renewable logging, both of which have the potential to generate considerable income if managed effectively.

POTENTIAL VALUE OF SUSTAINABLE FORESTRY

At present, although sustainable forest management in specific buffer zones is feasible, such schemes are virtually absent from tropical forests in practice (Poore

continued

___ *continued* ___

1988). In many cases, the failing of such schemes is due to lack of organisation rather than inherent deficiencies in the concept of sustainable forestry. For example, a recent analysis of the economic failing of timber extraction in Kenya (Norton-Griffiths & Southey 1994) has revealed a number of important points:

- Estimated value of felled trees, fuel collected and standing timber was about $148 million
- The overall value of the scheme in practice constituted a net loss of $15 million
- Much of the leakage was due to inefficient monitoring and lack of organisation. For example, the Kenya Forestry Department collected only 30 per cent of the royalty fees intended to cover the costs of re-establishing felled trees
- If the collection of stumpage fees were more efficient, it is estimated that sustainable forestry could generate up to about $4 million in annual profits
- The authors estimate that a figure of $14 million is plausible for the total annual returns generated by forestry lands in Kenya—including benefits from gathering fuelwood and medicinal plants in gazetted areas

POTENTIAL VALUE OF SUSTAINABLE TOURISM

One activity which has featured prominently in the debate on sustainable development has been that of promoting wildlife tourism within conservation areas and game reserves. And since the advent of high-speed, wide-bodied aeroplanes has facilitated mass tourist travel into a number of developing nations, this concept has appealed to many governments as a means of gaining access to foreign currency and investment, improved infrastructure and enhanced prestige (Pleumarom 1994). Since the early 1980s, concerns over the adverse social and environmental effects of mass tourism led to the emergence of *alternative tourism,* which now includes *ecotourism, low-impact* and *sustainable tourism,* which together encapsulate the idea of travel with the specific intention of enjoying natural habitats and wildlife, at the same time fulfilling economic, social and aesthetic needs of local populations, and maintaining both cultural and ecological integrity (Jenner & Smith 1992). Hence, it is envisaged that environmentally friendly tourism could act as a positive force for conservation, able to satisfy both present and future demands of foreign visitors wishing to experience wildlife *in situ,* while introducing a source of foreign currency to developing areas.

However, net returns due to the existence of wildlife areas can be difficult to quantify, and again this is discussed in relation to Kenya by Norton-Griffiths and Southey (1994). For example, in 1989, the typical overseas visitor spent about 14 days in Kenya, spending an average of 6.1 nights in a coastal hotel, 1.9 nights in Nairobi, 1.1 nights in a gamepark, and 4.9 nights elsewhere. While it is difficult to tell from such information to what extent tourism is directly or indirectly due to the presence of wildlife, the authors estimate that overall, at least 50 per cent of the tourist revenues generated nationwide are attributable to the wildlife parks and reserves.

___ *continued* ___

continued

Net benefits generated by wildlife tourism in Kenya (1989)	
Gross revenues from tourism (1989)	$419.0 million
Revenues attributed to wildlife sector (50%)	$209.5 million
Estimated costs of providing services and investment	$215.8 million
Net returns to wildlife tourism sector (excluding foreign exchange premium)	−$6.3 million
Net returns to wildlife tourism sector (including foreign exchange premium)	$27.2 million

Data modified from Norton-Griffiths and Southey (1994).

The indirect benefits of bioculture reserves include both the provision of environmental services and the conservation of _future_ economic potential in terms of useful chemicals or genetic materials which have yet to be identified by Western scientists. Environmental functions are fulfilled as a consequence of plants' integral roles in climatic, nutrient and hydrological cycles, and these can have repercussions at both local and global levels. At a local level, trees play a fundamental part in water and nutrient cycling, regulating watershed activity and providing protection against soil erosion (e.g. Clark 1992). On a global level, tropical forests play a vital part in storing carbon and therefore in the regulation of atmospheric levels of _carbon dioxide_ (CO_2)—an important _greenhouse gas_ which is released from the burning of fossil fuels. At present, the trees of the tropical forests and woodlands store about 55 per cent (188 billion tons) of the world's organic carbon, while providing a net sink for CO_2 during periods of active growth (Panayotou 1992). Any reduction in forest area would therefore have twofold repercussions, for while any carbon sequestered in existing trees would be released as CO_2, potential sinks for the free gas would be reduced.

In terms of the future potential which may result from biodiversity conservation, as biotechnological advances continue to improve the likelihood of gene transfer between unrelated species, the demand for novel genetic material looks set to increase (Reid _et al._ 1993). Equally, even as pharmaceutical companies remain ambivalent about their interest in natural products research, several large companies including the US National Cancer Institute, the biotech company Monsanto, and the pharmaceutical giants Smith Kline, Merck and Glaxo have revived plant screening programmes in the last 5 years; in addition, two new US companies Affymax and Shaman Pharmaceuticals Inc. have been set up recently with the specific aim of developing drugs based solely on natural products (Pearce *et al.* nd). Hence, as a consequence of this increasing interest in *biodiversity prospecting*, it is anticipated that both ethno-directed, and more random plant collection could provide additional income to local peoples, assuming that mechanisms for capturing the financial benefits from these *option values* or *future use-values* are developed (Box 12.2).

Box 12.2   Indirect benefits of standing forests. In addition to the value of direct benefits from standing forests, factors such as the environmental services they currently provide, and the commercial potential they may realise in the future can also be valued by environmental economists. In some cases, these hidden benefits may exceed more direct benefits—particularly where the effective value of carbon sequestration is considered (see Box 12.3).

Indirect benefits of standing forests are related both to the environmental services they provide, and to their *potential* uses. For example, a given plant which currently has no known commercial value, may, in the future be found to exhibit important medicinal or other useful properties. The economic value of these benefits can therefore be estimated either on the basis of the costs of repairing environmental damage, or through estimating the future value of plant products which are yet to be discovered.

ENVIRONMENTAL SERVICES PROVIDED BY TROPICAL FORESTS

Trees fulfil a range of vital environmental functions through their roles in climatic, nutrient and hydrological cycles. And while many of these functions are particularly significant at a local level, others have profound significance on a global scale. As with many of the values which are used in effective environmental accounting, many of these figures are difficult to estimate; that nature of some of these are outlined in the table below:

Local level	Global level
Water and nutrient recycling— Clark (1992) estimates that the replacement of tropical forests with grassland could reduce annual evapotranspiration by about 300 mm and rainfall by about 650–800 mm; losses sustained as a result of these changes can, therefore, be used in estimating the economic value of these services.	*Carbon sequestration—*the effective value of carbon sequestration can be estimated on the basis of current taxes payable on carbon emissions; in some areas, 1 ha of forest can balance emissions from the energy sector of a developed country, equivalent to between $1250 and $3500 carbon tax *per annum* (Norton-Griffiths & Southey 1994).

— *continued* —

_____ *continued* _____

Local level	Global level
Protection of watersheds and against soil erosion—damage due to silt accumulation has been estimated at between $75 and $637 per hectare of land cleared (see Panayotou 1992).	*Biodiversity conservation*—while the commercial potential of biological diversity is clear, the economic value of biodiversity itself remains difficult to quantify, depending very much on the future role of plants in pharmaceutical, agrochemical and other industries (Panayotou 1992). However, using current values of tropical plants used in the manufacture of orthodox pharmaceuticals, Pearce *et al.* (nd) estimate that annual revenue from novel plant-based drugs could exceed $8.8 billion in the USA alone.
Biodiversity conservation—where species diversity is lost, local sources of income may be lost as useful species become rare or disappear completely	

INCOME FROM BIODIVERSITY PROSPECTING

As we have seen, in addition to the sale of specific plant materials, some income may also be generated by granting bioprospecting rights to institutions with an interest in screening plants and other organisms for novel compounds or genetic materials. Already a number of contracts have been set up betwen industrial organisations, academic researchers and local communities, some of which are outlined below (see Laird 1993).

- in 1991 the pharmaceutical giant Merck paid $1.135 million to the NGO INBio, for rights to biodiversity exploration in Costa Rica.
- in 1992 Eli Lilly & Co. agreed to make a $4 million investment in Shaman Pharmaceuticals, in return for an option to obtain exclusive worldwide marketing rights for products identified and screened by Shaman; Shaman in turn were to return a portion of its receipts to people and governments in countries where they worked
- in general collaborating in-country collectors usually receive between one-fifth and one-half of the sample fee paid by companies to contracted plant collectors—often botanical gardens and universities—which is between $50 and $200 per kg of plant material
- in recent years, royalties, rather than advanced payments have become a major point in discussions about payment for biodiversity prospecting; royalties are based on net sales and depend upon each parties' relative risks and contributions to the development of the final product. While royalties have

_____ *continued* _____

___ *continued* ___

been included in collecting contracts only recently, several organisations have produced such contracts including the US NCI, the New York Botanical Garden, the Royal Botanic Gardens, Kew, and the for-profit brokerage company Biotics Ltd, based in the UK

Finally, standing forests also provide a number of *non-use benefits*, which include factors such as the social and aesthetic values associated with the existence of certain species or habitats. While the value of these benefits is currently almost impossible to quantify directly, the concept of an *existence value* is used to describe the cultural or moral value attached to the existence of particular species, habitats or cultures. These benefits are therefore estimated using contingency valuation as outlined in Box 11.6.

Of course, in determining the overall benefits of any development project, it is important to consider the total costs associated with the scheme. These include any *marginal costs* associated with producing a marketable product and the cost of repairing any environmental damage incurred (see Chapter 11). In addition, the *opportunity costs* of the project must also be considered. For example, where tropical nations do continue to maintain conservation areas, the potential income which could have been generated from land clearance is considered as a lost opportunity. Opportunity costs are calculated on the basis of a number of criteria including the quality and total area of the land itself, the gross revenues from equivalent land which has already been converted, and any costs such as those associated with rectifying environmental damage. Once the overall costs and benefits of alternative development projects have been established, these figures may then be used to carry out a cost/benefit analysis, in which the value of standing forest is compared with the net income which could be generated through land conversion (Box 12.3). On the basis of such calculations, it is evident that a range of non-destructive activities in conjunction with effective habitat conservation could, theoretically, prove a viable alternative to forest clearance.

Box 12.3 Cost/benefit coefficient for biocultural conservation. Using the values estimated for net benefits of standing forest, compared with the net benefits of forest clearance, it is apparent that in almost all cases, the economic potential of conservation far exceeds that generated through the latter. Theoretically speaking then, providing that standing forests are managed sustainably, they should prove economically viable in the long-term.

In order to determine the relative economic advantage of conservation compared with land clearance, the potential benefits of both must be considered. A recent paper by Panayotou (1992) presents such data for a range of developing nations, indicating that theoretical cost/benefit coefficients often favour conservation. However, in practice, many of the potential benefits of conservation are not realised in financial terms, such that immediate incentives are generally in favour of the short-term strategy of forest clearance for forestry and agricultural activities.

POTENTIAL VALUES OF DIFFERENT LAND-USE STRATEGIES

While the values reported in the following table are probably gross underestimates of current and potential values of standing forests, it is clear that the value of carbon sequestration far exceeds any revenue generated by sustainable harvesting, and is comparable with some very lucrative land uses such as the production of rubber in Malaysia

				Net present values (US$ per hectare)			
Region	Conversion	Timber production	NTPPs	Soil and water conservation	Recreation and tourism	Carbon	Diversity
Brazil	−605 (ranging)	ns	1100	ns	ns	1625–3500	85
Peru	2960–3814 (ranching/ plantation)	490–1001	6330	ns	ns	1625–3500	ns
Cameroon	1248 (cocoa)	10–101	33	75	17	2520	6–92
Malaysia	3370–5660 (rubber)	1541	228–595	ns	ns	2950–3682	ns
Philippines	875 (farming)	1415	ns	312–637	ns	2520	ns
Thailand	1082–2575 (eucalyptus)	ns	ns	367	203–410	3342	220
Mean	1850	1541	412	475	307	3000	220

Data taken from Panayotou (1992); original sources cited therein.

While neither strictly comparable with each other nor representative of all forest types in each country, these estimated values give some idea of the net present values generated by different forest uses. *Conversion values* (the value obtained

_ continued _

_ continued _

from clearing forest for alternative land use) are clearly not additive with other values, except for a single harvesting of timber on-site; all other values are mutually compatible (see Panayotou 1992).

TOTAL ECONOMIC POTENTIAL OF BIOCULTURE RESERVES

On the basis of the rather fragmentary data provided in the table above, it is possible to estimate cost/benefit coefficients for conservation compared to other landuse strategies:

Region	Cost/benefit ratios	
	Best case[*] (+ timber)	Worst case[†] (–timber)
Brazil	−0.13:1	−0.22:1
Peru	0.27:1	0.56:1
Cameroon	0.44:1	0.53:1
Malaysia	0.58:1	1.71:1
Philippines	0.18:1	0.81:1
Thailand	0.25:1	0.62:1

Values calculated from Panayotou (1992); original sources cited therein. Where a range of net present value is given: *lowest value for conversion, and highest values for conservation are used, †highest values for conversion and lowest values for conservation are used (see previous table).

Options for Improving Net Benefits of Biocultural Reserves

In theory then, the potential benefits of forest conservation may far exceed the benefits obtained through land clearance. However, in practice, only a small proportion of the benefits of standing forest are actually captured by local populations. For example, Principe (1991) has estimated that the value for a given plant species of widespread medicinal value could be worth up to US$390 million _per annum_, and that the annual revenue from plant-based drugs could equal about US$15.5 billion each year in the USA alone. However, considering that the gross national product (GNP) produced in the whole of Brazilian Amazonia is only about US$18 billion, returns from medicinal plants to producing countries are very limited, as prices paid for raw plant materials remain low (Cunningham 1993a). Equally, while the potential economic value of carbon sequestration accounts for a considerable proportion of the benefits attributed to forest conservation (see Box 12.2), these benefits are not realised currently in financial terms. Hence, while the world as a whole may ultimately benefit from the existence of tropical rainforests and other wildlife areas, these benefits are currently paid for largely by the tropical nations themselves. As a result, in real terms, economic incentives often favour development practices which are incompatible with forest conservation.

In order to address this fundamental problem of economic viability, it is important that any benefits from conservation are realised both nationally and locally. In some cases, this may require the development of local industries which add value to raw materials at source. For example, in cases where *royalty payments* have been agreed in contracts concerning plant collection, if the supplier is involved only in the early stages of research, the royalties agreed are typically between only and 1 and 5 per cent of any profits made; where suppliers provide additional pre-clinical data, royalties increase to between 5 and 10 per cent; where an active product is actually identified and data regarding efficacy are supplied, royalties can reach between 10 and 15 per cent (Laird 1993). In cases of pharmaceuticals, or specialised crafts which may ultimately yield considerable profits, the increase in local income as a result of *in situ* product development can be highly significant (see Clay 1992; Pearce *et al.* nd).

However, while value-adding at source may have some effect on local income, at present, much of the unrealised economic potential from forest conservation is related to global enviromental benefits—and in particular to carbon sequestration. In order to capture these *external benefits* locally, it will be necessary to develop appropriate legislation to ensure the payment and equitable distribution of, for example, taxes paid on carbon emissions. The nature and current status of some of these legislative solutions are detailed in Panayotou (1992).

Problems of Sustainability

Although the conservation implications of tropical hardwood logging have been the focus of global attention for more than a decade, the implications of harvesting minor forest products have been considered only recently (Cunningham 1993a). Nevertheless, there exists considerable evidence which warns of the adverse effects of excessive plant harvesting as a consequence of commercial success. Indeed, Maurice Iwu, of the Phytotherapy Research Laboratory at the University of Nigeria, maintains that in declaring a plant as useful—particularly in terms of its having medicinal value—one is 'more or less signing the death warrant of the plant' (Iwu 1994). Examples of overharvesting are evident throughout the world and include the collection of several medicinal plants in Africa (Cunningham 1993a), of medicinal plants and woody species in Mexico (Felger & Moser 1985; Lozoya 1994), of rubber in Amazonia (Nepstad *et al.* 1992) and of palm fronds used in craft products in Ecuador (Borgtoft Pedersen 1994). It is imperative therefore, that precautions are taken to assess and ensure the sustainability of the multiple activities carried out within biocultural reserves.

Commercialisation and Resource Depletion: A Case Study of Prunus Africana

One of the most well documented examples of the overharvesting of a useful plant, is that of *Prunus africana*—a medicinal plant in Cameroon, whose bark extract has become a major pharmaceutical in recent years. The use of this hardwood tree in the treatment of benign prostatic hyperplasia (BPH) has increased enormously since 1972, when its processed extracts were initially exported to Europe (primarily France and Italy) on a large scale. Today, bark extracts are used in Europe to prepare two drug products sold under the brand names of *Tadenan* (France) and *Pygenil* (Italy).

Under normal circumstances, *Prunus africana* exhibits a remarkable ability to withstand bark removal, and until now, all bark has been taken from wild populations in the Afromontane forests of Cameroon, Zaire, Kenya and Madagascar. However, in spite of this species' relative hardiness, and despite the conservation efforts of the companies involved in bark collection, loss of trees has been observed quite frequently in recent years—even in high-priority conservation sites. In response rural communities, traditional healers and government departments have expressed increasing concern over the sustainability of this international trade (Cunningham & Mbenkum 1993). In response to such concerns, a working paper produced as a part of the People and Plants initiative, reports on recent research into the extent of this problem and makes a number of recommendations regarding the possibility of creating artificial plantations in order to limit any further losses in the wild (Box 12.4).

Box 12.4   Commercialisation and sustainability—harvesting *Prunus africana* bark in Cameroon. It is clear from the study outlined below, that the establishment of an international market for *Prunus* bark has led to the rapid erosion of wild populations of this multiple-use species. It seems likely that cultivation offers the most feasible solution to this problem, although problems of ensuring stable markets, and increasing prices to encourage cultivation must be addressed.

Prunus africana is a hardwood tree found in the Afromontane forests of mainland Africa and Madagascar. A multiple-use species, its durable timber is favoured for

_ continued _

_____ *continued* _____

domestic use while its bark is widely used in traditional medicine to treat malaria, stomach ache and fever (Cunningham & Mbenkum 1993). About 30 years ago, extracts of *Prunus* bark were patented as active in the treatment of benign prostatic hypertrophy (BPH), and have since been marketed in Europe— particularly in Austria, France, Italy and Switzerland. As an internationally traded forest product harvested from the wild, *Prunus africana* provides a useful case study of the problems of NTPP commercialisation, revealing practical implications for policy development on the harvesting and sale of forest products in the future.

EFFECTS OF *PRUNUS AFRICANA* TRADE

Internationally, Cameroon is the most important source of *Prunus africana*, exporting an average of 1923 tons of bark each year. All the bark harvested here is processed and exported by Plantecam Medicam, a subsidiary of the French company, Laboratoires Debat. Plantecam Medicam first started harvesting bark from the forests of Mount Cameroon, later expanding its collection area into northwestern regions of the country.

* In 1972 Plantecam Medicam recruited at least 180 workers in western Cameroon
* For 15 years, the company employed its own workers, and harvesting of *Prunus africana* was controlled, with most trees surviving
* In 1985 an additional 50 collecting licences were provided to Cameroon entrepreneurs, ending the monopoly of Plantecam Medicam, although the company remained the sole exporter of the bark
* While all Special Permit holders play a Regeneration Tax and a Transformation Tax which are supposed to cover costs of forest regeneration, licensing of local contractors has encouraged overharvesting of *Prunus* bark
* Despite the ability of *Prunus* to regenerate its bark, and the efforts made by Plantecam Medicam to ensure sustainability, there are many cases where the company's guidelines for harvesting are not followed. This is resulting in significant tree loss from wild stocks, particularly in the NW Province
* In 1992, concern about the killing of *Prunus* trees was expressed by traditional leaders and many expressed an interest in cultivating trees
* In addition to concerns about the conservation status of *Prunus*, loss of trees has also affected populations of frugivorous animals, including several endangered bird species, and a number of primates

ALTERNATIVES TO HARVESTING FROM WILD POPULATIONS

Under their current licence, it is specified that Plantecam Medicam should plant 5 ha of *Prunus africana* per year for 5 years from 1993. However, there is little doubt that this requirement has not been fulfilled. Equally, it is clear that even if 5 ha of trees were planted each year, this would not be sufficient to replace the trees harvested from the wild. Based on data from the cultivation of *Acacia*

_____ *continued* _____

_____ *continued* _____

mearnsii and *Cinchona* spp. both of which are cultivated for their bark, a number of conclusions regarding the possible cultivation of *Prunus africana* have been reached.

- If 12-year trees were felled and stripped as in the case of *Acacia* (the trunks sold as timber or fuel), a 12 year rotation of 68.4 ha of trees would require 820.2 ha of trees in total. Hence even if Plantecam Medicam fulfilled their obligation to plant 5 ha per year, this would be totally inadequate to supply the existing demand for bark
- Cultivation of *Prunus* is easier than other forest species with slower growth rates and more specific habitat requirements
- *Prunus* can be propagated from fresh seed or from cuttings; rooting could be improved significantly by the use of a commercial auxin-based rooting powder
- Project failure due to inadequate market assessment or problems of land tenure must be avoided
- A para-statal corporation should be set up to monitor and control the production and export of medicinal plants
- Core conservation areas of viable wild populations should be designated for conservation
- Field gene banks for *ex situ* conservation should also be established
- Field assessment of bark damage, and detailed analysis of population structure should be carried out

Assessing the Environmental Impact of Extractive Activities

While commercialisation can have a considerable impact on the harvesting of valuable plant species, Hall and Bawa (1993) argue that over-harvesting is not restricted to products which are collected for commercial markets. For example, in South America, several cases of NTPP depletion have been noted even where products are used primarily by local communities. However, while there exists considerable evidence of the overharvesting of NTPPs—whether for commercial or domestic use—few quantitative analyses of the effects of extraction have been carried out. Hence, as it is difficult to determine the effects of harvesting on resource availability, it currently remains difficult to design truly effective conservation and management plans.

In ecological terms, the extraction of NTPPs can be considered sustainable, if the harvest has no long-term deleterious effect on the reproduction and regeneration of plant populations, or on any other species within a given ecosystem (Hall & Bawa 1993). These effects can be determined only by comparing the diachronic behaviour of those populations which are harvested, with those which are not, as the effects of harvesting are superimposed upon natural *population dynamics.* Hence, comparative sampling of natural populations, and of populations which are subject to varying degrees of harvesting intensity must be undertaken in order to investigate the sustainability of different harvesting regimes. Details of methods involved in the assessment of the environmental impact of extractive activities can be found in Hall and Bawa (1993) and in Martin (1995).

DEVELOPING PRACTICAL SOLUTIONS: LEGAL MECHANISMS AND ETHICAL CODES

It is clear from this discussion that there are a number of significant problems associated with the practical application of TBK. Hence, while the notion of using ethnobotanical data to combine the conservation with economic development appears desirable in the long-term, these problems must be addressed if the strategy is to succeed. In many cases, solutions to these difficulties will depend on the development of suitable legislation which facilitates the equitable sharing of both the costs and benefits of conservation; which provides protection for traditional people's *intellectual, scientific and cultural property*; which ensures the long-term conservation of valuable plant species. In the meantime, however, it is vital that ethnobotanists are encouraged to behave in an ethical manner, which neither exploits nor endangers the local communities with whom they work. As the fields of environmental law and professional ethics are vast, only the main points relating specifically to ethnobotanical research are outlined here. However, interested students might wish to consult more detailed texts such as the World Resources Institute's *Biodiversity Prospecting* (Reid *et al.* 1993), the Crucible Group's *People, Plants and Patents* (1994) and Birnie and Boyle's *International Law and the Environment* (1992).

The Role of Environmental Law

Throughout this century, the problem of ownership over biological organisms has provided a significant focus for the activities of the legal profession. For example, in 1922 an international group of patent lawyers met in London to discuss the possibility of patenting plant varieties; by 1930, the USA had adopted the *Plant Patent Act* for fruits and ornamentals (Crucible Group 1994). While these patents apply specifically to domesticated plants developed by commercial breeders, ownership over wild resources has also formed a centre for debate. Indeed, the past perception of biological materials as *common global property*, with no restrictions concerning the collection and export of raw materials, has had serious consequences for many developing nations. For example, while current sales of the African violet (*Saintpaulia ionantha*) generate about US$30 million *per annum*, none of this income is invested in the conservation of Tanzania—the country from which it was originally derived (Lovett 1988 cited in Cunningham 1993b). Fortunately, since the 1970s, the world has seen the emergence of International Environmental Law as a key component in the areas of plant protection and sustainable development, and in recent years several important international agreements have been developed to address problems of both species conservation and the equitable distribution of the benefits associated with biodiversity conservation.

Legislating for Conservation

Until recently, world trade in biological materials was essentially unrestricted, leading both to problems of overharvesting (Cunningham 1990; Elisabetsky 1991) and to the spread of pest and disease organisms. This situation arose largely due to the fact that in historical terms, unimproved genetic materials were regarded as global property. However, as questions concerning the ownership of such materials were raised in the international arena during the mid-1970s, a number of significant legal developments have taken place. These developments include the signing of the *Convention on International Trade in Endangered Species of Wild Fauna and Flora* (CITES) in 1973 and of the *Convention on the Conservation of European Wildlife and Natural Habitats* in 1979 (Table 12.2), and have culminated recently in the UN *Convention on Biological Diversity* (1992). Significantly, within this last agreement, the rights of producing countries over their biological materials are made explicit, the conservation implications of which are outlined in Box 12.5.

Legislating for the Equitable Distribution of Revenue from Biodiversity Conservation and Traditional Botanical Knowledge

On a national basis, the issue of ensuring fair returns to producing countries has been addressed to some extent by the granting of *national sovereignty* over biological resources. For example, the Rio *Convention on Biological Diversity* states that 'producing nations will have the ability to negotiate agreements with those who seek access to genetic resources, to ensure that it receives a share of any resulting profits' (Box 12.6). However, as current legislation stands, there are a number of problems remaining (Barton 1994):

Table 12.2 Developing legislation towards the conservation of useful plants. The development of legislation to protect biological resources has been complex, involving a range of governmental and non-governmental organisations. With the negotiation of the *Convention on Biological Diversity* a state's authority over its genetic resources is formalised; however, these rights are prospective only, and nations have no right to seek compensation for resources which have already been exported from the country. In terms of species which are already endangered, CITES currently provides the only international instrument for listing species of plants and animals which are considered to be endangered to the extent that commercial trade must either be monitored and controlled, or prohibited entirely (see Birnie & Boyle 1992; Lewington 1993; Barton 1994)

Year	Official comments and legislation
1973	*Convention on International Trade in Endangered Species of Wild Fauna and Flora* (CITES); originally signed by 21 countries to control and monitor international trade in animals and plants considered to be threatened, or likely to become threatened due to commercial exploitation.
1975	CITES takes effect; by 1993 over 120 countries were party to the convention.
1976	In its stated aim of 'health for all by the year 2000' the World Health Organisation (WHO) promote the evaluation and conservation of medicinal plants.
1979	*Convention on the Conservation of European Wildlife and Natural Habitats*; took force in 1982.
1988	WHO state that 'the importance of conservation is recognised by WHO and its Member States'; conservation is consequently considered to be an essential feature of national programmes on traditional medicine.
1991	IUCN, WWF and WHO declare jointly that they regard medicinal plants as a major, but neglected group of plants for which conservation is a priority.
1992	At its 44th World Health Assembly, WHO recognise that 'many species of medicinal plants are threatened by ecological and environmental changes'.
1992	At the UN Conference on Environment and Development (Earth Summit) in Rio de Janeiro, more than 150 Member States sign the *Convention on Biological Diversity*, which recognises the sovereign rights of states over the collection and export of their nation's biological resources.

- There is no protection against competition; if a researcher suspects that a plant may be valuable, there is no means of preventing that researcher from obtaining the same plant from another nation whose terms of contract are less severe
- The rights of the source nation are prospective only; there is no right to obtain royalties for genetic materials which have already left the nation, and there is no right to claim royalties for folk knowledge at all

Box 12.5 Conservation implications of the UN Earth Summit. At the United Nations Earth Summit held in June 1992, the majority of Member States in attendance were signatories to two important pieces of legislation, both of which have considerable significance in terms of the conservation of biological and cultural diversity. *Agenda 21* is concerned with sustainable development and the safe disposal of hazardous wastes; the *Convention on Biological Diversity* is specifically concerned with ensuring the sustainable use of biological resources, and the equitable distribution of benefits from that use. Some of the main points of these agreements are outlined here (see Reid *et al.* 1993; UN 1993).

The notion of generating economic growth while promoting environmental protection formed the main subject of debate during the 1992 United Nations Conference on Environment and Development in Rio de Janeiro, Brazil, from which emerged the international agreement, Agenda 21. This agreement adopted by 153 of the countries who attended this first *Earth Summit*, outlines a programme of co-operative research aimed at combining strategies for the efficient use of resources and proper waste management alongside the conservation of biological and cultural diversity (UN 1993).

AGENDA 21

Agenda 21 constitutes a set of integrated strategies to halt and reverse the effects of environmental degradation and to promote the integration of economic development and conservation. As a result of this Agenda, a new UN Commission on Sustainable Development has been mandated to examine the progress and implementation of such strategies. The major aims and proposed action encapsulated in Agenda 21 are outlined below.

Aims

- Combating deforestation
- Managing fragile ecosystems
- Promoting sustainable agriculture
- Maintaining water quality
- Ensuring the rights of participation of indigenous peoples

Action

- To strengthen laws to promote sustainable agriculture and rural development
- To press for early entry into force of the *Convention on Biological Diversity*
- To promote effective economic development
- To promote economic diversification
- To promote integrated management and local participation
- To encourage realistic pricing of natural resource commodities
- To promote afforestation and the provision of a sustainable energy supply

continued

___ continued ___

THE CONVENTION ON BIOLOGICAL DIVERSITY

The *Convention on Biological Diversity* represents an integration of diverse concepts concerning the environment, trade and intellectual property law. Its main aims include the conservation of biodiversity, the sustainable development of genetic resources and the fair and equitable distribution of the resulting benefits (Gollin 1993). Among its 17 articles, the convention addresses a number of important problems facing developing nations.

- Recognises the need for reciprocity between access to genetic resources and transfer of relevant technologies
- Defines the basic terms of biodiversity prospecting
- Recognises the sovereign right to exploit domestic resources
- Promotes the importance of *in situ* conservation while recognising the continued need for *ex situ* strategies
- Promotes the use of biological resources 'in accordance with traditional cultural practices'
- Promotes the idea of providing incentives to conserve by protecting the intellectual property of producing nations
- Addresses the need for information and technology transfer between nations

Box 12.6 INBio—negotiating for equitable distribution of revenue from biodiversity prospecting. When the agreement between INBio and Merck was first announced in 1991, it was widely discussed in both the scientific and popular press. Attitudes to the agreement range from those who regard it as a 'start' in the process of ensuring an equitable global system for biodiversity conservation, to those who regard it as no more than a 'rip-off'. Those of the latter persuasion argue that as Merck's sales in 1991 were $8.6 billion, compared with Costa Rica's GNP of $5.2 billion; and as pharmaceutical companies spend an average of $231 million on researching each new drug, then, even accounting for the fact that most of the costs of research go on proving safety and efficacy, the payment to Costa Rica of only $113 per sample is exploitative (Crucible Group 1994).

One of the main substantive provisions from the UN *Convention on Biological Diversity* is the recognition of nations' obligations to ensure that both the countries supplying biodiversity, and those using it should receive an equitable share of the financial benefits which result. This important issue has resulted in the development of various contracts to facilitate access of bioprospectors to genetic and biochemical resources in exchange for a proportion of any later profits. However, even before the signing of the convention, several countries had set up organisations aimed at discovering sustainable uses for their biological diversity.

___ continued ___

___ *continued* ___

COSTA RICA'S INBio

INBio, is a private, non-profit organisation which was specifically established to facilitate the conservation and sustainable use of biodiversity. INBio's basic goals are 'to identify and locate biodiversity, to save representative samples of such diversity in protected wild lands, and to use it non-destructively for the public good' (Reid *et al.* 1993). As such, in addition to biodiversity prospecting itself, one of INBio's roles is to function as an intermediary between external bio-prospectors and Costa Rica, in order to ensure that the country receives appropriate returns from any prospecting activities. It is hoped, as a result of this initiative, that biodiversity prospecting will help to fund conservation, while demonstrating the economic value of biological diversity. This in turn should help to convince policy makers that biodiversity conservation should figure centrally in all development planning.

- INBio's biodiversity prospecting is focused on Costa Rica's conservation areas and the wild species in them, and is committed to ensuring the survival of these areas
- INBio aims to find non-destructive uses for these wild species
- It supports a taxonomic and inventory process aimed at thorough collection and documentation of replicable samples of each species
- Samples are to be screened for chemicals, micro-organisms and genes
- It is committed to developing contracts with national and international pharmaceutical, biotech and agribusiness industries, that insure direct compensation for costs of research, conservation and development of wild land biodiversity
- It is also committed to training and otherwise helping Costa Rica to take charge of all of these activities

THE INBio-MERCK CONTRACT

One of the most famous activities of Costa Rica's INBio, was the agreement it negotiated with the pharmaceutical giant Merck, in 1991. While many of the details of this agreement have not been made public, a number of significant factors are outlined below:

- INBio's contractual arrangements include 8 items:
    Direct payments in cash or barter (equipment, training, technology) to enable INBio to develop
    Payment of a significant percentage of INBio's initial budget plus royalties to cover costs of conservation
    A significant, fair royalty on net sales from the commercialisation of Costa Rica's biodiversity
    Help in moving drug research and development to the source country
    Minimal exclusivity (in relation to the involvement of industrial competitors)
    Agreement on sample ownership and patent ownership
    The use of chemical synthesis and domestication to protect wild sources
    Protective legal mechanisms

___ *continued* ___

_____ *continued* _____

- The Merck–INBio Bioprospecting Contract involved a 2-year agreement announced in 1991
- Merck paid US$1.135 million to INBio for rights to biodiversity exploration in Costa Rica
- INBio was to provide 10 000 biosamples from Costa Rica's nature parks which Merck was to screen for bioactivity
- If any profitable drug resulted from this material, Merck would have sole rights to market it, and an (undisclosed) percentage of the royalties would be shared with INBio

---

- There is no protection with respect to derived knowledge; that is to say that an ethnobotanical investigation could lead to the publication of a scientific paper, which in turn leads to the development of new pharmaceuticals by companies outside the scope of the research agreement
- At present, the rights under the *Convention on Biological Diversity* belong to the producing nation, and there is little legal pressure to share any compensation with its indigenous peoples

This last point is particularly significant for two reasons. First, as deforestation for timber and agriculture frequently results from the lack of any alternative source of income, unless local peoples receive financial benefits from conservation, they will have no economic incentive to conserve. Secondly, the historical role of indigenous and other traditional peoples in generating and protecting biodiversity over the past millennia, surely suggests that they should receive an equitable share of its economic benefits. This problem of ensuring returns to local people is closely linked with the broader issue of the world's treatment of its indigenous peoples, and both have received considerable attention during the second half of this century (Table 12.3), beginning with the International Labour Organisation's negotiation of the *Convention Concerning the Protection and Integration of Indigenous and Other Tribal and Semi-tribal Populations in Independent Countries* (1957) and culminating so far in the establishment of the Working Group on Traditional Intellectual, Cultural and Scientific Resources, based in Oxford, UK (Posey 1994). This working group is particularly concerned with exploring the options within existing laws on *intellectual property rights* (IPRs) as a means of ensuring local people's rights over the exploitation of their biological resources and their traditional knowledge (Box 12.7).

Professional Codes of Practice for Ethnobotanical Researchers

It is clear from the discussion above, that the legal mechanisms currently in place to protect the rights of indigenous people over their

Table 12.3   Developing legislation towards the protection of indigenous people's scientific and cultural resources. Over the last 50 years, considerable attention has focused on the rights of indigenous peoples, most recently in relation to rights of ownership over both natural resources and traditional knowledge. At present the mechanisms available for their protection and the legislation in force are inadequate. Hence, many ethnobiologists, including Darell Posey—who currently holds a fellowship at Oxford University—are working to improve methods for the protection, compensation and empowerment of traditional communities

Date	Events
1957	*Convention Concerning the Protection and Integration of Indigenous and Other Tribal and Semi-tribal Populations in Independent Countries* (International Labour Organisation).
1981	A UN Working Group on Indigenous Populations was created following an initiative by the Sub-commission on Prevention of Discrimination and Protection of Minorities. This Working Group has been working to produce a *Draft Universal Declaration on the Rights of Indigenous Peoples.*
1987	IUCN Working Group on Traditional Ecological Knowledge—working to establish recognition for the value of indigenous knowledge and the need for genuinely equitable partnerships.
1989	Since the 1957 convention became regarded as too assimilationist, the convention was updated to the *Convention Concerning Indigenous and Tribal Peoples in Independent Countries*; this granted indigenous people very limited rights over natural resources, although no rights to royalties on ethnobotanical materials.
1989	The International Society of Ethnobiology called for mechanisms to compensate indigenous peoples for use of their specialist knowledge.
1989	UNESCO pass a resolution on the *Safeguarding of Traditional Culture and Folklore.*
1992	Agenda 21 advocates the active involvement of indigenous peoples in projects concerned with conservation and sustainable development.
1993	An updated version of the *Draft Declaration on the Rights of Indigenous Peoples* goes much further than any other document in recognising indigenous rights over genetic resources; however, this does not yet form a part of international law
1994	Working Group on Traditional Intellectual, Cultural and Scientific Resources, set up in Oxford, UK to discuss mechanisms for the protection and recognition of indigenous knowledge and natural resources.

See Birnie & Boyle 1992; Cunningham 1993b; Posey 1994b.

cultural and scientific resources are inadequate. Hence, despite increased national control over the export and exploitation of their plant genetic resources, there is commonly little evidence of effective 'trickle down' of financial returns from national governments to local communities. At present then, it is largely the responsibility of ethnobotanical researchers—whether they are based in academic or commercial organisations—to ensure a fair deal for the communities among whom they work. In recent years, several NGOs and academic societies have been looking both at the ethical issues involved in ethnobotanical research generally, and, more specifically, at those involved in the commercial applications of ethnobotanical data.

Box 12.7   Towards genuinely equitable distribution of returns from biodiversity. At present there are few legally binding agreements which can assist in the protection of indigenous people's natural resources and intellectual property, and arguments regarding the best methods for achieving this are extremely complex. However, many ethnobotanists agree that a uniform agreement which addresses the relative rights of both indigenous peoples and their governments should be developed by NGOs, and several have become involved in groups which are attempting to develop such agreements (see Barton 1994; Posey 1994).

As we have seen, the recognition of national sovereignty has given developing nations the ability to develop agreements which grant prospecting rights to external organisations in exchange for payment—either in cash or in terms of training and technology transfer. However, it is rare that a nation's indigenous peoples— the historical custodians of biodiversity, and the keepers of much of the knowledge about its use—themselves benefit from such agreements. There are three fundamental issues to consider here—the question of ownership of unimproved natural resources (which is closely linked with recognising traditional land rights); the ownership of natural resources which have been improved by traditional communities; and the ownership of traditional knowledge relating to the potential uses of natural resources. The latter of these two fundamental issues are considered briefly below.

FARMER BREEDERS' RIGHTS OVER IMPROVED GERMPLASM

While this century has seen the granting of patent rights to commercial plant breeders for the development of genetically improved organisms, the patent rights of traditional farmer breeders have been much less well established. For example, while it was not until 1987 that farmers' rights were recognised at all, in recent years the rights of farmers have been challenged again (White 1994):

Commercial breeders' rights	Farmer breeders' rights
1922—a meeting of patent lawyers in London consider the possibility of protection for plant varieties. 1930—US adopts the *Plant Patent Act* for fruits and ornamentals.	1987—Commission on Plant Genetic Resources accepted legitimacy of IPR protection for formal breeders in exchange for recognition of farmer breeders' rights.

_____ *continued* __|

_____ *continued* _____

Commercial breeders' rights	Farmer breeders' rights
1961—Union for the protection of New Varieties of Plants (UPOV) is established; this is later modified and strengthened in 1972 and again in 1978.	1987—Commission on Plant Genetic Resources established the *Fund for Plant Genetic Resources* to fulfil obligations inherent in the concept of farmers' rights.
1991—*UPOV Convention* strengthened to prevent farmers from replanting patent-protected varieties.	1991—the FAO Commission on Plant Genetic Resources draft the *International Code of Conduct for Plant Germplasm Collecting and Transfer.* This is a voluntary code of
1992—'species' patent granted on genetically modified cotton.	conduct which recognises farmer breeders' rights over their
1994—'species' patent granted on soya bean.	germplasm, and sets guidelines for germplasm exchange.
	1994—under a bill currently before the US Senate Subcommittee on Agricultural Research, indigenous and small farmers stand to lose their rights over traditional seed varieties; large US seed companies have lobbied heavily for the bill.

TRADITIONAL BOTANICAL KNOWLEDGE AS INTELLECTUAL PROPERTY

The concept of applying intellectual property law to the protection of traditional knowledge has been a subject of debate for some years—particularly as the IUCN identified the need to develop equitable partnerships for the commercialisation of plant products based on traditional botanical knowledge. However, a number of important problems associated with contemporary intellectual property law have led to the exploration of alternative avenues in the protection of traditional people's intellectual property (see Barton 1994).

- Contemporary intellectual property law permits only the patenting of an identified active principle from a plant, not the plant or the folk information pertaining to it
- The most significant rights of indigenous peoples are those deriving from physical control over plants and TBK; this can provide the basis for *trade secret protection*; such agreements are currently enforceable in developed nations
- Intellectual property provisions of the Uruguay Round of the General Agreement on Tariffs and Trade (1993) will require all participating nations to enforce trade secret agreements in the future
- Despite recent efforts to strengthen indigenous people's rights over genetic resources and folk knowledge, the most far-reaching of these are not yet a part of international law

Ethical Aspects of Ethnobotanical Research

There are several ethical issues associated with ethnobotanical research *per se*, including issues of acculturation and exploitation which are common to all anthropological studies. In addition there are perhaps two major concerns associated specifically with more applied research: first, that researchers should be totally honest about the aims of their research, and must involve local people in project design and execution; secondly, that researchers should negotiate equitable partnerships with local people to ensure the fair distribution of any financial returns or academic recognition which result from the work. This second point is particularly pertinent, for while most academics are under considerable pressure to publish their research in order to achieve both peer approval, and financial support for further research, the premature publication of such findings can seriously jeopardise traditional people's rights over their resources and knowledge. Indeed, Barton (1994) cites a case where a patent to a traditional poison was invalidated on the basis of previous academic publications describing indigenous uses of the plant. This situation arises due to the fact that published information is already considered to be in the *public domain*, and is therefore excluded from any subsequent claims regarding novelty or intellectual property.

The Development of Professional Codes of Practice

The opposing considerations relating to the publication of data clearly put academic researchers in a difficult position. Indeed, one ethnobotanist of my acquaintance submitted a paper to the *Journal of Ethnopharmacology* in which he purposely omitted species' names of the medicinal plants he described. The paper was rejected on the grounds that his data were incomplete, thus highlighting the difficulties of fulfilling both professional and ethical demands. It seems likely that the only reasonable solution to this problem therefore, is to ensure from the outset, that legally binding agreements are drawn up between academics, commercial organisations and traditional peoples. To facilitate the smooth operation of this difficult, yet crucial aspect of ethnobotanical research, a number of professional organisations have begun to develop codes of profession conduct, the main points of which are outlined in Table 12.4.

Table 12.4 Summary of issues relating to new natural products development, covered in the resolutions or statements of various organisations. While formal international legislation regarding the protection of indigenous people's intellectual and cultural resources may remain undeveloped, many professional bodies have drawn up codes of conduct to which their members are expected to adhere. Many of these address the ethical need to establish equitable partnerships between researchers and local communities—in both an intellectual and a financial sense

Organisation	Summary of main issues covered					
	Equitable partnerships	Training/ technotransfer	Health	Sustainable resource use	Survey species and TBK	National sovereignty
International Society of Ethnobiology (1988)	✓	✓	✓	✓	✓	✓
International Society of Chemical Ecology (1991)	✓			✓	✓	
Society for Economic Botany (1991)	✓	✓			✓	
NIH/NCI workshop of Drug Development, Biological Diversity & Economic Growth (1991)	✓	✓	✓	✓		
Global Biodiversity Strategy (WRI/IUCN/UNEP) (1992)	✓	✓	✓	✓	✓	
ASOMPS (Asian Symposium for Medicinal Plants, Spices and other Natural Products) (1992)	✓	✓	✓	✓	✓	✓
American Society of Pharmacognosy (1992)	✓	✓	✓	✓	✓	✓

Data taken from Cunningham (1993b).

SUMMARY

While the potential benefits of combining conservation with sustainable economic development are considerable, and while the role of ethnobotanical information in such ventures is clear, a number of significant economic and ecological problems currently limit the practical success of such projects. First, despite recent legislation granting national sovereignty over unimproved genetic materials, local communities often see little of the financial returns from conservation, such that immediate economic incentives often favour more destructive activities. On the other hand, where plants do become a commercial success, this can lead to problems of overharvesting and the eventual loss of local resources. Only by addressing these, and other related issues, can the many direct and indirect benefits of habitat conservation be enjoyed both globally and at a local level. The solutions to these problems lie largely in the hands of environmental economists—who are able to draw up a realistic picture of the long-term costs and benefits of a given development project—and of environmental lawyers, who are involved in the development of legislation aimed at protecting the rights and resources of all parties concerned. However, one of the fundamental problems which remains, is that of protecting traditional people themselves—their rights of access to essential resources, and their rights to benefit from the knowledge and genetic diversity which they have developed throughout millennia. At present, legislation to protect indigenous communities is far from adequate, and it is up to ethnobotanists to ensure that this situation improves.

Bibliography

Abbiw DK (1990) *Useful Plants of Ghana*. Intermediate Technology Publications Ltd/RBG Kew, London.

Acevedo-Rodríguez P (1990) The occurrence of piscicides and stupefactants in the plant kingdom. In: Prance GT, Balick MJ (eds) *New Directions in the Study of Plants and People (Advances in Economic Botany 8)*. New York Botanical Garden, New York, pp. 1–23.

Adesida GA, Adesogan EK, Okorie DA, Taylor DAH, Styles BT (1971) The limonoid chemistry of the genus *Khaya* (Meliaceae). *Phytochemistry* **10**: 1845–1853.

Alcorn JB (1981) Haustic noncrop resource management: implications for prehistoric rainforest management. *Human Ecology* **9**: 395–417.

Alcorn JB (1989) Process as resource: the traditional agricultural ideology of Bora and Huastec resource management and its implications for research. In: Posey DA, Balée W (eds) *Resource Management in Amazonia: Indigenous and Folk Strategies (Advances in Economic Botany 7)*. New York Botanical Garden, New York, pp. 63–77.

Alkire BH, Tucher AO, Maciarello MJ (1994) Tipo, *Minthostachys mollis* (Lamiaceae) an Ecuadorian mint. *Economic Botany* **48**: 60–64.

Allaby M (1992) *The Concise Oxford Dictionary of Botany*. Oxford University Press, Oxford.

Allan W (1965) *The African Husbandman*. Oliver & Boyd, Edinburgh.

Altman JC (1987) *Hunter-gatherers Today: An Aboriginal Economy in North Australia*. Australian Institute of Aboriginal Studies, Canberra.

Anderson *et al.* (1946) in O de M 1990.

Anderson AB (1991) Forest management strategies by rural inhabitants in the Amazon estuary. In: Gómez-Pompa A, Whitmore TC, Hadley M (eds) *Rainforest Regeneration and Management*. UNESCO, Paris, pp. 351–360.

Anderson AB, Posey DA (1989) Management of a tropical scrub savanna by the Gorotire Kayapó Indians. In: Posey DA, Balée W (eds) *Resource Management in Amazonia: Indigenous and Folk Strategies (Advances in Economic Botany 7)*. New York Botanical Garden, New York, pp. 174–188.

Anderson R (1992) The efficacy of ethnomedicine: research methods in trouble. In: Nichter M (ed) *Anthropological Approaches to the Study of Ethnomedicine*. Gordon & Breach, Reading, pp. 1–18.

Anon (1988) Just what the witch doctor ordered. *The Economist* **307(7544)**: 75–76.

Anon (1994) Nontire exceeds tire demand. *Recycling Today* **May**: 20.

Appasamy P (1993) Role of non-timber forest products in a subsistence economy: the case of a joint forestry project in India. *Economic Botany* **47**: 258–267.

Arenas P (1987) Medicine and magic among the Maka Indians of the Paraguayan Chaco. *Journal of Ethnopharmacology* **21**: 279–296.

Arnold T, Wells JH, Wehmeyer AS (1985) Khoisan food plants: taxa with potential for future economic exploitation. In: Wickens FE, Goodin JR, Field DV (eds) *Plants for Arid Lands*. Unwin Hyman, London, pp. 69–86.

Avé W (1988) Small-scale utilisation of rattan by a Semai community in West Malaysia. *Economic Botany* **42**: 105–119.

Baines G, IIviding E (1992) Traditional environmental knowledge from the Marovo area of the Solomon Islands. In: Johnson M (ed) *Lore: Capturing Traditional Environmental Knowledge*. Dene Cultural Institute, Fort Hay, Canada, pp. 91–110.

Baker L, Woenne-Green S, Mutitjulu Community (1992) The role of Aboriginal ecological knowledge in ecosystem management. In: Birckhead J, de Lacy T, Smith L (eds) *Aboriginal Involvement in Parks and Protected Areas*. Australian Institute of Aboriginal and Torres Strait Islander Studies, Canberra, pp. 65–74.

Balée W, Daly DC (1990) Resin classification by the Ka'apor Indians. In: Prance GT, Balick MJ (eds) *New Directions in the Study of Plants and People (Advances in Economic Botany 8)*. New York Botanical Garden, New York, pp. 24–34.

Balée W, Gély A (1989) Managed forest succession in Amazonia: the Ka'apor case. In: Posey DA, Balée W (eds) *Resource Management in Amazonia: Indigenous and Folk Strategies (Advances in Economic Botany 7)*. New York Botanical Garden, New York, pp. 129–158.

Balick MJ (1984) Ethnobotany of palms in the neotropics. In: Prance GT, Kallunki JA (eds) *Ethnobotany in the Neotropics (Advances in Economic Botany 1)*. New York Botanical Garden, New York, pp. 24–33.

Balick MJ (1990) Ethnobotany and the identification of therapeutic agents from the rainforest. In: Chadwick DJ, Marsh J (eds) *Bioactive Compounds from Plants* (Ciba Foundation Symposium No. 154). Wiley, Chichester, pp. 22–32.

Balick MJ (1994) Ethnobotany, drug development and biodiversity conservation—exploring the linkages. In: Chadwick DJ, Marsh J (eds) *Ethnobotany and the Search for New Drugs* (Ciba Foundation Symposium 185). Wiley, Chichester, pp. 4–18.

Balick MJ, Mendelsohn R (1992) Assessing the economic value of traditional medicines from tropical rain forests. *Conservation Biology* **6**: 128–130.

Banack SA (1991) Plants and Polynesian voyaging. In: Cox PA, Banack SA (eds) *Islands, Plants and Polynesians: An Introduction to Polynesian Ethnobotany*. Dioscorides Press, Portland, Oregon, pp. 25–40.

Bancroft J (1878) *Further Remarks on the Pituri Group of Plants*. Government Press, Brisbane.

Barbera G, Carimi F, Inglese P (1992) Past and present role of the Indian-fig prickly pear (*Opuntia ficus indica*. L Miller) Cactaceae, in the agriculture of Sicily *Economic Botany* **46**: 10–20.

Barker R, Cross N (1992) Documenting oral history in the African Sahel. In: Johnson M (ed) *Lore: Capturing Traditional Environmental Knowledge*. Dene Cultural Institute, Fort Hay, Canada, pp. 113–135.

Barrows D (1931) Prehistoric Pueblo foods. *Museum of Arizona Notes* **4**: 1–4.

Bartlett HH (1929) *Colour Nomenclature in Batak and Malay* (Papers of the Michigan Academy of Science Arts and Letters) **10**: 1–52.

Barton JH (1994) Ethnobotany and intellectual property rights. In: Chadwick DJ, Marsh J (eds) *Ethnobotany and the Search for New Drugs* (Ciba Foundation Symposium 185). Wiley, Chichester, pp. 214–221.

Beattie J (1966) *Other Cultures: Aims, Methods and Achievements in Social Anthropology.* Routledge, London.

Beck W (1992) Aboriginal preparation of *Cycas* seeds in Australia. *Economic Botany* 46: 133–147.

Bedini F, Masera D (1994) Local farmers innovate in irrigation: development of new low-cost sprinklers in Kenya. *Indigenous Knowledge & Development Monitor* 2: 19–21.

Beecher CWW, Gyllenhaal C (1993) MEDFLOR an ethnobiological database. *Journal of Ethnopharmacology* 39: 223–229.

Begley S (1988) Zombies and other mysteries; ethnobotanists seek magical, medicinal plants. *Newsweek* 111(8): 79.

Begossi A, Leitão-Filho HF, Richerson PJ (1993) Plant uses in a Brazilian coastal community (Búzios Island). *Journal of Ethnobiology* 13: 233–256.

Bell EA (1980) Plant phenolics. In: Bell EA, Charlwood BV (eds) *Secondary Plant Products (Encyclopedia of Plant Physiology, New Series* Vol. 8). Springer-Verlag, New York, pp. 11–22.

Bell WH, Castetter EF (1937) *The Utilisation of Mesquite and Screwbean by the Aborigines in the American Southwest.* University of New Mexico Bulletin No. 314, University of New Mexico Press, Albuquerque.

Bellamy R (1993) *Ethnobiology in Tropical Forests: Expedition Field Techniques.* Royal Geographic Society, London.

Bellon MR, Brush SB (1994) Keepers of Maize in Chiapas, Mexico. *Economic Botany* 48: 196–209.

Bellon MR, Taylor JE (1993) Farmer soil taxonomy and technology adoption. *Economic Development and Cultural Change* 41: 764–786.

Bennett BC (1992) Plants and people of the Amazonian rainforests: the role of ethnobotany in sustainable development. *BioScience* 42: 599–607.

Bennett BC, Alarcón R (1994) *Osteophloeum platyspermum* and *Virola duckei* (Myristicaceae): newly reported as hallucinogens from Amazonian Ecuador. *Economic Botany* 48: 152–158.

Bennett BC, Alarcón R, Cerón C (1992) The ethnobotany of *Carludovica palmata* Ruiz & Pavón (Cyclantaceae) in Amazonian Ecuador. *Economic Botany* 46: 233–240.

Berlin B (1970) *A Preliminary Ethnobotanical Survey of the Aguaruna Region of the Upper Marañón River Valley, Amazonas, Peru.* Report to the Wenner-Green Foundation for anthropological research, Washington DC.

Berlin B (1984) Contributions of Native American collectors to the ethnobotany of the neotropics. In: Prance GT, Kallunki JA (eds) *Ethnobotany in the Neotropics (Advances in Economic Botany* 1). New York Botanical Garden, New York, pp. 34–40.

Berlin B (1992) *Ethnobiological Classification: Principles of Categorization of Plants and Animals in Traditional Societies.* Princeton University Press, Princeton.

Berlin B, Berlin EA (1994) Anthropological issues in medical ethnobotany. In: Chadwick DJ, Marsh J (eds) *Ethnobotany and the Search for New Drugs* (Ciba Foundation Symposium 185). Wiley, Chichester, pp. 240–259).

Berlin B, Breedlove DE, Raven PH (1973) General principles of classification and nomenclature in folk biology. *American Anthropology* **75**: 214–242.

Berlin B, Breedlove DE, Raven PH (1974) *Principles of Tzeltol Plant Classification*. Academic Press, New York.

Best J, Baxter P (eds) (1994) *The Rural Extension Bulletin*, No. 4. Agricultural Extension and Rural Development Department, University of Reading.

Biggam CP (1993) Haewenhnydele: an Anglo-Saxon medicinal plant. Poster presentation, *Botanical Society of Scotland 'Plants and People' symposium*, 24–27 September 1993.

Birckhead J, Smith L (1992) Introduction: conservation and country—a reassessment. In: Birckhead J, de Lacy T, Smith L (eds) *Aboriginal Involvement in Parks and Protected Areas*. Australian Institute of Aboriginal and Torres Strait Islander Studies, Canberra, pp. 1–6.

Birnie PW, Boyle AE (1992) *International Law and the Environment*, Clarendon Press, Oxford.

Birnie M, de Koning J, Cooper V (1992) Joint management of the Kakadu National Park. In: Birckhead J, de Lacy T, Smith L (eds) *Aboriginal Involvement in Parks and Protected Areas*. Australian Institute of Aboriginal and Torres Strait Islander Studies, Canberra, pp. 263–264.

Bohrer (1986) Guideposts in Ethnobotany. *Journal of Ethnobiology* **6**: 27–43.

Bold HC, Alexopoulos CJ, Delevoryas T (1987) *Morphology of Plants and Fungi* (5th edition). Harper & Row Publishers, New York.

Boom BM (1989) Use of plant resources by the Chácobo. In: Posey DA, Balée W (eds) *Resource Management in Amazonia: Indigenous and Folk Strategies* (*Advances in Economic Botany* 7). New York Botanical Garden, New York, pp. 78–96.

Borgtoft Pedersen H (1994) Mocora palm fibres: use and management of *Astocaryum standleyanum* (Arecaceae) in Ecuador. *Economic Botany* **48**: 310–325.

Borlaug N (1992) Small-scale agriculture in Africa: the myths and realities. *Feeding the Future* (Newsletter of the Sasakawa Africa Association) **4**: 2.

Boster JS (1984) Classification, cultivation and selection of Aguaruna cultivars of *Manihot esculenta* (Euphorbiaceae). In: Prance GT, Kallunki JA (eds) Ethnobotany in the Neotropics. (*Advances in Economic Botany* 1) New York Botanical Garden, New York, pp. 34–47.

Boster JS (1985) Requiem for the omniscient informant: there's life in the old girl yet. In: Dougherty J (ed) *Directions in Cognitive Anthropology*. University of Illinois Press, Champaign, Illinois, pp. 177–198.

Box L (1989) Virgilio's theorem: a method for adaptive agricultural research. In: Chambers R, Pacey A, Thrupp LA (eds) *Farmer First: Farmer Innovation and Agricultural Research*. Intermediate Technology Publications Ltd., London, 61–68.

Bradfield M (1971) The changing pattern of Hopi agriculture. *Royal Anthropological Institute Occasional Paper*, No. 30. Royal Anthropological Institute of Great Britain and Ireland, London.

Brady NC (1974) *The Nature and Properties of Soils,* 8th edn. Macmillan Publishing, New York.

Brand JC, Cherikoff V (1985) The nutritional composition of Australian Aboriginal food plants of the desert regions. In: Wickens GE, Goodin JR, Field DV (eds) *Plants for Arid Lands*. Unwin Hyman, London, pp. 53–68.

Bridson D, Forman L (1992) *The Herbarium Handbook* (revised edition). Royal

Botanic Gardens, Kew.

Brown CH (1985) Mode of subsistence and folk biological taxonomy. *Current Anthropology* **26**: 43–53.

Brown WL, Anderson EG, Tuchawena JR (1952) Observations on three varieties of Hopi maize. *American Journal of Botany* **39**: 597–609.

Brush SB (1991) A farmer-based approach to conserving crop germplasm. *Economic Botany* **45**: 153–165.

Bryant VM (1989) Pollen: nature's fingerprints of plants. In: *1990 Yearbook of Science and the Future*. Encyclopedia Britannica Inc., Chicago, pp. 92–111.

Bryant VM (1993) Phytolith research: a look toward the future. In: Pearsall DM, Piperno DR (eds) *Current Research in Phytolith Analysis: Applications in Archaeology and Paleoecology* (MASCA Research Papers in Science and Archaeology, **10**). University Museum of Archaeology and Anthropology, University of Pennsylvania, Philadelphia, pp. 175–199.

Bryant VM (1994) Callen's Legacy. In: Sobolik KD (ed.) *Palaeonutrition: The Diet and Health of Prehistoric Americans* (Center for Archaeological Investigation Occasional Paper, No 22). Board of Trustees, South Illinois University, pp. 151–160.

Bryant VM, Holloway RG (1983) The role of palynology in archaeology. In: Schiffer MS (ed.) *Advances in Archaeological Method and Theory*, Vol. 6. Academic Press, New York, pp. 191–223.

Bryce JH, Hill SA (1993) Energy production in plant cells. In: Lea PJ, Leegood RC (eds) *Plant Biochemistry and Molecular Biology*. Wiley, Chichester, pp. 1–26.

Bryson W (1994) *Made in America*. QPD, London.

Bunch R (1989) Encouraging farmers' experiments. In: Chambers R, Pacey A, Thrupp LA (eds) *Farmer First: Farmer Innovation and Agricultural Research*. Intermediate Technology Publications, London, pp. 55–60.

Burch ES, Ellanna LJ (1994) Introduction. In: Burch ES, Ellanna LJ (eds) *Key Issues in Hunter-Gatherer Research*. Berg Publishers Inc., Oxford, pp. 1–8.

Burger J (1990) *The Gaia Atlas of First Peoples: A Future for the Indigenous World*. Robertson McCarta Ltd, London.

Butzer KW (1982) *Archaeology as Human Ecology*. Cambridge University Press, Cambridge.

Campbell DG, Daly DC, Prance GT, Maciel UN (1986) Quantitative ecological inventory on *terra firme* and *várzea* tropical forest on the Rio Xingu, Brazilian Amazon. *Britonnia* **38**: 369–393.

Cane S (1989) Australian Aboriginal seed grindling and its archaeological record: a case study from the Western Desert. In: Harris DR, Hillman GC (eds) *Foraging and Farming: The Evolution of Plant Exploitation*. One World Archaeology 13, Unwin Hyman, London, pp. 99–119.

Carroll MP (1992) Allomotifs and the psychoanalytic study of folk narratives. *Folklore* **103**: 225–234.

Castelló i Vidal JI, Lillo AL (1993) Co-ordination mechanisms in the Spanish biosphere reserves. *Nature and Resources* **29**: 12–16.

Castetter EF (1944) The domain of ethnobiology. *American Naturalist* **78**: 158–170.

Castetter EF, Bell WH (1951) *Yuman Indian Agriculture*. University of New Mexico Press, Albuquerque.

Chambers R, Pacey A, Thrupp LA (eds) (1989) *Farmer First: Farmer*

Innovation   and   Agricultural   Research.   Intermediate   Technology
Publications, London.

Chase AK (1989) Domestication and domiculture in northern Australia: a social
perspective. In: Harris DR, Hillman GC (eds) *Foraging and Farming: The
Evolution of Plant Exploitation (One World Archaeology* 13). Unwin Hyman,
London, pp. 42–54.

Cherry M (1985) The needs of the people. In: Wickens GE, Goodin JR, Field DV
(eds) *Plants for Arid Lands.* Unwin Hyman, London, pp. 1–8.

Chhabra SC, Mahunnah RLA (1994) Plants used in traditional medicine by
Hayas of the Kagera region, Tanzania. *Economic Botany* 48: 121–129.

Chickwendu VE, Okezie CEA (1989) Factors responsible for the ennoblement of
African yams: inferences from experiments in yam domestication. In: Harris
DR, Hillman GC (eds) *Foraging and Farming: The Evolution of Plant
Exploitation (One World Archaeology* 13). Unwin Hyman, London, pp.
344–357.

CIRAN (1993) *Indigenous Knowledge and Development Monitor* 1(1)

Clark C (1992) Empirical evidence for the effect of tropical deforestation on climatic change. *Environmental Conservation* 19: 39–47.

Clark CMH (1962) *History of Australia* (Vol I). Melbourne University Press,
Carlton, Victoria.

Clay J (1992) Some general principles and strategies for developing markets in
North America and Europe for non-timber forest products: lessons from
Cultural Survival Enterprises 1989–1990. In: Nepstad DC, Schwartzman S
(eds) *Non-timber Products from Tropical Forests: Evaluation of a Conservation
and Development Strategy (Advances in Economic Botany* 9). New York
Botanical Garden, New York, pp. 101–106.

Cohen JH, Tokheim CH (1994) *Shaman Pharmaceuticals: Ethnobotany,
Biodiversity and Drug Discovery.* Smith Barney Special Situations Research.
28 June.

Concar D (1994) How to burn a wilderness. *New Scientist* 144 (1949): 48–52.

Conklin HC (1954a) An ethnoecological approach to shifting agriculture.
Transactions of the New York Academy of Science 17: 133–142.

Conklin HC (1954b) *The Relation of Hanunóo Culture to the Plant World* (see
Conklin 1974).

Conklin HC (1957) *Hanunóo Agriculture, a Report on an Integral System of
Shifting Cultivation in the Philippines.* Food and Agriculture Organisation of
the United Nations, Rome.

Conklin HC (1974) *The Relation of Hanunóo Culture to the Plant World* (Yale
University PhD, 1955). University Microfilms Ltd, High Wycombe.

Conway GR (1989) Diagrams for farmers. In: Chambers R, Pacey A, Thrupp LA
(eds) *Farmer   First: Farmer   Innovation   and   Agricultural   Research.*
Intermediate Technology Publications, London, pp. 77–86.

Cordy-Collins A (1982) Psychoactive painted Peruvian plants: the Shamanism
Textile. *Journal of Ethnobiology* 2: 144–153.

Cortella AR, Pochettino ML (1994) Starch grain analysis as a microscopic diagnostic feature in the identification of plant material. *Economic Botany* 48:
171–181.

Costantini ES (1975) El uso de alucinógenos de orígen vegetal por las tribus
indígenas del Paraguay actual. *Cuadernos Científicos CEMEF* 4: 35–48.

Cotterell A (ed) *The Penguin Encyclopedia of Ancient Civilisations.* Penguin

Books, London.

Cotton C, Hodgson D (1994) Development of a predictive database for ecological management. Poster presentation, *35th Annual Meeting of the Society for Economic Botany,* Mexico, 20–26 June 1994.

Cotton CM, Evans LV and Gramshaw JW (1991a) The accumulation of volatile oils in whole plants and cell cultures of tarragon (*Artemisia dracunculus*). *Journal of Experimental Botany* **42**: 365–375.

Cotton CM, Gramshaw JW, Evans LV (1991b) The effect of α-naphthalene acetic acid (NAA) and benzylaminopurine (BAP) on the accumulation of volatile oil components in cell cultures of tarragon (*Artemisia dracunculus*). *Journal of Experimental Botany* **42**: 377–386.

Coughenour MB, Ellis JE, Swift DM, Coppock KL, Galvin K, McCabe JT, Hart TC (1985a) Energy extraction and use in a nomadic pastoral ecosystem. *Science* **230**: 619–625.

Coughenour M, McNaughton S, Wallace L (1985b) Responses of an African graminoid (*Themeda triandra* Forsk) to frequent defoliation, nitrogen and water: a limit of adaption to herbivory. *Oecologia* **68**: 105–110.

Cowan J (1992) *Mysteries of the Dream-time: The Spiritual Life of Australian Aborigines* (revised edition). Prism Press, Bridport, Dorset.

Cox PA (1990) Ethnopharmacology and the search for new drugs. In: Chadwick DJ, Marsh J (eds) *Bioactive Compounds from Plants* (Ciba Foundation Symposium No. 154). Wiley, Chichester, pp. 40–55.

Cox PA (1994) The ethnobotanical approach to drug discovery: strengths and limitations. In: Chadwick DJ, Marsh J (eds) *Ethnobotany and the Search for New Drugs* (Ciba Foundation Symposium 185). Wiley, Chichester, pp. 25–36.

Cox PA, Balick MJ (1994) The ethnobotanical approach to drug discovery. *Scientific American* **270(6)**: 60–65.

Crane E (1985) Bees and honey in the exploitation of arid land resources. In: Wickens GE, Goodin JR, Field DV (eds) *Plants for Arid Lands.* Unwin Hyman, London, pp 163–176.

Craveiro AA, Machado MIL, Alencar JW, Matos FJA (1994) Natural product chemistry in northeastern Brazil. In: Chadwick DJ, Marsh J (eds) *Ethnobotany and the Search for New Drugs* (Ciba Foundation Symposium 185). Wiley, Chichester, pp. 95–102.

Crawford GW (1992). Prehistoric plant domestication in East Asia. In: Cowan CW, Watson PJ (eds) *The Origins of Agriculture: An International Perspective.* Smithsonian Institution Press, London, pp. 7–38.

Croom EM (1983) Documenting and evaluating herbal remedies. *Economic Botany* **37**: 13–27.

Crucible Group (1994) *People, Plants and Patents.* International Development Research Centre, Ottawa.

Cunliffe B (ed) (1994) *The Oxford Illustrated Prehistory of Europe.* BCA, London.

Cunningham AB (1990) *African Medicinal Plants: Setting Priorities at the Interface between Conservation and Primary Healthcare.* (People and Plants Working Paper No 1) UNESCO, Paris.

Cunningham AB (1993) *Ethics, Ethnobiological Research and Biodiversity.* WWF. Gland.

Cunningham AB, Mbenkum FT (1993) *Medicinal Bark in International trade: A Case Study of the Afromontane Tree* Prunus africana. Report to WWF

International.

Darvill T (1987) *Prehistoric Britain*. Batsford, London.

Davis TA, Johnson DV (1987) Current utilisation and further development of the palmyra palm (*Borassus flabellifer* L. Arecaceae) in Tamil Nadu State, India *Economic Botany* **41**: 247–266.

Deans SG, Waterman PG (1993) Biological activity of volatile oils. In: Hay RKM, Waterman PG (eds) *Volatile Oil Crops: Their Biology, Biochemistry and Production*. Longman Scientific and Technical, Harlow, UK, pp. 97–112.

de Boef W, Amanor K, Wellard K (1993) *Cultivating Knowledge: Genetic Diversity, Farmer Experimentation and Crop Research*. Intermediate Technology Publications, London.

Decker DS, Wilson HD (1986) Numerical analysis of seed morphology in *Cucurbita pepo. Systematic Botany* **11**: 595–607.

de Lacy T (1992) The evolution of a truly Australian national park. In: Birckhead J, de Lacy T, Smith L (eds) *Aboriginal Involvement in Parks and Protected Areas*. Australian Institute of Aboriginal and Torres Strait Islander Studies, Canberra, pp. 383–390.

Delfeld MA (1994) *Datura* species in the Aztec herbal. Presentation, *35th Annual Meeting of the Society for Economic Botany*. Mexico City, Mexico, 20–26 June.

Denevan WM, Padoch C (1988) Introduction: the Bora Agroforestry Project. In: Denevan WM, Padoch C (eds) *Swidden-fallow Agroforestry in the Peruvian Amazon* (*Advances in Economic Botany* **5**). New York Botanical Garden, New York, pp. 1–7.

Densmore F (1928) How Indians use Plants for Foods, Medicine and Crafts (see Densmore 1974).

Densmore F (1974) *How Indians use Wild Plants for Food, Medicine and Crafts*. Dover Publications Inc., New York.

Desmond A, Moore J (1992) *Darwin*. Penguin Books, London.

Dialla BE (1993) The Mossi indigenous soil classification in Burkina Faso. *Indigenous Knowledge & Development Monitor* **1(3)**: 17–18.

Dialla BE (1994) The adoption of soil conservation practices in Burkina Faso. *Indigenous Knowledge & Development Monitor* **2(1)**: 10–12.

Dimbleby GW (1963) Pollen analyses from two Cornish barrows. *Journal of the Royal Institution of Cornwall* (New Series) **4**: 364–375.

Dinwoodie JM (1981) *Timber, its Nature and Behaviour*. Van Nostrand Reinhold, New York.

Douglas JD (ed) (1988) *New Bible Dictionary*. Intervarsity Press, Leicester.

Dove MR (1993) Smallholder rubber and swidden agriculture in Borneo: a sustainable adaptation to ecology and economy of tropical forest. *Economic Botany* **47**: 136–147.

Dove MR (1994) Transition from native forest rubbers to *Hevea brasiliensis* (Euphorbiaceae) among tribal smallholders in Borneo. *Economic Botany* **48**: 382–396.

Drinkwater M (1994) Knowledge, consciousness and prejudice: adaptive agricultural research in Zambia. In: Scoones I, Thompson J (eds) *Beyond Farmer First: Rural People's Knowledge, Agricultural Research and Extension Practice*. Intermediate Technology Publications, London, pp. 32–41.

Driver HE (1969) *Indians of North America*, 2nd edn (revised). University of Chicago Press, London.

Dumond DE (1987) *The Eskimos and Aleuts* (revised edition). Thames and Hudson, London.

Dyson-Hudson R, Dyson-Hudson N (1980) Nomadic pastoralism. *Annual Review of Anthropology* 9: 15–62.

Eaton SB, Shostak M. Konner M (1988) *The Paleolithic Prescription*. Harper & Row, London.

Edwards DA, Josephson SC, Brenner Coltrain J (1994) Burkina Faso herdsmen and Optimum Foraging Theory: a reconsideration. *Human Ecology* 22: 213–217.

Efron DH, Holmstedt B, Kline NS (eds) (1967) *Ethnopharmacologic Search for Psychoactive Drugs*. Raven Press, New York.

Elizabetsky E (1991) Sociopolitical, economic and ethical issues in medicinal plant research. *Journal of Ethnopharmacology* 32: 235–239.

Ellen RF (1993) Modes of subsistence: hunting and gathering to agriculture and pastoralism. In: Ingold T (ed.) *Companion Encyclopaedia of Anthropology: Humanity, Culture and Social Life*. Routledge, London.

Ellen RF (1994) Putting plants in their place: anthropological approaches to understanding the ethnobotanical knowledge of rainforest populations. Presentation, UBD-RGS Conference.

Ellen RF, Fukui K (eds) (1994) *Beyond Nature and Culture: Cognition, Ecology and Domestication*. Berg, London.

Estes RD, Small R (1981) The large herbivore populations of Ngorongoro Crater. *East African Wildlife Journal* 19: 175–186.

Evans D (1984) The Doctrine of Signatures as the explanation of some puzzling names and uses of plants. In: Vickery R (ed) *Plant-lore Studies (The Folklore Society Mistletoe Series* Vol 18). The Folklore Society, London, pp. 66–74.

Fahn A (1979) *Secretory Tissues in Plants*. Academic Press, London.

Fahn A (1990) *Plant Anatomy* (4th edition). Pergamon Press, Oxford.

Farnsworth NR (1990) The role of ethnopharmacology in drug development. In: Chadwick DJ, Marsh J (eds) *Bioactive Compounds from Plants* (Ciba Foundation Symposium 154). Wiley, Chichester, pp. 2–21.

Farnsworth NR (1994) Ethnopharmacology and drug development. In: Chadwick DJ, Marsh J (eds) *Ethnobotany and the Search for New Drugs (Ciba Foundation Symposium* 185). Wiley, Chichester, pp. 42–51.

Farnsworth NR, Loub WD, Soejarto DD, Cordell GA, Quinn ML, Mulholland K (1981) Computer services for research on plants for fertility regulation. *Korean Journal of Pharmacognosy* 12: 98–109.

FEB (1993) *The Foundation for Ethnobiology*. Media Launch & Symposium, 2 February.

Feeny P (1976) Plant apparency and chemical defence. *Recent Advances in Phytochemistry* 10: 1–40.

Felger RS, Moser MB (1985) *People of the Desert and Sea: Ethnobotany of the Seri Indians*. University of Arizona Press, Tuscon.

Fernandez PG (1994) Indigenous seed practices for sustainable agriculture. *Indigenous Knowledge & Development Monitor* 2: 9–12.

Fieldhouse P (1986) *Food and Nutrition: Customs and Culture*. Chapman & Hall, London.

Fitzgerald R (1982) *A History of Queensland: From the Dreaming to 1915*. University of Queensland Press, London.

Flaster P (1994) SEB Classes. *Plants & People: Society for Economic Botany*

Newsletter. **Spring**: 2.

Fleisher A, Fleisher Z (1988) Identification of Biblical hyssop and origin of the traditional use of oregano-group herbs in the Mediterranean region. *Economic Botany* **42**: 232–241.

Flenley JR, King ASM, Jackson J, Chew C, Geller JT, Prentice ME (1991) The late Quaternary vegetational and climatic history of Easter Island. *Journal of Quaternary Science* **6**: 85–115.

Flood J (1983) *Archaeology of the Dreamtime: The Story of Prehistoric Australia and her People.* University of Hawaii Press, Honolulu.

Fong FW (1992) Perspectives for sustainable resource utilisation and management of Nipa. *Economic Botany* **46**: 45–54.

Ford RI (1978) Ethnobotany: historical diversity and synthesis. In: Ford RI (ed) *The Nature and Status of Ethnobotany* (Anthropological papers, Museum of Anthropology, University of Michigan No 67). Ann Arbor, Michigan, pp. 33–50.

Frechione J, Posey DA, Francelino da Silva (1989) The perception of ecological zones and natural resources in the Brazilian Amazon: an ethnoecology of Lake Coari. In: Posey DA, Balée W (eds) *Resource Management in Amazonia: Indigenous and Folk Strategies. (Advances in Economic Botany* **7**) New York Botanic Garden, New York pp. 260–282.

Fujisaka S, Jayson E, Dapsula A (1993) 'Recommendation domain' and a farmer's upland rice technology. *Indigenous Knowledge & Development Monitor* **1(3)**: 8–10.

Futuyama D (1976) Food plant specialisation and environmental predictability in Lepidoptera. *American Naturalist* **110**: 285–292.

Gaillard MJ, Birks HJB, Emanuelsson U, Berglund BE (1992) Modern pollen/land-use relationships as an aid in the reconstruction of past land-uses and cultural landscapes: an example from south Sweden. *Vegetation History and Archaeobotany* **1**: 3–17.

GELA (1994) *Directorio Latinoamericano de Etnobotánicos.* Autonomous University of Mexico, Mexico.

George EF, Sherrington PD (1984) *Plant Propagation by Tissue Culture: Handbook and Directory of Commercial Laboratories.* Eastern Press, Reading.

Gifford EM, Foster AS (1989) *Morphology and Evolution of Vascular Plants* (3rd edition). WH Freeman, New York.

Gilmore MR (1932) Importance of ethnobotanical investigation. *American Anthropologist* **34**: 320–327.

Gilmore MR (1919) *Uses of Plants by the Indians of the Missouri River Region* (see Gilmore 1991).

Gilmore MR (1991) *Uses of Plants by the Indians of the Missouri River Region.* University of Nebraska Press, London.

Godoy RA, Lubowski R, Markandaya A (1993) A method for the economic valuation of non-timber tropical forest products. *Economic Botany* **47**: 220–233.

Goldstein A & Kalant H (1990) Drug policy: striking the right balance. *Science* **249**: 1513–1521.

Goodale JC (1982) Production and reproduction of key resources among the Tiwi of north Australia. In: Williams NM, Hunn ES (eds) *Resource Managers: North American and Australian Hunter-gatherers.* Australian Institute of Aboriginal Studies, Canberra, pp. 197–210.

Goodman AH et al. (1984) Indications of stress from bone and teeth. In: Cohen

MN, Armelagos GJ (eds) *Palaeopathology at the Origins of Agriculture.* Academic Press, New York, pp. 13–49.

Goodwin TW, Mercer EI (1983) *Introduction to Plant Biochemistry* (2nd edition). Pergamon Press, Oxford.

Goody J (1993) *The Culture of Flowers.* Cambridge University Press, Cambridge.

Goonatilake S (1991) *The Evolution of Information: Lineages in Gene, Culture and Artefact.* Pinter Publishers, London.

Gottesfeld CMJ (1992) The importance of bark products in the aboriginal communities of New British Columbia, Canada. *Economic Botany* **46**: 148–157.

Gould RA (1982) To have and have not: the ecology of sharing among hunter-gatherers. In: Williams NM, Hunn ES (eds) *Resource Managers: North American and Australian Hunter-gatherers.* Australian Institute of Aboriginal Studies, Canberra, pp. 69–92.

Grayzel JA (1990) Markets and migration: a Fulbe pastoral system in Mali. In: Galaty JG, Johnston DL (eds) *The World of Pastoralism.* Belhaven Press, London, pp. 35–68.

Gremillion KJ (1993) The evolution of seed morphology in domesticated *Chenopodium*: an archaeological case study. *Journal of Ethnobiology* **13**: 149–170.

Griggs B (1981) *Green Pharmacy.* Healing Arts Press, Rochester.

Groube L (1989) The taming of the rainforests: a model for Late Pleistocene forest exploitation in New Guinea. In: Harris DR, Hillman GC (eds) *Foraging and Farming: the Evolution of Plant Exploitation (One World Archaeology* **13**). Unwin Hyman, London, pp. 292–304.

Guillet DW, Furbee L, Sandor J, Benfer R (1995) The Lari Soils Project in Peru: a methodology for combining cognitive and behavioural research. In: Warren DM, Slikkerveer LJ, Brokensha D (eds) *The Cultural Dimension of Development: Indigenous Knowledge Systems.* Intermediate Technology Publications, London, pp. 71–82.

Gupta AK, IDS Workshop (1989) Maps drawn by farmers and extensionists. In: Chambers R, Pacey A, Thrupp LA (eds) *Farmer First: Farmer Innovation and Agricultural Research.* Intermediate Technology Publications, London, pp. 86–93.

Hall A (1993) A dyeing art revealed: evidence for plants used in dyeing and mordanting at 9th–11th Century Coppergate, York. Presentation, *Botanical Society of Scotland, Plants and People Symposium.* Royal Botanic Garden, Edinburgh, 24–27 September.

Hall AL (1991) *Developing Amazonia: Deforestation and Social Conflict in Brazil's Carajás Programme.* Manchester University Press, Manchester.

Hall P, Bawa KS (1993) Methods to assess the impact of extraction of non-timber forest products on plant populations. *Economic Botany* **47**(3): 234–247.

Hall R (1986) *Egyptian Textiles.* Shire Egyptology, Aylesbury.

Hallam SJ (1989) Plant usage and management in southwest Australian Aboriginal societies. In: Harris DR, Hillman GC (eds) *Foraging and Farming: The Evolution of Plant Exploitation (One World Archaeology* **13**). Unwin Hyman, London: 136–151.

Haman O (1991) The joint IUCN–WWF conservation programme and its interest in medicinal plants. In: Akerele O. Heywood V, Synge H (eds) *Conservation of Medicinal Plants.* WHO, IUCN, WWF, Cambridge University Press, Cambridge, pp. 13–22.

Hamilton MB (1994) *Ex situ* conservation of wild plant species: time to reassess the genetic assumptions and implications of seed banks. *Conservation Biology* 8: 39–49.

Hammond N (1978) The myth of the milpa: agricultural expansion in the Maya lowlands. In: Harrison PD, Turner BL (eds) *Pre-Hispanic Maya Agriculture.* University of New Mexico Press, Albuquerque, pp. 23–24.

Harborne JB (1980) Plant Phenolics. In: Bell EA, Charwood BV (eds) *Secondary Plant Products* (Encyclopedia of plant physiology New Series Vol 8). Springer-Verlag, New York, pp. 329–402.

Harborne JB (1984) *Phytochemical Methods: A Guide to Modern Techniques of Plant Analysis,* 2nd edn. Chapman & Hall, London.

Harborne JB (1988) *Introduction to Ecological Biochemistry,* 3rd edn. Academic Press, London.

Harborne JB (ed) (1989) *Plant Phenolics* (Methods in Plant Biochemistry 1). Academic Press, London.

Harborne JB, Turner BL (1984) *Plant Chemosystematics.* Academic Press, London.

Harlan JR (1989) Wild-grass seed harvesting in the Sahara and Sub-Sahara of Africa. In: Harris DR, Hillman GC (eds) *Foraging and Farming: The Evolution of Plant Exploitation (One World Archaeology* 13). Unwin Hyman, London pp. 79–98.

Harlan JR (1992) Indigenous African agriculture. In: Cowan CW, Watson PJ (eds) *The Origins of Agriculture: An International Perspective.* Smithsonian Institution Press, London, pp. 59–70.

Harris DR (1989) An evolutionary continuum of people–plant interaction. In: Harris DR, Hillman GC (eds) *Foraging and Farming: The Evolution of Plant Exploitation (One World Archaeology* 13). Unwin Hyman, London, pp. 11–26.

Harris SR, Hillman GC (eds) (1989a) *Foraging and Farming: The Evolution of Plant Exploitation (One World Archaeology* 13). Unwin Hyman, London.

Harris SR, Hillman GC (1989b) Introduction. In: Harris SR, Hillman GC (eds) *Foraging and Farming: The Evolution of Plant Exploitation (One World Archaeology* 13). Unwin Hyman, London, pp. 1–8.

Harshberger JW (1896) The purposes of ethnobotany. *Botanical Gazette* 21: 146–154.

Haslam E (1986) Milestones in plant tissue culture *Natural Products Reports* 3; 217–249.

Hastorf CA, DeNiro MG (1985) Reconstruction of prehistoric plant production and cooking practices by a new isotopic method. *Nature* 315: 489–491.

Hay RKM, Svoboda KP (1993) Botany. In: Hay RKM, Waterman PG (eds) *Volatile Oil Crops: Their Biology, Biochemistry and Production.* Longman Scientific & Technical, Harlow, UK, pp. 5–22.

Hay RKM, Waterman PG (1993) Introduction. In: Hay RKM, Waterman PG (eds).

Hecht SB, Posey DA (1989) Preliminary results on soil management techniques of the Kayapó Indians. In: Posey DA, Balée W (eds) *Resource Management in Amazonia: Indigenous and Folk Strategies (Advances in Economic Botany* 7). New York Botanical Garden, New York, pp.174–188.

Helbaek H (1960) The paleo-ethnobotany of the Near East and Europe. In: Braidwood RJ, Howe B (eds) *Prehistoric Investigations in Iraqui Kurdistan (Studies in Ancient Oriental Civilisations* 31). Chicago University Press,

Chicago.

Hellemans A, Bunch B (1988) *The Timetables of Science: A Chronology of the Most Important People and Events in the History of Science.* Simon & Schuster, London.

Hendry G (1993) Plant Pigments. In: Lea PJ, Leegood RC (eds) *Plant Biochemistry and Molecular Biology.* Wiley, Chichester, pp. 181–196.

Hepburn G (1989) Pesticides and drugs from the neem tree. *The Ecologist* **19**: 31–32.

Hill HE, Evans J (1989) Crops of the Pacific: new evidence from the chemical analysis of organic residues in pottery. In: Harris DR, Hillman GC (eds) *Foraging and Farming: the Evolution of Plant Exploitation (One World Archaeology* **13**). Unwin Hyman, London, pp. 418–425.

Hillman GC, Colledge SM, Harris DR (1989) Plant-food economy during the Epipalaeolithic period at Tel Abu Hureyra, Syria: dietary diversity, seasonality, and modes of exploitation. In: Harris DR, Hillman GC (eds) *Foraging and Farming: The Evolution of Plant Exploration (One World Archaeology* **13**). Unwin Hyman, London, pp. 240–268.

Hobhouse H (1992) *Seeds of Change: Five Plants that Transformed Mankind.* Papermac, London.

Holland BK (1994) Prospecting for drugs in ancient texts. *Nature* **369**: 702.

Homewood KM, Rodgers WA (1991) *Maasailand Ecology: Pastoralist Development and Wildlife Conservation in Ngorongoro, Tanzania.* Cambridge University Press, Cambridge.

Hormaza JI, Dollo L, Polito VS (1994) Determination of relatedness and geographical movements of *Pistacia vera* (Pistachio: Anacardiaceae) germplasm by RAPD analysis. *Economic Botany* **48**: 349–358.

Howe HF, Westley LC (1988) *Ecological Relationships of Plants and Animals.* Oxford University Press, Oxford.

Hunn E, French D (1984) Alternatives to taxonomic hierarchy: the Sahaptin case. *Journal of Ethnobiology* **3**: 73–92.

Hunn ES, Williams NM (1982) Introduction. In: Williams NM, Hunn ES (eds) *Resource Managers: North American and Australian Hunter-gatherers.* Australian Institute of Aboriginal Studies, Canberra, pp. 1–16.

ICCO (1995) Cocoa. *World Commodity Forecasts—Food, Feedstuffs and Beverages* August, USDA: 4.1 Cocoa.

IDS Workshop (1989) Interactions for local innovation. In: Chambers R, Pacy A, Thrupp LA (eds) *Farmer First: Farmer Innovation and Agricultural Research.* Intermediate Technology Publications, London, pp. 43–51.

Ingrouille M (1992) *Diversity and Evolution of Land Plants.* Chapman & Hall, London.

Irvine D (1989) Succession management and resource distribution in an Amazonian rainforest. In: Posey DA, Balée W (eds) *Resource Management in Amazonia: Indigenous and Folk Strategies (Advances in Economic Botany* **7**). New York Botanical Garden, New York, pp. 223–237.

Isman MB (1994) Botanical Insecticides. *Pesticide Outlook* **5**: 26–31.

Iverson J (1941) Land occupation in Denmark's stoneage. *Danmarks Geologiske Undersogelse* **2(66)**: 1–67.

Iwu MM (1994) African medicinal plants and the search for new drugs based on ethnobotanical leads. In: Chadwick DJ, Marsh J (eds) *Ethnobotany and the Search for New Drugs (Ciba Foundation Symposium* 185). Wiley, Chichester, pp. 116–126.

Jain SK (ed) (1981) *Glimpses of Indian Ethnobotany*. Oxford & IBH, New Delhi.

Jain SK (ed) (1987) *A Manual of Ethnobotany*. Scientific Publishers, Jodphur.

Jain SK (ed) (1989) *Methods and Approaches in Ethnobotany*. Proceedings of the 2nd training course in ethnobotany, Lucknow. Surya Publications, Dehradun.

Jain SK (1994) Ethnobotany and research on medicinal plants in India. In: Chadwick DJ, Marsh J (eds) *Ethnobotany and the Search for New Drugs (Ciba Foundation Symposium* 185). Wiley, Chichester, pp. 153–164.

Jain SK, Minnis P, Shah NC (1986) *World Directory of Ethnobotanists*. Surya Publications, Dehradun, India.

Jenner P, Smith C (1992) *The Tourism Industry and the Environment* (Economist Intelligence Unit Special Report No 2453). Economist Intelligence Unit, London.

Johannes RE (1989) Introduction. In: Johannes RE (ed.) *Traditional Ecological Knowledge: A Collection of Essays*. IUCN, Cambridge, pp. 5–6.

Johns T (1989) A chemical–ecological model of root and tuber domestication in the Andes. In: Harris DR, Hillman GC (eds) *Foraging and Farming: The Evolution of Plant Exploitation (One World Archaeology* 13). Unwin Hyman, London, pp. 504–522.

Johnson A (1975) Time allocation in a Machiguenga community. *Ethnology* **14**: 301–310.

Johnson M (ed.) (1992) Research on traditional environmental knowledge: its development and its role. In: *Lore: Capturing Traditional Environmental Knowledge*. Dene Cultural Institute, Fort Hay, Canada, pp. 3–22.

Johnson M, Ruttan RA (1992) Traditional environmental knowledge of the Dene: a pilot project. In: Johnson M (ed.) *Lore: Capturing Traditional Environmental Knowledge*. Dene Cultural Institute, Fort Hay, Canada, pp. 35–63.

Jones R, Meehan B (1989) Plant foods of the Gidjingali: ethnographic and archaeological perspectives from northern Australia on tuber and seed exploitation. In: Harris DR, Hillman GC (eds) *Foraging and Farming: the Evolution of Plant Exploitation (One World Archaeology* 13). Unwin Hyman, London, pp. 120–135.

Joshi CP, Nguyen HT (1993) RAPD (random amplified polymorphic DNA) analysis-based intervarietal genetic relationships among hexaploid wheats. *Plant Science* **93**: 95–103.

Juniper BE, Jeffree CE (1983) *Plant Surfaces*. Edward Arnold, London.

Karim WJB (1981) *Ma'Betisék Concepts of Living Things*. The Athlone Press, Humanities Press, New Jersey.

Kean S (1988) Developing a partnership between farmers and scientists: the example of Zambia's Adaptive Research Planning Team. *Experimental Agriculture* **24**: 289–299.

Keeley LH, Toth N (1981) Microwear polishes on early stone tools from Koobi Fora, Kenya. *Nature* **293**: 464–465.

Keesing RM (1981) *Cultural Anthropology*, 2nd edn. Holt, Rinehart & Winston, Inc., London.

Khanna KR, Shukla S (1991) Genetics of secondary plant products and breeding for their improved content and modified quality. In: Khanna KR (ed.) *Biochemical Aspects of Crop Improvement*. CRC Press, Boston, pp. 283–326.

Khanna KR, Singh SP (1991) Genetics of fatty acid composition and breeding for modified seed oils. In: Khanna KR (ed) *Biochemical Aspects of Crop Improvement*. CRC Press, Boston, pp. 255–282.

King SR, Tempesta MS (1994) From shaman to human clinical trials: the role of industry in ethnobotany, conservation and community reciprocity. In: Chadwick DJ, Marsh J (eds) *Ethnobotany and the Search for New Drugs* (Ciba Foundation Symposium 185). Wiley, Chichester, pp. 197–206.

Kleiman R, Plattner RD, Wisleder D (1988) *Journal of Natural Products* (Lloydia) **51**: 249–256.

Kuhnlein HV, Turner NJ (1991) *Traditional Plant Foods of Canadian Indigenous Peoples: Nutrition, Botany and Use*. Gordon & Breach, Reading.

Laird SA (1993) Contracts for biodiversity prospecting. In: Reid WV *et al.* (eds) *Biodiversity Prospecting: Using Genetic Resources for Sustainable Development*. World Resources Institute Publications, Washington DC, pp. 99–130.

Lamprey HF (1983) Pastoralism yesterday and today: the overgrazing problem. In: Bourlier F (ed.) *Tropical Savannas: Ecosystems of the World*, Vol. 13. Elsevier, Amsterdam, pp. 643–666.

Layton R (1989) Introduction: who needs the past? In: Layton R (ed) *Who needs the past? Indigenous Values and Archaeology (One World Archaeology* 5). Unwin Hyman, London, pp. 1–20.

Lebot V (1991) Kava (*Piper methysticum* Forst. f). The Polynesian dispersal of an Oceanian plant. In: Cox PA, Banack SA (eds) *Islands, Plants and Polynesians: An Introduction to Polynesian Ethnobotany*. Dioscorides Press, Portland, Oregon, pp. 169–201.

Lenneberg EH (1953) Cognition in ethnolinguistics. *Language* **29**: 463–471.

Lentz DL (1993) Medicinal and other economic plants of the Paya of Honduras. *Economic Botany* **47**: 358–370.

Lévi-Strauss C (1963) *Totemism* (English translation). Beacon Press, Boston.

Lévi-Strauss C (1966) *The Savage Mind* (English translation). Weidenfeld and Nicolson, London.

Lewin R (1993) *Human Evolution: An Illustrated Introduction* (3rd edition). Blackwell Scientific Publishers, Oxford.

Lewington A (1990) *Plants for People*. The Natural History Museum Publications, London.

Lewington A (1993) *Medicinal Plants and Plant Extracts: A Review of Their Importation into Europe* (TRAFFIC network report). Traffic International, Cambridge.

Lewis HT (1982) Fire technology and resource management in aboriginal North America and Australia. In: Williams NM, Hunn ES (eds) *Resource Managers: North American and Australian Hunter-gatherers*. Australian Institute of Aboriginal Studies, Canberra, pp. 45–68.

Lewis HT (1989) A parable of fire: hunter-gatherers in Canada and Australia. In: Johannes RE (ed.) *Traditional Ecological Knowledge: A Collection of Essays*. IUCN, Cambridge, pp. 7–20.

Lewis HT (1992) The technology and ecology of nature's custodians: anthropological perspectives on Aborigines and national parks. In: Birckhead J, de Lacy T, Smith L (eds) *Aboriginal Involvement in Parks and Protected Areas*. Australian Institute of Aboriginal and Torres Strait Islander Studies, Canberra, pp. 15–28.

Lewis W, Elvin-Lewis M (1984) Plants and dental care among the Jivaro of the upper Amazon basin. In: Prance GT, Kallunki JA (eds) *Ethnobotany in the Neotropics (Advances in Economic Botany* 1). New York Botanical Garden, New York, pp. 53–61.

Lewis W, Elvin-Lewis M (1994) Basic, quantitative and experimental research phases of future ethnobotany with reference to the medicinal plants of South America. In: Chadwick DJ, Marsh J (eds) *Ethnobotany and the Search for New Drugs (Ciba Foundation Symposium* 185). Wiley, Chichester, pp. 60–72.

Lewis-Williams JD, Dowson TA (1988) The signs of all times: entoptic phenomena in Upper Palaeolithic art. *Current Anthropology* **34**: 201–245.

Ley SV (1990) Synthesis of antifeedants for insects: novel behaviour-modifying chemicals from plants. In: Chadwick DJ, Marsh J (eds) *Bioactive Compounds from Plants (Ciba Foundation Symposium* 154). Wiley, Chichester, pp. 80–87.

Lightfoot C, Guia O de, Aliman A, Ocado F (1989) Systems diagrams to help farmers decide in on-farm research. In: *Farmer First: Farmer Innovation and Agricultural Research*. Intermediate Technology Publications, London, pp. 93–100.

Lipp FJ (1989) Methods for ethnopharmacological field work. *Journal of Ethnopharmacology* **25**: 139–150.

Loub WD, Farnsworth NR, Soejarto DD, Quinn ML (1985) NAPRALERT: computer handing of natural product research data. *Journal of Chemical Information and Computer Science* **25**: 99–103.

Lozoya X (1994) Two decades of Mexican ethnobotany and research on plant-derived drugs. In: Chadwick DJ, Marsh J (eds) *Ethnobotany and the Search for New Drugs (Ciba Foundation Symposium* 185). Wiley, Chichester, pp. 130–140.

Ludvico LR, Bennett IM, Beckerman S (1991) Risk-sensitive feeding behaviour among the Barí. *Human Ecology* **19**: 509–516.

MacDonald N (1991) *Brazil: A Mask called Progress*. Oxfam, Oxford.

Maikhuri RK, Gangwar AK (1993) Ethnobiological notes on the Khasi and Garo tribes of Meghalaya, northeast India. *Economic Botany* **47**: 345–357.

Mann J (1994) *Murder, Magic and Medicine*. Oxford University Press, Oxford.

Marschner H (1986) *Mineral Nutrition in Higher Plants*. Academic Press, London.

Martin GJ (1995) *Ethnobotany: A Conservation Manual*. Chapman & Hall, London.

Maurya DM (1989) The innovative approach of Indian farmers. In: Chambers R. Pacey A, Thrupp LA (eds) *Farmer First: Farmer Innovation and Agricultural Research*. Intermediate Technology Publications, London, pp. 9–14.

May CE, Reddy P, Clarke BC, Appels R (1991) Recent advances in analyzing chromosome structure in cereals and their impact on breeding programs. In: Khanna KR (ed) *Biochemical Aspects of Crop Improvement*. CRC Press, Boston, pp. 70–107.

Maybury-Lewis D (1992) *Millennium: Tribal Wisdom and the Modern World*. Viking Penguin, New York.

McClatchey WC, Cox P (1992) Use of the sago palm *Metroxylon warburgii* in the Polynesian island, Rotuma. *Economic Botany* **46**: 305–309.

McClaughlin S, Schuck SM (1991) Fiber properties of several species of Agavaceae from southwestern United States and northern Mexico. *Economic Botany* **45**: 480–486.

McDonald Fleming M (1992) Reindeer management in Canada's Belcher Islands: documenting and using traditional environmental knowledge. In

Johnson M (ed) *Lore: Capturing Traditional Environmental Knowledge.* Dene Cultural Institute, Fort Hay, Canada, pp. 69–87.

McGrew WC (1992) *Chimpanzee Material Culture: Implications for Human Evolution.* Cambridge University Press, Cambridge.

McIntosh Baring (1993) Czarnikow Sugar estimate for the 1993–4 crop year. *Sugar Market Update* **1 September**: 14.

McMeekin D (1992) Representations on pre-Columbian spindle whorls of the floral and fruit structures of economic plants. *Economic Botany* **46**: 171–180.

McMeekin D (1994) Domesticated plants, fungi and art. Presentation, *35th Annual Meeting of the Society for Economic Botany.* Mexico City, Mexico, 20–26 June.

McNeely JA (1994) Lessons from the past: forests and biodiversity. *Biodiversity and Conservation* **3**: 3–20.

McRae M (1994) Creature cures. *Equinox* **75**: 47–55.

Medley KE (1993) Extractive forest resources of the Tana River National Primate Reserve, Kenya. *Economic Botany* **47**: 171–183.

Mellars P (1994) The upper palaeolithic revolution. In: Cunliffe B (ed.) *The Oxford Illustrated Prehistory of Europe.* BCA, London, pp. 42–78.

Merculieff I (1994) Western society's linear systems and aboriginal cultures: the need for two-way exchanges for the sake of survival. In: Burch ES, Ellanna LJ (eds) *Key Issues in Hunter-gatherer Research.* Berg Publishers Inc. Oxford, pp. 405–415.

Messer AC (1990) Traditional and chemical techniques for the stimulation of *Shorea Javanica* (Dipterocarpaceae) resin exudation in Sumatra. *Economic Botany* **44**: 463–469.

Millar D (1993) Farmer experimentation and the cosmovision paradigm. In: de Boef W, Amanor K, Wellard K *Cultivating Knowledge: Genetic Diversity, Farmer Experimentation and Crop Research.* Intermediate Technology Publications, London, pp. 44–50.

Miller M, Taube K (1993) *The Gods and Symbols of Ancient Mexico and the Maya: An Illustrated Dictionary of Mesoamerican Religion.* Thames & Hudson, London.

Miller NF (1992) The origins of plant cultivation in the Near East. In: Cowan CW, Watson PJ (eds) *The Origins of Agriculture: An International Perspective.* Smithsonian Institution Press, London, pp. 39–58.

Milliken W, Miller RP, Pollard SR, Vandelli EV (1992) *Ethnobotany of the Waimiri Atroari Indians of Brazil,* Royal Botanic Gardens, Kew.

Mione T (1994) Taxonomy and ethnobotany of Mexican and Central American *Jaltomata* (Solanaceae). Presentation, *35th Annual Meeting of the Society for Economic Botany,* Mexico City, Mexico, 20–26 June.

Mithen SJ (1994) The mesolithic age. In: Cunliffe B (ed.) *The Oxford Illustrated Prehistory of Europe.* BCA, London, pp. 79–135.

Moore PD, Webb JA (1978) *An Illustrated Guide to Pollen Analysis.* Hodder & Stoughton, London.

Morakinyo A (1994) The commercial rattan trade in Nigeria. *Forests, Trees and People Newsletter* **25**: 12–14.

Morris B (1976) Whither the savage mind? Notes on the natural taxonomies of a hunting and gathering people. *Man* **11**: 542–557.

Myers N (1992) *The Primary Source: Tropical Forests and our Future.* WW Norton and Co., London.

Nabhan GP (1985) Native crop diversity in aridoamerica: conservation of regional gene pools. *Economic Botany* **39**: 387–399.

Nazarea-Sandoval VD (1995) Indigenous decision-making in agriculture: a reflection of gender and socioeconomic status in the Philippines. In: Warren DM, Slikkerveer LJ, Brokensha D (eds) *The Cultural Dimension of Development: Indigenous Knowledge Systems*. Intermediate Technological Publications, London, pp. 155–174.

Nelson EC, Stalley RA (1993) The earliest botanical carvings in northwestern Europe: 1205–1210, Corcomroe Abbey, the Burren, County Clare, Ireland. Poster presentation, *Botanical Society of Scotland 'Plants and People' Symposium*. 24–27 September 1993.

Nepstad DC, Brown IF, Luz L, Alechandra A, Viana V (1992) Biotic impoverishment of Amazonian forests by tappers, loggers and cattle ranchers. In: Nepstad DC, Schwartzman S (eds) *Non-timber Products from Tropical Forests: Evaluation of a Conservation and Development Strategy (Advances in Economic Botany* **9**). New York Botanical Garden, New York, pp.1–14.

Nequatewa E (1967) *Truth of a Hopi: Stories Related to the Origin, Myths and Cland Histories of the Hopi*. Northland Press, Flagstaff.

Nicholson MS, Arzeni CB (1993) The market medicinal plants of Monterrey, Nuevo León, México. *Economic Botany* **47**: 184–192.

Norton-Griffiths M (1979) The influence of grazing, browsing and fire on vegetation dynamics of the Serengeti. in: Sinclair ARE, Norton-Griffiths M (eds) *Serengeti: Dynamics of an Ecosystem*. University of Chicago Press. Chicago, pp. 310–352.

Norton-Griffiths M (1995) Economic incentives to develop the rangelands of the Serengeti: implications for wildlife conservation. In: Sinclair ARE, Arcese P (eds) *Serengeti II: Research, Management and Conservation of an Ecosystem*. University of Chicago Press, Chicago.

Odin GS (ed.) (1982) *Numerical Dating in Stratigraphy*. Wiley, Chichester.

Orlove BS (1980) Biological anthropology. In: Siegel BJ (ed) *Annual Review of Anthropology* **9**, 235–274.

Ortiz de Montellano BR (1990) *Aztec Medicine, Wealth and Nutrition*. Rutgers University Press, London.

Ott J (1993) *Pharmacotheon: Entheogenic Drugs, their Plant Sources and History*. Natural Products Co., Kennewick, WA.

Padoch C (1988) The economic importance and marketing of forest and fallow products in Iquitos region. In: Denevan WM, Padoch C (eds) *Swidden-fallow Agroforestry in the Peruvian Amazon (Advances in Economic Botany* **5**). New York Botanical Garden, New York, pp. 74–89.

Padoch C, Inuma JC, de Jong W, Unruh J (1988) Market-oriented agroforestry at Tamshiyacu. In: Denevan WM, Padoch C (eds) *Swidden-fallow Agroforestry in the Peruvian Amazon (Advances in Economic Botany* **5**). New York Botanical Garden, New York, pp. 90–96.

Panayotou T (1992) *Protecting Tropical Forests*. Presentation, Annual Meeting of the American Economic Association, New Orleans, 3–5 January.

Parrish (1994) Indigenous post-harvest knowledge in an Egyptian oasis. *Indigenous Knowledge & Development Monitor* **2**(1): 7–9.

Pearce D, Moran D, Fripp E DATE *The Economic Value of Biological and Cultural Diversity, Main Report: A Report to the World Conservation Union*. CSERGE, London.

Pearsall DM (1989) *Paleoethnobotany: A Handbook of Procedures*. Academic Press, London.

Peeters JP, Galway NW (1988) Germplasm collections and breeding needs in Europe. *Economic Botany* **42**: 503–521.

Peluso NL (1992) The rattan trade in East Kalimantan, Indonesia. In: Nepstad DC, Schwartzman S (eds) *Non-timber Products from Tropical Forests: Evaluation of a Conservation and Development Strategy (Advances in Economic Botany* **9***)*. New York Botanical Garden, New York, pp. 115–127.

Peters CA, Gentry A, Mendelsohn R (1989) Valuation of an Amazonian rainforest. *Nature* **339**: 655–656.

Petersen D, McGinnes BS (1979) Vegetation of south Maasailand, Tanzania: a range habitat classification. *East African Agriculture and Forestry Journal* **44**: 252–271.

Phillips O, Gentry AH (1993a) The useful plants of Tambopata, Peru: I. Statistical hypotheses tests with a new quantitative technique. *Economic Botany* **47**: 15–32.

Phillips O, Gentry AH (1993b) The useful plants of Tambopata, Peru: II. Additional hypothesis testing in quantitative ethnobotany. *Economic Botany* **47**: 33–43.

Pickersgill B (1989) Cytological and genetical evidence on the domestication and diffusion of crops within the Americas. In: Harris DR, Hillman GC (eds) *Foraging and Farming: The Evolution of Plant Exploitation (One World Archaeology* **13***)*. Unwin Hyman, London, pp. 426–439.

Pleumaron A (1994) The political economy of tourism. *The Ecologist* **24**: 142–148.

Plotkin MJ (1988) The outlook for new agricultural and industrial products from the tropics. In: Wilson EO (ed) *Biodiversity*. National Academy Press, Washington DC, pp. 106–118.

Poore D (1988) *Natural Forest Management for Sustainable Timber Production*. Report for the International Tropical Timbers Organisation.

Porter D, Allen B, Thompson G (1991) *Development in Practice: Paved with Good Intentions*. Routledge, London.

Posey DA (1983) Indigenous knowledge and development: an ideological bridge to the future. *Cientia y Cultura* **35**: 877–894.

Posey DA (1984) A preliminary report on diversified management of tropical forest by the Kayapó Indians of the Brazilian Amazon. In: Prance GT, Kallunki JA (eds) *Ethnobotany in the Neotropics (Advances in Economic Botany* **1***)*. New York Botanic Garden, New York, pp. 112–126.

Posey DA (1990) Intellectual property rights and just compensation for indigenous knowledge. *Anthropology Today* **6**: 13–16.

Posey DA (1994) Ethnobiology and commercial ethics. Presentation, *Intecol: VI International Congress of Ecology*. Manchester, 21–26 August.

Prance GT (1990) Fruits of the rainforest. *New Scientist* **13 January**: 42–45.

PT Data Consult Inc (1994) Price of rubber up due to poor supply. *Indonesian Commercial Newsletter*. **April**: 22.

Puleston DE (1978) Terracing, raised fields and tree cropping in the Maya lowlands: a new perspective on the geography of power. In: Harrison PD, Turner BL (eds) *Pre-Hispanic Maya Agriculture*. University of New Mexico Press, Albuquerque, pp. 225–246.

Quansah N (1994) *Forest Conservation and Healthcare Development, Madagascar*. Presentation, *People and Plants Initiative Annual Reporting and Planning Meeting*, Royal Botanic Gardens, Kew, April.

Raharijaona V (1989) Archaeology and oral traditions in the Mitongoa-Andrainjato area (Betsileo region) of Madagascar. In: Layton R (ed) *Who Needs the Past? Indigenous Values and Archaeology (One World Archaeology* 5). Unwin Hyman, London, pp. 189–194.

Rajan, Sethuraman (1993) *Indigenous Knowledge & Development Monitor.*

Rajasekaran B, Warren DM (1994) Indigenous knowledge for socioeconomic development and biodiversity conservation: the Kolli hills. *Indigenous Knowledge & Development Monitor* 2(2): 13–17.

Rappaport RA (1968) *Pigs for the Ancestors: Ritual in the Ecological of a New Guinea People* (see Rappaport 1984).

Rappaport RA (1984) *Pigs for the Ancestors: Ritual in the Ecology of a New Guinea People* (new, enlarged edition). Yale University Press, London.

Reeves C (1992) *Egyptian Medicine.* Shire Publications Ltd, Princes Risborough.

Reid WV, Laird SA, Gámez R, Sittenfeld A, Janzen DH, Gollin MA, Juma C (1993) A new lease on life. In: *Biodiversity Prospecting: Using Genetic Resources for Sustainable Development.* World Resources Institute, Washington DC, pp. 1–52.

Reinhard KJ, Bryant VM (1992) Coprolite analysis: a biological perspective on archaeology. In: Schiffer MB (ed.) *Archaeological Method and Theory,* Vol. 4. The University of Arizona Press, Tuscon, pp. 245–288.

Renfrew JM (1973) *Palaeoethnobotany: The Prehistoric Food Plants of the Near East and Europe.* Methuen & Co., Ltd, London.

Rhoades R (1989) The role of farmers in the creation of agricultural technology. In: Chambers R, Pacey A, Thrupp LA (eds) *Farmer First: Farmer Innovation and Agricultural Research.* Intermediate Technology Publications, London, pp. 3–9.

Ribeiro BG, Kenhíri T (1989) Rainy seasons and constellations: the Desâna economic calendar. In: Posey DA, Balée W (eds) *Resource Management in Amazonia: Indigenous and Folk Strategies (Advances in Economic Botany* 7). New York Botanical Garden, New York, pp. 97–114.

Richards P (1985) *Indigenous Agricultural Revolution.* Unwin Hyman, London.

Richards P (1989) Agriculture as a performance. In: Chambers R, Pacey A, Thrupp LA (eds) *Farmer First: Farmer Innovation and Agricultural Research.* Intermediate Technology Publications Ltd, London, pp. 39–42.

Richards P (1994) Local knowledge formation and validation: the case of rice production in central Sierra Leone. In: Scoones I, Thompson J (eds) *Beyond Farmer First: Rural People's Knowledge, Agricultural Research and Extension Practice.* Intermediate Technology Publications, London, pp. 165–170.

Rocheleau D, Wachira K, Malaret L, Wanjohi BM (1989) Local knowledge for agroforestry and native plants. In: Chambers R, Pacey A, Thrupp LA (eds) *Farmer First: Farmer Innovation and Agricultural Research.* Intermediate Technology Publications Ltd, London, pp. 14–23.

Rollo F, Venanzi FM, Amici A (1991) Nucleic acids in mummified seeds: biochemistry and molecular genetics of pre-Columbian maize. *Genetics Research* 58: 193–201.

Rose JC, Burnett BA, Nassaney MS, Blaeuer MW (1984) Paleopathology and the origins of maize agriculture in the lower Mississippi valley and Caddoan Culture areas. In: Cohen MN, Armelagos GJ (eds) *Paleopathology at the Origins of Agriculture.* Academic Press Inc., New York, pp. 393–424.

Rudgley R, (1993) *The Alchemy of Culture: Intoxicants in Society*. British Museum Press, London.

Saint-Pierre C, Bingrong O (1994) Lac host trees and the balance of agroecosystems in South Yunnan, China. *Economic Botany* **48**: 21–28.

Salas MA (1994) 'The technicians only believe in science and cannot read the sky.' The cultural dimension of the knowledge conflict in the Andes. In: Scoones I, Thompson J (eds) *Beyond Farmer First: Rural People's Knowledge, Agricultural Research and Extension Practice*. Intermediate Technology Publications, London, pp. 57–69.

Salick J (1989) Ecological basis of Amuesha agriculture, Peruvian upper Amazon. In: Posey DA, Balée W (eds) *Resource Management in Amazonia: Indigenous and Folk Strategies (Advances in Economic Botany 7)*. New York Botanical Garden, New York, pp. 189–212.

Salick J, Lundberg M (1990) Variation and change in Amuesha agriculture in the Peruvian Amazon. In: Prance GT, Balick MJ (eds) *New Directions in the Study of Plants and People (Advances in Economic Botany 8)*. New York Botanic Garden, New York, pp. 189–198.

Salisbury FB, Ross CW (1978) *Plant Physiology* (2nd edition). Wadsworth Publishing Company, Belmont, California.

Sampson HC, Crowther EM (1943) Crop production and soil fertility problems. *The West Africa Commission 1938–39: Technical Reports* (Part 1). Leverhulme Trust, London.

Sánchez GJJ, Goodman MM, Rawlings JO (1993) Appropriate characters for racial classification in maize. *Economic Botany* **47**: 44–59.

Schmeda-Hirschmann G (1994) Plant resources used by the Ayoreo of the Paraguayan Chaco. *Economic Botany* **48**: 252–259.

Schultes RE (1983) Richard Spruce: an early ethnobotanist and explorer of the northwest Amazon and northern Andes. *Journal of Ethnobiology* **3**: 139–147.

Scoones I, Thompson J (1994) Knowledge, power and agriculture—towards a theoretical understanding. In: Scoones I, Thompson J (eds) *Beyond Farmer First: Rural People's Knowledge, Agricultural Research and Extension Practice*. Intermediate Technology Publications, London, pp. 16–31.

Seymour-Smith C (1986) *Macmillan Dictionary of Anthropology*. Macmillan Press Ltd, London.

Shaman Pharmaceuticals Inc (1994) *Shaman begins Provir*[TM] Phase II Clinical Trial. Press release, 28 January.

Shewry PR, Kreis M (1991) Genetics of fatty acid composition and breeding for modified seed oils. In: Khanna KR (ed.) *Biochemical Aspects of Crop Improvement*. CRC Press, Boston, pp. 225–282.

Shoup J (1990) Middle Eastern sheep pastoralism and the Hima system. In: Galaty J, Johnson D (eds) *The World Pastoralism: Herding Systems in Comparative Perspective*. The Guilford Press, New York, pp. 195–215.

Sidick (1994) The current status of *jamu* and suggestions for further research and development. *Indigenous Knowledge & Development Monitor* **2(1)**:13–15.

Silberbauer GB (1994) A sense of place. In: Burch ES, Ellanna LJ (eds) *Key Issues in Hunter-Gatherer Research*. Berg Publishers Inc., Oxford, pp. 119–143.

Simpson BB, Conner-Ogorzaly MC (1986) *Economic Botany*. McGraw Hill, London.

Simpson J (1994) Margaret Murray: who believed her and why? *Folklore* **105**:

89–96.

Sindiga I (1984) Land and population problems in Kajiado and Narok, Kenya. *African Studies Review* **27**: 23–39.

Sindiga I (1994) Indigenous (medical) knowledge of the Maasai. *Indigenous Knowledge & Development Monitor* **2**: 16–18.

Smith AB (1992) *Pastoralism in Africa: Origins and Development Ecology*. Hurst & Company, London.

Smith BD (1992) Prehistoric plant husbandry in eastern North America. In: Cowan CW, Watson PJ (eds) *The Origins of Agriculture: An International Perspective*. Smithsonian Institution Press, London, pp. 101–120.

Smith M, Kalotas AC (1985) Bardi plants: an annotated list of plants and their use by the Bardi Aborigines of Dampierland in northwestern Australia. *Records of the Western Australia Museum* **12**: 317–359.

Smith MA (1989) Seed gathering in inland Australia: current evidence from seed-grinders on the antiquity of the ethnohistorical pattern of exploitation. In: Harris DR, Hillman GC (eds) *Foraging and Farming: The Evolution of Plant Exploitation (One World Archaeology* **13**). Unwin, Hyman, London, pp. 305–317.

Smith PM (1976) *The Chemotaxonomy of Plants*. Edward Arnold, London.

Smole WJ (1989) Yanoama horticulture in the Parima Highlands of Venezuela and Brazil. In: Posey DA, Balée W (eds) *Resource Management in Amazonia: Indigenous and Folk Strategies (Advances in Economic Botany* **7**). New York Botanical Garden, New York, pp. 115–128.

Soleri D (1989) Hopi gardens. *Arid Lands Newsletter* **29**: 11–14.

Soleri D, Cleveland DA (1993) Hopi crop diversity and change. *Journal of Ethnobiology* **13**: 203–232.

Soleri D, Smith SE (1994) Morphological, phenological and genetic comparisons of two maize varieties conserved *in situ* and *ex situ*. Presentation, *Annual Meeting of the Society for Economic Botany*, Mexico City, 20–26 June.

Soleri D, Smith SE (1995) Morphological and phenological comparisons of two Hopi maize varieties conserved *in situ* and *ex situ*. *Economic Botany* **49**: 56–77.

Srivastava VK (1992) Should anthropologists pay their respondents? *Anthropology Today* **8**: 16–20.

Stahl AB (1989) Plant–food processing: implications for dietary quality. In: *Foraging and Farming: The Evolution of Plant Exploitation (One World Agriculture* **13**). Unwin Hyman, London, pp. 171–196.

Stocks A (1983) Candoshi and Cocamilla swiddens in eastern Peru. *Human Ecology* **11**: 69–84.

Stoffle RW, Halmo DB, Olmsted JE, Evans MJ (1990) *Native American Cultural Resource Studies at Yucca Mountain, Nevada*. Institute for Social Research, University of Michigan, Ann Arbor.

Sullivan S (1992) Aboriginal site management in national parks and protected areas. In: Birckhead J, de Lacy T, Smith L (eds) *Aboriginal Involvement in Parks and Protected Areas*. Australian Institute of Aboriginal and Torres Strait Islander Studies, Canberra, pp. 169–178.

Sutton MQ (1993) Midden and coprolite derived subsistence evidence: an analysis of data from the La Quinta site, Salton Basin, California. *Journal of Ethnobiology* **13**: 1–14.

Sutton P, Rigsby B (1982) People with 'politics': management of land and personnel on Australia's Cape York peninsula. In: Williams NM, Hunn ES (eds) *Resource Managers: North American and Australian Hunter-gatherers*.

Australian Institute of Aboriginal Studies, Canberra, pp. 155–172.

Tabor JA, Hutchison CF (1994) Using indigenous knowledge, remote sensing and GIS for sustainable development. *Indigenous Knowledge & Development Monitor* **2(1)**: 2–6.

Tainton NM, Groves RH, Nash RC (1977) Time of mowing and burning veld: short-term effects on production and tiller development. *Proceedings of the Grassland Society of South Africa* **12**: 59–64.

Taube K (1993) *The Legendary Past: Aztec and Maya Myths*. British Museum Press, London.

Thompson JES (1974) Canals of the Rio Candelaria Basin, Campeche, Mexico. In: Hammond N (ed.) *Mesoamerican Archaeology: New Approaches*. University of Texas Press, Austin, pp. 197–302.

Thursday Plantation nd *Tea tree oil*. Thursday Plantation, Wollongbar, Australia.

Titilola T (1994) Indigenous knowledge systems and sustainable development in agriculture: essential linkages. *Indigenous Knowledge & Development Monitor* **2(2)**: 18–21.

Tjamiwa T (1992) Tjunguringkula Waakaripai: joint management of the Uluru National Park. In: Birckhead J, de Lacy T, Smith L (eds) *Aboriginal Involvement in Parks and Protected Areas*. Australian Institute of Aboriginal and Torres Strait Islander Studies, Canberra, pp. 7–14.

Torres CM, Repke DM, Chan K, McKenna D, Llagostera A, Schultes RE (1991) Snuff powders from pre-hispanic San Pedro de Atacama: chemical and contextual analysis. *Current Anthropology* **32**: 640–649.

Turner BL (1978) The development and demise of the swidden thesis of Maya agriculture. In: Harrison PD, Turner BL (eds) *Pre-Hispanic Maya Agriculture*. University of New Mexico Press, Albuquerque, pp. 13–22.

Turner NJ (1988) Ethnobotany of coniferous trees in Thompsons and Lillooet Interior Salish of British Columbia. *Economic Botany* **46**: 177–194.

Turner NT, Davis A (1993) 'When everything was scarce': the role of plants as famine foods in northwestern North America. *Journal of Ethnobiology* **13**: 171–201.

Turner RK, Pearce D, Bateman I (1994) *Environmental Economics: An Elementary Introduction*. Harvester Wheatsheaf, London.

Ucko PJ, Dimbleby GW (1969) *The Domestication and Exploitation of Plants and Animals*. Duckworth, London.

Ulluwishewa R (1993) Indigenous Knowledge, national IK resource centres and sustainable development. *Indigenous Knowledge & Development Monitor* **1(3)**: 11–13.

UN (1993) *The Global Partnership for Environment and Development: A Guide to Agenda 21*. UN, New York.

Unruh J, Flores Paitán S (1987) Relative abundance of the useful component in old managed fallows at Brillo Nuevo. In: Denevan WM, Padoch C (eds) *Swidden-fallow Agroforestry in the Peruvian Amazon (Advances in Economic Botany 5)*. New York Botanical Garden, New York, pp. 67–73.

USDA (1994a) *Research Studies-USDA FAS World Tobacco*. **February**: 9.

USDA (1994b) *Research Studies-USDA FAS World Tobacco*. **January**: 4.

USDA (1994c) *Tropical Products: World Markets and Trade*. **December**: 15.

USDA (1995a) *World Commodity Forecasts—Food, Feedstuffs and Beverages*. **August**: 4 Cocoa.

USDA (1995b) *World Commodity Forecasts—Food, Feedstuffs and Beverages.* **August**: 6 Maize.

USDA (1995c) *World Commodity Forecasts—Food, Feedstuffs and Beverages.* **August**: 9 Tea.

van den Berg ME (1984) Ver-o-Peso: the ethnobotany of an Amazonian market. In: Prance GT, Kallunki JA (eds) *Ethnobotany in the Neotropics (Advances in Economic Botany* 1). New York Botanic Garden, New York, pp. 140–149.

Vaughan JG (1991) Biochemical characterisation of cultivated species, cultivars and their wild relations. In: Khanna KR (ed) *Biochemical Aspects of Crop Improvement.* CRC Press, Boston, pp. 37–58.

Verlet N (1993) Commercial aspects. In: Hay RKM, Waterman PG (eds) *Volatile Oil Crops: Their Biology, Biochemistry and Production.* Longman Scientific & Technical, Harlow, UK, pp. 137–174.

Vickery R (1985) *Unlucky Plants: A Folklore Survey.* The Folklore Society, London.

Vidal J (1993) Whose new lease on life? *The Guardian.* 21 May.

Vilarem G, Périneau F, Gaset A (1992) Exploitation of the molecular potential of plants: *Equisetum arvense* (Equisetaceae). *Economic Botany* **46**: 401–407.

von Geusan LA, Wongprasert S, Trakansupakon P (1992) Documenting and applying traditional environmental knowledge in Northern Thailand. In: Johnson M (ed) *Lore: Capturing Traditional Environmental Knowledge.* Dene Cultural Institute, Fort Hay, Canada, pp. 164–174.

Vorren K-D (1986) The impact of early agriculture on the vegetation of northern Norway: a discussion of anthropogenic indicators in biostratigraphical data. in: Behre KE (ed) *Anthropogenic Indicators in Pollen Diagrams.* AA Balkema, Rotterdam, pp. 1–18.

Wallace J *et al.* (1992) Aboriginal involvement in national parks: Aboriginal rangers' perspectives. In: Birckhead J, de Lacy T, Smith L (eds) *Aboriginal Involvement in Parks and Protected Areas.* Australian Institute of Aboriginal and Torres Strait Islander Studies, Canberra, pp. 29–38.

Walters M, Hamilton A (1993) *The Vital Wealth of Plants.* WWF, Gland.

Warren DM, Slikkerveer LJ, Brokensha D (eds) (1995) *The Cultural Dimension of Development: Indigenous Knowledge Systems.* Intermediate Technology Publications, London.

Waterman PG (1993) The chemistry of volatile oils. In: Hay RKM, Waterman PG (eds) *Volatile Oil Crops: Their Biology, Biochemistry and Production.* Longman Scientific & Technical, Harlow, UK, pp. 47–62.

Weller SC, Ruebush TK, Klein RE (1992) An epidemiological description of folk illness: a study of *empacho* in Guatemala. In: Nichter M (ed) *Anthropological Approaches to the Study of Ethnomedicine.* Gordon & Breach, Reading, pp. 19–32.

Wenke RJ (1990) *Patterns in Prehistory: Humankind's First Three Million Years* (3rd edition). Oxford University Press, Oxford.

Western D, Dunne T (1979) Environmental aspects of settlement site decisions among pastoral Maasai. *Human Ecology* **7**: 75–98.

Whistler WA (1988) Ethnobotany of the Tokelau: the plants, their Tokelau names and their uses. *Economic Botany* **42**: 155–176.

White DJ (1994) Whose seeds are they anyway? *Cultural Survival Quarterly* **18** (2/3): 8–10.

Whiting AF (1939) *Ethnobotany of the Hopi* (Museum of Northern Arizona

Bulletin No. 15). Northern Arizona Society of Science and Art, Flagstaff.

Wickens GE (1990) What is economic botany? *Economic Botany* **44**: 12–28.

Wickens GE, Goodin JR, Field DV (eds) (1985) *Plants for Arid Lands*. Unwin Hyman, London.

Wilcox WW, Botsai EE, Kubler J (1991) *Wood as a Building Material: A Guide for Designers and Builders*. Wiley, Chichester.

Wilken GC (1987) *Good Farmers: Traditional Agricultural Resource Management in Mexico and Central America*. University of California Press, Berkeley.

Wilkes G (1989) Maize: domestication, racial evolution and spread. In: Harris DR, Hillman GC (eds) *Foraging and Farming: The Evolution of Plant Exploitation (One World Archaeology* **13***)*. Unwin Hyman, London, pp. 440–455.

Williams DE (1993) *Lycianthes moziniana* (Solanaceae): an underutilized Mexican plant food with 'new' crop potential. *Economic Botany* **47**: 387–400.

Williamson E (1992) *The Penguin History of Latin America*. Penguin Books Ltd, London.

Wilson EO (1988) The current state of biological diversity. In: Wilson EO, Peter FM (eds) *Biodiversity*. National Academy Press, Washington DC, pp. 3–18.

Wilson EO (1992) *The Diversity of Life*. The Belknap Press of Harvard University Press, Cambridge, MA.

Wolf EC (1986) *Beyond the Green Revolution: New Approaches for Third World Agriculture* (Worldwatch Paper 73). Worldwatch Institute, Washington.

Womersley JS (1981) *Plant Collecting and Herbarium Development*. FAO Plant production and Protection Paper 33. Rome.

WWF/UNESCO/RBG, Kew (1993) *People and Plants: Ethnobotany and the Sustainable use of Plant Resources*. Information Pack, RBG Kew.

Xiao PG (1994) Ethnopharmacological investigation of Chinese medicinal plants. In: Chadwick DJ, Marsh J (eds) *Ethnobotany and the Search for New Drugs (Ciba Foundation Symposium* 185), Wiley, Chichester, pp. 169–173.

Yen DE (1989) The domestication of environment. In: Harris DR, Hillman GC (eds) *Foraging and Farming: The Evolution of Plant Exploitation (One World Archaeology* **13***)*. Unwin Hyman, London, pp. 55–78.

Yen DE (1993) The origins of subsistence agriculture in Oceania and the potentials for future tropical food crops. *Economic Botany* **47**: 3–14.

Yesner DR (1977). Resource diversity and population stability among hunter-gatherers. *Western Canadian Journal of Anthropology* **7**: 18–59.

Yesner DR (1994) Seasonality and resource 'stress' among hunter-gatherers: archaeological signatures. In: Burch ES, Ellanna LJ (eds) *Key Issues in Hunter-Gatherer Research*. Berg Publishers Inc., Oxford, pp. 151–168.

Zigmond ML (1941) *Ethnobotanical Studies among California and Great Basin Shoshoneans* (PhD thesis). Yale University, New Haven.

Zingg RM (1934) American plants in Philippine ethnobotany. *Philippine Journal of Science* **54**: 221–274.

Zohary D (1989) Domestication of the southwest Asian neolithic crop assemblage of cereals, pulses and flax: the evidence from the living plants. In: Harris DR, Hillman GC (eds) *Foraging and Farming: The Evolution of Plant Exploitation (One World Archaeology* **13***)*. Unwin Hyman, London, pp. 358–373.

Postscript

[The Pilgrim Fathers] packed as if they had misunderstood the purpose of the trip. They found room for sundials and candle snuffers, a drum, a trumpet, and a complete history of Turkey. One William Mullins packed 126 pairs of shoes and 13 pairs of boots. Yet between them they failed to bring a single cow or horse or plough or fishing line ...

They were, in short, dangerously unprepared for the rigours ahead, and they demonstrated their manifest incompetence in the most dramatic possible way: by dying in droves ...

For two months they tried to make contact with the natives, but every time they spotted any, the Indians ran off. Then one day in February a young brave of friendly mien approached a party of Pilgrims on the beach. His name was Samoset and he was a stranger in the region himself, but he had a friend named Tisquantum from the local Wampanoag tribe, to whom he introduced them. They showed them how to plant corn and catch wildfowl. ... Before long, as every schoolchild knows, the Pilgrims were thriving and Indians and settlers were sitting down to a cordial Thanksgiving feast.

Bill Bryson 1994, in *Made in America*

Index

aboriginal botany 6
Aborigines (*see also* Bardi,
    indigenous peoples)
  early studies 9, 13
  extant groups 130, 131, 135, 150
  in conservation of national parks
    325–326
  in-law avoidance 75
  numbers of 147, 150
  resource management 154–157,
    325–326
  subsistence mode 128
  use of fire 154–157
  use of pituri 13
  wild plant foods 130, 135–136,
    149–152
abortifacients 39
Acacia spp 54–55, 65
Adansonia digitata 134
Africa (*see also* pastoralists,
    rangelands)
  extant groups 128
  pastoralists 144–149
  traditional knowledge and rural
    development 2
  traditional rangeland management
    13
Afzelia africana 195
Agave fourcroydes 46
Agave sisalana 46, 51
Agave spp 46
Agave subsimplex 205
Agelaea trifolia 200
agricultural science 124
agricultural scripts (*see* behavioural
    scripts)
agriculture (*see also* low-input
    agriculture)
  origins of 12, 310–311

agroforestry (*see also* forests, shifting
    cultivation, swidden
    agriculture)
  definition of 178
  changing attitudes towards 180
Aguaruna (*see also* indigenous
    peoples)
  as shifting cultivators 180
  ethno-ornithology 77
  manioc cultivation 77, 78, 165
Alaska 80–81
Aleut
  traditional ecological knowledge
    80–81
  environmental perception 252
alkaloids (*see also* arrow poisons,
    drugs, fish poisons, medicinal
    plants)
  as arrow poisons 221
  functions in plants 54
  nature and activity 40, 221
  numbers of 41
  sources of 221
allelopathy 38
Allium cepa (onion) 21, 22, 33, 38,
    41, 203
allspice (*Pimenta dioica*) 4
Aloë arborescens 33
Amaranthus cruentus 175
Amazonia (*see also* indigeneous
    peoples, shifting cultivation)
  development programmes 88
  early studies 5, 6
  indigenous peoples 49, 72, 137, 180
America (*see* Amazonia, Central
    America, North America,
    South America)
Amuesha (*see also* indigenous
    peoples)

as shifting cultivators 180
 farming practices 162, 163
 land classification 163
Andes 4, 5
Andropogon greenwayii 156
animal behaviour (*see*
 zoopharmacognosy)
anise 230
Annona cherimola 112
Anseranas semipalmata (magpie
 goose) 157
Anthriscus sylvestris (cow parsley)
 67
anthropogenic effects (*see also*
 domestication, low-input
 agriculture)
 on plant genotype 57
 on plant phenotype 293–294,
 309
 on vegetation 13, 292–293, 308
anthropological field methods
 91–106, 107
 analytical tools 95–96, 103–106
 data verification 95–96, 101–103
 triadic and paired comparisons
 102–103
 ecological mappping 137
 economic calendars 137, 141
 elicitation methods 94–95
 interview techniques and
 questionnaires 92–93
 numerical indices 95–101
 direct matrix ranking 98
 preference ranking 95, 97–98, 102
 use-values 97, 98, 99
 participant choice 100, 103
 choosing researchers 106
 indices of agreement 105–106
 influence of sociological
 variables 103
 methods of 103
 relative use-value 103–104
 sample size 100, 101
 participant observation 91
 pile sorting 102
 problems in 91–92
 quantitative and qualitative
 methods 92, 93
 structured surveys 91
 summary of methods 107

anthropology 8
 in ethnobotanical studies 1, 6, 8
 in 'rescue operations' 15
antibacterial plants 9
antiviral plants 16, 321–322 (*see also*
 HIV)
apple (*Malus pumila*) 20, 21
applied anthropology 124
applied ethnobotany 18, 313–374
 (*see also* applied
 palaeoethnobotany,
 environmental economics,
 ethno-directed research,
 cultural survival, sustainable
 development)
 interest in 313
 potential benefits of 18, 314–345
 in economic development 18,
 314, 315, 331, 334–336
 in germplasm conservation
 327–329
 in habitat conservation 18, 314,
 323–326, 331–332
 in identifying 'novel' crops
 315–319, 320–322
 in improved healthcare 82, 83,
 329–330
 problems of 347–373
 ethical considerations 368–373
 realising economic potential
 349–358
 rationale behind 16
applied palaeoethnobotany
 in conservation 345
 in drug discovery 342–345
archaeobotanical methods 119–122
 (*see also* archaeological record,
 palaeoethnobotany, pollen
 analysis)
 archaeochemical analysis 287–289
 chemical spectra of Pacific food
 plants 288–289
 collecting samples 119–120, 122
 coprolite analysis 285
 dating methods 122, 123, 280
 identifying fossils 121
 interpreting the evidence 283–284
 recovering macrofossils 120
 recovering microfossils 120,
 285–286

role of palaeopathology 285, 286
use-wear analysis 286, 287
archaeobotany (*see also* archaeobotanical methods, pollen analysis)
definition of 11
archaeochemistry (*see* archaeobotanical methods)
archaeological record
of the American continent 3
interpretation of
archaeobotanical data 283–284
artistic evidence 294–300
evidence from animal behaviour 303–304
role of ethnoarchaeological evidence 302–303
role of ethnographic evidence 301–303
role of ethnohistoric evidence 302–303
role of linguistic evidence 301–302
role of modern plants 284, 289–294
role of pollen analysis in 13
Arctic 128
Aristida sp 147
Aristida pungens 149
Arnhemland (*see also* Aborigines, indigenous peoples)
food processing 135
resource management 154–155
arrow poisons
curare 219, 221
mode of activity 221, 222
sources and processing 219–223
Strychnos nux vomica 41
art (*see also* art and technology, documentary evidence)
and the use of psychoactive plants 295, 297–298
as a source of ethnobotanical data 76, 124
in palaeoethnobotany 294–298
plants represented in, 13, 106
symbolic analysis of plant motifs 295, 296–298
Palaeolithic art 295–298

plants used in 203–208
shamanism textile 298, 301
art and technology (*see also* extracts and exudates)
factors affecting use 206
plant processing 210–215
plants used in 203 208
as pigments 203
as sealants and adhesives 203, 226–227
in basketry 203, 212–213
in tanning 220, 224, 226
in tools and utensils 203, 204, 205
in toys 203, 205
numbers of plants used 205
resource management 209–210
suitability of tools 166–167
Artemisia dracunculus (tarragon) 30, 31
Arum 55
Asia
ethnobotanical projects in 2, 13, 76
indigenous peoples 128
Atrocaryum jauari 140
Atropa belladonna (deadly nightshade) 56, 343
Australasia 13
Australia (*see also* Aborigines)
discovery of tea tree 9
ethnobotanical projects in 3
land allocation 68
myths 75
traditional foods 65
autotoxicity 31
Avicienna marina 141
Azadirachta indica (neem) 56, 65
azadirachtin 65
Aztec
early studies 4
herbal 297
human sacrifice 75

Bahamas 4
Baja California (*see* Seri)
Balick, Mike 106
bamboos (Bambusoidae)
flowering in 23
in material culture 46, 201

bananas (*Musa* spp) 127, 170
Bancroft, Joseph 13
Banisteriopsis caapi 4
Banks, Joseph 135
Bardi (*see also* Aborigines)
  environmental indicators 141
  material culture 205, 208
  medicinal plants 239
  numbers of 150
  resource units 139
  wild plant foods 131
bark (*see also* plant processing)
  in material culture 47, 199–201,
    211–212
  in medicine 83
  processing of 211–212
  structure and development 35,
    202
  variation in 47, 201, 202
barley 305
Barrows, David 7
basil 39
bay (*Laurus nobilis*) 31
beans
  growth in artificial mounds 61
  growth requirements 163, 183
  Hopi varieties 177
behavioural scripts 72
Bertholletia excelsa 141
Betula papyrifera 200
Betula spp 47
biocultural reserves
  Belize Ethnobotany Project 341,
    349
    Terra Nova Rain Forest Reserve
    342, 350
  Beni Biosphere reserve, Bolivia 83
  cost/benefit analysis 349–355
  economic potential 349, 357
  improving net benefits 357–358
  Tana River National Primate
    Reserve (TRNPR) 213
biodiversity (*see also* conservation,
    genetic diversity)
  estimated numbers of plant species
    20, 21
  prospecting 16, 315–319
biological control (*see also* food
    storage, low input agriculture,
    resource management)

of pathogens 53
of pests 56, 65, 167, 168
of weeds 51
rediscovered by Western science
  80
role of volatile oils 229–231
Bixa orellana (anatto) 113, 184
black pepper (*Piper nigrum*) 8, 230
Boehmeria nivea 28, 29
Bora
  as shifting cultivators 209
  Bora Agroforestry Project
    209–210
  flavours and fragrances 230
Boster, James 78, 103, 105
botanical methods (*see also*
    domestication, molecular
    biology, plant names)
  collection of voucher specimens
    113–114, 115
  endangered species 113, 165
  cytogenetics 289, 292–293
  morphometric analysis 114
  phytochemistry 124
  plant classification 116, 117–118
  plant identification 113, 114,
    116–117
  taxonomic skills 15
    chemotaxonomy 116, 118
botany (*see also* aboriginal botany,
    economic botany, ethnobotany)
  in the nineteenth century 5
Brachiaria deflexa 149
Brassica 116
Brazilnut tree 331
Bridelia ferrugina 200
brittlebush 204, 232
bronchitis 10
broom (*Sarothamnus scoparius*) 68
Brosimum utile 197
Burkina Faso 259
Bursera microphylla 138
Bursera spp 47, 138

cabbage 41
cabbage white butterfly (*Pieris
    brassicae*) 42
caboclo 59, 139 (*see also* Lake Coari)
cacti 22, 55 (*see also* plant
    structures and functions)

Caesalpinia pulcherrima　118
caffeine　41
Calamus spp (rattan)　46
Callitris tropicana (cypress pine)　157
Calotropis procera　200
Camellia sinensis (tea)　9
Cameroon　83
Canada　128
Canarium schweinfurthii　195
Canarium spp　151
cancer　83
Cannabis sativa (hemp)　46
Capsicum annuum　175
Carajás Development Programme　88
Carapra procera　199–200
cardiac glycosides　10, 39
Carribbean　83
Carludovica palmata (Panama hat
　palm)　46
carrot (*Daucas carota*)　22, 42
carrying capacity (*see also* cultural
　ecology)
　limiting factor theory (LFT)　74
　modification by humans　74
cashew　331
Cassia　315
Cassia fisulosa (golden shower tree)
　55
Catoblepas gnu (wildebeest)　155
cedar　209, 211
Ceiba pentrandra (kapok)　49
Celtis spp　138
Cenchrus biflorus　149
Cenchrus ciliaris　147
Central America　128
ceremony (*see* ritual and ceremony)
Chácobo
　as shifting cultivators　180
　material culture　191
Chagas' disease　75
Chapman & Hall Chemical database
　(*see* information systems)
Chenopodium belandieri　294
Chicago　7
chimpanzees　303
CHIMPP (Group for Chemo-ethology
　of Hominoid Interactions with
　Medicinal Plants and
　Parasites)　304
China　9

Chippewa Indians
　material culture　211–212
　uses of plants　63
　wild plant foods　131, 133
Chloris spp　147
Chrysanthemum spp　56
Cinnamomum zeylanica (cinnamon)
　8
CIRAN　14
Citrullus colocynthis　65
Citrullus lanatus　134
Citrus spp　31
Clathotropis macrocarpa　135
clinical trials　16, 322
Coca　242
Coccinia adoensis　134
cocoa (*Theobroma cacao*)　4, 5, 185
coconut (*Cocos nucifera*)　33, 49
Coffea arabica (coffee)　9, 51
cognates (*see also* linguistics)
　and cultural significance　276–277
　construction of ancestral terms
　　112
　detection of　112
　in human dispersal　112
　in palaeoethnobotany　302
cognitive ethnobotany (*see also*
　　ethnotaxonomic systems,
　　environmental perception)
　as a field of ethnobotany　15
　factors affecting cognition
　　248–253
　in the interpretation of
　　ethnobotanical data　66,
　　245–247
　methods of study　254–277
　　role of ethnotaxonomy　255–256,
　　269–277
　nature of　64
Colobus monkey　304
Colorado　6
Columbus, Christopher　3, 4, 9
commercialisation (*see also* applied
　　ethnobotany, economic
　　botany)
coniine　41
Conium maculatum (hemlock)　41
Conklin, Harold　69–71 (*see also*
　　Hanunóo)
conservation (*see also* applied

ethnobotany, genetic diversity, sustainable development)
biology   16
in national parks   155–157, 322–326
of habitats   16
of useful germplasm   83, 327–329
strategies   327–329
construction materials (*see also* bamboos, bark, rattan, timber, wood)
factors affecting use   199–203
from animals   199
plant parts used   199–203
roofing materials   199–200
tying materials and sealants   203
Cook, Captain James   9
coprolites (*see* archaeobotanical methods, plant fossils)
Corchorus capsularis (jute)   46
coriander   230
cork (*see also* bark, wood)
in material culture   47
structure and development   27, 33, 35
corn (*Zea mays*) (*see also* Hopi)   4, 5, 78
corozo palm   75
cosmetics   228
cotton (*Gossypium* spp)   4, 49, 291
Cox, Paul   64
Craterispermum laurinum   195
creosote bush   204, 214
crop repertoires (*see* Hopi, Ka'apor, native crops)
cropping systems (*see also* land-use, low-input agriculture)
experimentation   168
types of multicropping   169
Western perceptions   85
crops (*see also* Hopi, native crops)
distribution of   182
evolution of   291
germplasm   166, 168–169, 174, 179
farmer breeders' rights   370
management   179, 327
indigenous innovations in   13
management of   166–171, 179, 181–184
protection of   167

storage   167
yields   171
Cuba   4
Cucumis africanus   134
Cucurbita spp   175
Culpepper, Nicholas   10
cultural diversity (*see also* cultural survival, indigenous peoples)
conservation of   83, 88
erosion of   88
cultural ecology   64, 73 (*see also* carrying capacity, optimal foraging theory)
criticisms of   73, 74
cultural ecological interpretations
of drug preparation   76
of human sacrifice   75
of in-law avoidance   75
of Ka'apor burial   73
of male supremacy   75
of pollution taboos   75
of sacred groves   75
cultural materialism   74
determinism   73, 74
development of theories   74–75
nature of   64
of Tsembaga horticulture   8
cultural impact assessment   82, 339–340
cultural survival   329–340 (*see also* applied ethnobotany, cultural diversity)
Cultural Survival   84, 335–336
cupuaçu fruits   331
curare   219
cutin   31
cyanogenic glycosides   54
cycads   135–136
Cycas spp (cycads)   135–136, 151–152
cycasin   135–136
Cydonia oblonga (quince)   112
Cymbopogon excavatus   147
Cynodon spp   147
Cyperus spp   152
cypress pine   157
cytogenetics (*see* botanical methods, domestication)

Dactyloctenium sp   149
Dacus dorsalis   55

Darwin, Charles 5
databases (*see* information systems)
Datura stramonium 343
de Sagahún, Bernardino 62
de Torquemada, Juan 62
decision-making
  influence of the perceived
      environment 2, 245 247
  methods of study 254–255
defence mechanisms in plants 53, 54
Denmark
  development of pollen analysis 12
  origins of agriculture 12
Desâna
  as shifting cultivators 180
  economic calendar 164
desert tomato 136
digitalin 11
digitalis 10
Digitalis lanata (woolly foxglove) 39
Digitalis purpurea (purple foxglove)
  10
Digitaria macroblephara 147
digitoxin 11
Dioscorea spp (yams)
  in Aboriginal diet 149, 151–152
  lateral expansion in 33
  management of 153
  protection from fire 75
  toxins in 135
Dioscorea transversa (parsnip yam)
  135
Diospyros 195
Directorio Latinoamericano de
    Etnobotánicos 125
Doctrine of Signatures 251
documentary evidence
  developments in 299
  from Egypt 300
  New World 300
  in palaeoethnobotany 298–300
Dodecastigma integrifolium 186
domestication (*see also*
      anthropogenic effects,
      botanical methods)
  definition of domesticated plants
    165
  evolution of domesticated plants
    42, 308–310
  methods of study 289–294

cytogenetics 289, 292–293
  morphometric analysis 293–294
  phytogeography 290–291
  protected plants 165
  role of traditional agricultural
      practices 169
  semi-domesticates 165
dosage
  problems in traditional medicine
    82
  standardisation in foxglove 10
dropsy (*see* heart disease)
Drosophila pachea 55
drought resistance 52, 53, 149
drug prospecting (*see also* ethno-
      directed research)
  in ancient texts 106
drugs (*see also* medicinal plants)
  ethno–directed discovery 318–319
  from folklore 10, 342, 345
  from plants 10, 316
  preparation 76, 242–243
  Western *versus* indigenous
    318–319
Drypetes natalensis 197
Duboisia spp (pituri) 13
Duguetia cauliflora 198
Duguetia flagellaris 198
dyes 212, 224, 233

Echinochloa stagnina 149
Echinochloa utilis 293
ecological anthropology (*see* cultural
    ecology)
ecological methods
  ecological maps 137
  economic calendars 137
  environmental impact assessment
    361–362
  role in ethnobotany 124
ecological repercussions (*see also*
    cultural ecology, efficacy)
  in Africa 156–157
  in Australia 156–157
  of commercialisation 359–361
  of extractive activities 361–362
  of farming practices 166–167
  of ritual behaviour 73, 75, 144
  of swidden management 186, 187
economic botany (*see also* applied

ethnobotany, environmental
economics)
and conservation 340–342
and resource depletion 358–362
commercialisation
of non-timber products (*see*
NTPPs)
of Seri ironwood sculptures 214
economic significance
of Australian plants 9
of plants in European history
5, 9
of traditional plant uses 1, 9
in Brazil 83
undergraduate studies in 14
Economic Botany 3
education
graduate classes 14
in reviving traditional culture 88
role in ethnobotany 124
undergraduate classes 14
efficacy
of traditional plant uses
insect repellents 64, 64
nutritional quality 64, 65
remedies 10, 64, 65, 79, 239–240,
243
of traditional methods
drug preparation 66, 76, 243
farming practices 79
food processing 65, 66, 135–136
resource management 80, 143,
156–157, 323–326
research into 66, 67
Ehrenberg, Christian 11
Eleocharis spp 151–152
elephant tree 203
Eleusine spp 156
Ellen, Roy 9
emergency foods (*see* famine foods)
England 10
entopic phenomena 295, 297–298
environmental economics
cost/benefit analysis 349–358
biodiversity prospecting
354–355
environmental services 353–354
existence values 355
for different land-use strategies
356–357

marginal costs 355
non-use benefits 355
opportunity costs 355
sustainable forestry 350–351
tourism 351–352
improving net benefits 357–358
role in ethnobotany 124
environmental functions of plants 19
environmental services 353–354
exploitation of 51, 57, 160
environmental indicators 140–141
Australia 141
economic calendars 164
Lake Coari 139–141
Peru 163
environmental perception (*see also*
ethnotaxonomy, cognitive
ethnobotany)
factors affecting 247–253
biophysical characteristics of
plants 248–249
categories of 247
magical and religious beliefs
250–251
personal factors 253
social influences 68, 247,
249–253
methods of study 254–277
sources of evidence 112, 254–277
human behaviour 254–255
marginal comments 253
patterns of resource use 246,
249–252
plant-lore 67–68
prototypic evaluations 253
studies in human cognition 8, 15
enzymes (*see* plant biochemistry)
Eragrostis spp 148
Eschweilera rhododendrifolia 138
essential oils (*see also* extracts and
exudates)
biological properties of 38, 53
components of 39
ecological roles 52–55
sources of 31, 39, 221
uses of 229–230, 230–231
estragole (aka allylanisole,
methylchavicol) 39, 42, 43
ethical considerations (*see also*
legislation)

in ethnobotanical study   124
of ethnobotanical researchers
    368–370, 372–374
Ethiopia   9, 141
ethnobiology (*see also* organisations,
    publications, societies)
  fields of study   16
ethnobotany
  as a scientific discipline   6, 15
  criticisms of   85
  definitions of   1, 2
  economic potential of   1, 5, 9, 18
  fields of study   15, 16, 17
  first use of term   6
  history of   1, 13–16
    in America   3–8
    in Europe   9–13
Ethnobotany   17
ethno-directed research
  biodiversity prospecting   217, 320
    in drug discovery   243, 315–319,
      321–322
  conservation
    environmental management
      319–326
    germplasm   327–329
  efficiency of   316–319, 321–322
  ethical considerations   368–373
ethnopharmacology (*see also*
    traditional medicine,
    medicinal plants)
  definition of   10
  history of   10–11, 13
  in the study of traditional
    medicines   64, 65, 236–237
  medthods in   237
ethnoscience   15 (*see also*
    ethnobiology, ethnobotany,
    ethnopharmacology)
  ethnoecology   15
  ethnoentomology   16
  ethnohistory   302
  ethnomedicine   15, 16, 329–330
  ethnomineralogy   16
  ethnomycology   16
  ethno-ornithology   77
  ethnotaxonomy   15
  ethnozoology   16
ethnotaxonomic classification (*see
    also* environmental perception,

ethnotaxonomic nomenclature,
    ethnotaxonomic systems,
    ethnotaxonomies)
  and human cognition   8
  and scientific classification
    264–266
  categories   68–70, 262, 265
  general principles   260–262
  of disease   258–259
  of grasses   147
  of land use   161–162, 163
  of resins   267–268
  of soils   259
  special purpose   266–269
ethnotaxonomic nomenclature (*see
    also* ethnotaxonomic
    classification, plant names)
  and cultural significance   111–112,
    269–277
  and human cognition   8, 269–277
  attributive terms   69
  basic plant names   69
  general principles   263–264
ethnotaxonomic systems (*see also*
    ethnotaxonomic classification,
    ethnotaxonomic
    nomenclature)
  definition of   68
  history of study   69, 257
  in cognitive ethnobotany   68,
    255–256
  limitations of   70, 72
  methods of study (*see*
    anthropological field methods,
    linguistics)
  structural characteristics   165,
    259–277
  and subsistence mode   270–271
ethnotaxonomies
  Aguaruna   258, 272, 274
  Hanunóo   69–70
  Ka'apor   267–268
  Maasai   147
  Mossi   259
  Quechua   272
  Seri   273–274, 275
  Tzeltal Maya   68, 257–258, 258–259,
    272, 275, 276–277
  Tzoltzil Maya   258–259, 276–277
etymology

in ethnopharmacology 14
of Hopi plant names 7
Eucalyptus miniata 200
Eucalyptus spp 31, 47, 52, 156
Euurope 3–13
Euterpe oleraceae 139
extracts and exudates (*see also* lac, toxins)
  body products 227–229
  food additives and preservatives 56, 229
  fragrances and food additives 229–231
  functions in plants 218
  in material culture 208, 216–234
  in traditional medicine 234–243
  production and processing 231–233
    collecting resins 231–232
    collecting rubber 232
    extraction and processing 233
  sealants and adhesives 226–227
  sources 208, 217–231

famine foods
  categories of 132, 134
  in North America 134
  processing 135
Far East 9
farming practices (*see also* low-input agriculture, genetic diversity)
  artificial mounds 61–62
  chinampas 345
  milpas 76
Ferdinandusa sp 138
fertilisation (*see* low-input agriculture)
Fewkes 7
fibres (*see also* plant processing)
  in material culture 6, 46–47, 206–208, 212–213
  in plant organs 46–47, 49
  in textiles 49, 209, 216, 224, 298
  sources of 206–208
  structure and function 27, 28, 206
ficus nymphaefolia 197
Ficus spp 151
Fimbristylis oxytachya 154
fire resistance 52, 156
fire technology 153–157 (*see also*

Aborigines, resource management)
  ecology of 155, 156–157
  study of 153, 154–155
  uses of 153–154
    in conservation 156–157
    in disease control 156
    in farming 167
fish poisons 219–223 (*see also* saponins)
  mode of activity 222, 223
  sources and processing 219–223
  Strychnos nux-vomica 41
flavours (*see* extracts and exudates)
flowers (*see* reproductive structures)
fodder 37 (*see also* pasture, wild plant resources)
folk knowledge 2 (*see also* traditional knowledge)
folk taxonomy (*see* ethnotaxonomic systems)
folk varieties (*see* native crops)
folklore 108 (*see also* myth, witchcraft)
  British 67
  and drug discovery 342–345
food additives (*see* extracts and exudates)
food plants 3, 4, 6, 22 (*see also* crops, nutritional quality, wild plant foods)
food processing (*see also* cycads, *Cycas* spp, manioc)
  consequences of 65, 66, 135–136
  wild plant foods 132, 135–136
food storage (*see also* biological control)
  domesticated produce 167
  experiments in 78
  wild plant foods 132, 136
forests (*see also* agroforestry, environmental economies, non-timber plant products, shifting cultivation)
  benefits of 350
    environmental services 353–354
    products 335, 358
  conservation 83
  management 186, 188, 350–351
  sustainable forestry 350–351

Forests. Trees and People Newsletter 125

Foundation for Ethnobiology 14

Foxglove (*Digitalis purpurea*) 10

fragrances (*see* extracts and exudates)

France 4

Frankenia palmeri 232

fuel (*see also* wild plant resources)
 burning qualities 137, 138
 OXFAM project 138

game attractants 137 (*see also* plant–animal relationships—manipulation of)

Gazelle spp (gazelles) 155

genetic diversity (*see also* native crops)
 and traditional experimentation 78
 and traditional plant management 16, 57, 179, 186, 187
 conservation of 327–329

Genoma deversa 200

genotype 42 (*see also* anthropogenic effects, botanical methods)

Geographical Information Systems (GIS) 125

geophagy 304

geranium 39

Gerard, John 10

Ghana 137, 138

Gidjingali 150 (*see also* Aborigines)
 bush foods 150, 151–152
 cycad processing 135–136

Gilmore, Melvin 8

ginger 229

gorillas 303–304

Gossypium spp (cotton) 4

gourds (*Lagenaria siceraria*) 78

grape (*Vitis vinifera*) 20

grasses (*see also* pasture)
 as food 135, 145, 147–149
 in material culture 49, 200
 teaching among the Maasai 76

Greece 10

Green Revolution 13, 167

Guadua sp 140

Guarea scabra 198

Guatteria olivaceae 94

gymnosperms

as timber 47

reproductive behaviour 20

resins 33, 39

Gynerium satittatum 49

hairs (*see* trichomes)

hallucinogens (*see* psychoactive plants, psychoactive substances)

Hanunóo
 botanical terms 71
 plant classification 69–70
 rice cultivars 165
 subsistence activities 69, 128

Harshberger 1, 6

Hazzard collection 6

healthcare (*see* applied ethnobotany)

heart disease 10

Helianthus annuus (sunflower) 294

hemlock (*Conium maculatum*) 41

hemp (*Canabis sativa*) 46

henbane (*Hyoscyamus niger*) 56

henequen 46

herbal medicine (*see* traditional medicine)

herbals (*see also* pharmacopoeia) 11, 297, 299–300

Hevea brasiliensis (rubber) 4 (*see also* extracts and exudates)

HIV 65, 317, 318–319

Homalanthus nutans 65

Homo erectus 305

Homo sapiens 305, 307

Hopi
 agriculture 128, 171–178
 crop repertoires 165, 174–176, 177–178
 cultural significance 174
 germplasm management 176–178, 327–329
 cultural evolution 172–173
 etymology of plant names 7
 physicochemical environment 173–174

hormones in plants
 insect pheromones 55
 oestrogen-like 39

horticulture (*see* agriculture)

human behaviour (*see also* behavioural scripts, cognitive

ethnobotany, environmental
perception)
and cultural ecology   75
as a reflection of knowledge 62
human sacrifice (*see* cultural ecology)
Humirianthera rupestris   135
hunter-gatherers
  delayed-return   136
  extant groups   128–129
  immediate-return   136
Hyoscyamus niger   56

IATSIS (Institute of Aboriginal and
  Torres Strait Islander Studies)
  13
icthyothereol   222
Imperata cylindrica   51
INBio   366–368
Inca   4
index of saliency   95
India   13, 131
indices of agreement   103, 104
indigenous agricultural knowledge
  (IAK)   60 (*see also* traditional
  knowledge)
*Indigenous Knowledge and
  Development Monitor*   14
indigenous organisations   88
indigenous peoples 59–61 (*see also*
  traditional peoples)
  numbers of   127
indigo   233
Indonesia   76
infrared (IR) spectroscopy   121
information systems   123, 125,
  301–302
in-law avoidance (see cultural
  ecology)
insect repellents   56, 65 (*see also*
  biological control)
integrated healthcare systems   82
integrated knowledge systems (*see*
  knowledge systems)
intellectual property rights   368, 371
International Code of Botanical
  Nomenclature (ICBN)   117,
  118
International Phonetic Association
  (*see* linguistics)
Ipomoea sp   149

Iranthera juruensis   198
Iranthera paraensis   198
Iriartella setigera   204
ironwood   274
  sculptures   214
Iversen, Johannes   12

Jacaranda copaia   243
Jaltomata procumbens   118
Jatropha cuneata   212
Jessenia bataua   46, 51
jojoba   234
Josselyn, John   5, 62
Journal of Ethnobiology   17
Journal of Ethnopharmacology   3, 17
Juncus effusus   42
jute   46

Ka'apor
  as shifting cultivators   180
    crop repertoires   165
    manipulating species
      composition   185
    pest control   168
    recognition of resource units
      181
  burial   72
  resin classification   267–268
Kadavakurichi reserve   191
Kakadu National Park   325, 326
Kalanchoe spp   52
kapok (*Ceiba pentandra*)   49
kava (*Piper methysticum*) (*see also*
  psychoactive substances)
  myths   109
  properties of   110–111
  symbolic analysis of myths
    108–110, 343
Kayapó
  as shifting cultivators   128, 180
    ecological repercussions   186
    management techniques   186,
      188
  powerful plants   251
  use of *Bixa orellana*   113
Kenya (*see* Pokomo)
Khaya spp (African mahogany)
  116
khellin   316
knowledge (*see also* knowledge

systems, traditional
      knowledge)
   in human adaptation   59
knowledge systems (*see also*
      traditional knowledge)
   integrated (IKS)   61, 79, 82, 84
   structure of   15, 73
   Western   85, 86
Koobi Fora   287
Krameria grayi   212
Krugiodendron   196

lac 226–227
Lagenaria ciceraria (gourds)   78
Lake Coari   137, 168
   agriculture   161–162
   ecological knowledge   137, 139–141
   economic calendar   142–143
   resource units   139–141, 161–162
land use
   activity   254
   classification and planting patterns
      161–162, 163
   strategies   356
landraces (*see* native crops)
language
   revival of indigenous language   88
lanoxin   11
latexes (*see also* extracts and
      exudates)
   accumulation in laticifers   31
   sources of   221
laticifers (*see* latexes)
Laurus nobilis (sweet bay)   31
leaves (*see* material culture, plant
      structures and functions)
legislation (*see also* ethical
      considerations)
   Agenda   21, 365
   Convention on Biological Diversity
      366
   Declaration of American
      Independence   6
   Declaration of Indigenous Rights
      88
   for conservation   320–321, 323,
      362–363, 364
   for the protection of indigenous
      peoples   363, 364–368, 369,
      370–371

   equitable distribution of revenue
      363, 364, 366–368, 370–371
   farmer-breeders rights   370
   UN Convention on Environment
      and Development (Earth
      Summit)   88
   Wilderness Act 1964   322
   World Conservation Strategy   80
legumes   205
Leucopoa kingii   45
Lévi-Strauss,Claude   255
Lewington, Anna   14
Licania spp   50, 51
lignin (*see also* wood)
   and specific gravity   195
   distribution   23
   nature and functions   27, 40, 53,
      54, 56
Ligustrum ovafolium (privet)   67
lily
   Lilium spp   67
   Nymphaea spp   135
linguistics
   detection and uses of cognates
      112
   in ethnotaxonomic study   256,
      263–264
   International Phonetic Association
      (IPA)   112
   methods   106–108, 112
   phonological variation and
      synonymy   276
   role in ethnobotanical study   8,
      124
   role in palaeoethnobotanical study
      301–302
Linnaean classification
   nature of   68
   plant classification   116, 117–118
   plant identification   114, 116–117
   plant nomenclature   118
livestock (*see* pastoralists)
Livistonia humilis   151
Lophocereus schotti   55
low-input agriculture (*see also*
      shifting cultivation, farming
      practices, art and technology,
      Hopi, Kayapó)
   dependence on   128–129, 159
   efficiency   171

experimentation 78–79, 167–168
fertilisation 56, 72, 167
in the American Southwest
171–178
pest control 167, 168
productivity in 167–168, 170, 171
types of 160, 161
Lycianthes moziniana 65

Maasai
as pastoralists 128, 144
classification of grasses 147
rangeland conservation 324–325
Ma'Betisék 246
Macrozamia spp (cyads) 135–136
Madagascar 82, 83, 330
magpie goose 157
maize (*see also Zea mays*) 182, 291,
327, 329
Hopi varieties 165, 174–175, 177
growth requirements 163
male supremacy (*see* cultural
ecology)
malus pumila (apple) 20, 21
mandrake 343
mangroves 48
Manihot esculenta (*see* manioc)
manioc (*Manihot esculenta*) (*see also*
Aguaruna, native crops)
cultivation 77, 127
ethnotaxonomy 274
growth requirements 183
processing 66
yields 170
Marongarive Special Reserve 330
Marovo Project 84, 93
material culture (*see also* art and
technology, extracts and
exudates, fibres, timber)
categories of plant use 191
non-timber plant products 192
(*see also* non-timber plant
products)
uses of timber 192
cultivated plants 209
experiments in 213–215
resource management 209–210
role of plant structures
flowers 49
leaves 46

microscopic structures 50, 51
roots 49
seeds 49
stems 46–47
Mauritia carana 200
Mauritia flexuosa 139, 141
Maya (*see also* ethnotaxonomies)
early studies 4
medicinal plants 238
MEDFLOR 125
medicinal plants (*see also* drugs,
ethnopharmacology,
traditional medicine)
chemistry and pharmacology
238–240
conservation of 336–338
in Asia 2
in herbals 10
in South America 238
of the Chippewa Indians 63
medieval
economic plants 9
herbals 10
powerful plants 251
Melaleuca alternifolia (tea tree) 9,
65
Melaleuca dealbata 200
Metaleuca spp 47
Mentha citrata 68, 257
Mentha spicata 68, 257
Mesoamerica 4
Mesquite 204, 275
mestizo 191, 193–194
methyleugenol 55
Metroxylon warburii (sago palm)
46
Mexico 65, 72, 258
microscopy 116
scanning electron microscopy
(SEM) 116
transmission electron microscopy
(TEM) 116
Microseris scapigera (daisy yam)
135
Middle Ages (*see* medieval)
millet (*see also Panicum*) 135
Mimosa acaciaoides 242–243
Mimosa albida 112
Minquartia guianensis 194
mint 39, 229

molecular biology
  in plant taxonomy   118
  role in ethnobotany   15, 124
Monanthochloe   138
morphometric analysis (*see* botanical
    methods, domestication)
Morris, Brian   9
Mossi (*see* ethnotaxonomies)
Musa spp (bananas and plantains)
    127, 170, 183
Musa textilisi   46
myth (*see also* kava, symbolic
        analysis, witchcraft)
    and environmental perception   75,
        246
    as a source of ethnobotanical
        data
    representation of plants in   13,
        246, 76, 108, 110

NAPRALERT [National Products
    Alert)   66, 125, 238, 241
National Cancer Institute (NCI)   64,
    237
national parks (*see also* Kakadu,
        rangelands)
    conservation of   156–157
    in Africa   156
    in Australia   157, 325
    indigeneous involvement in
        325–326
    Uluru-Kata Tjuta National Park
        325
native crops (*see also* Hopi, Ka'apor)
    as a source of genetic diversity
        78, 165, 166, 168–169, 174, 178,
        179
    crop repertoires   165–166
    experiments in   13
    folk varieties   165, 175–176
    management   166–171, 179,
        181–184
    species   13, 174–177
    yields   171
neem   56
Neurolaena lobata   101
New England   5
New Guinea   8
New Mexico   14
New World   4, 300

Nicotiana tabacum (tobacco)   3, 4, 5,
    252
nicotine   41
Niger   137
nomadic (*see also* hunter-gatherers,
        pastoralists)
    agriculture   186
non-governmental organisations
    (NGO's)   84
non-timber plant products (NTPP's)
    (*see also* forests, environmental
        economics)
    developing markets   349
    economic potential   331–336
    environmental impact of
        extraction   361–362
    in sustainable development   341
    nature of   192
North America
    early settlers   6
    study of ethnobotany in   3, 6–8,
        14
Norway   12
nutrients
    required for plant growth   37
    synthesis in plants   37, 55
nutritional quality
    and plant defence   53
    of cereals   56
    of fodder plants   137
    of pasture   147
    of traditional foods   64, 65
        plant organs   134
        wild plants   130, 134, 135–136
    variation in   56
Nymphaea spp   135, 151–152

oak (*Quercus suber*)   47
Oceania   128
oedema (*see* heart disease)
oil palm (*Orbignya oleifera*)   56
Old World   9, 300
Olea spp (olive)   56
Oncoba spinosa   197
onion (*see* Allium cepa)
opium poppy (*Papava somniferum*)
    11, 31, 216
Optimal Foraging Theory (OFT)   74
    application to hunter-gatherer
        research   74

Marginal Value Theorem (MVT) 74
oral tradition 76
Orbignya oleifera (oil palm) 56
organ pipe cactus 275
organic chemistry
organisations (*see also* indigenous
    organisations and individual
    entries)
  ANPWS (Australian National
    Parks and Wildlife Service) 325
  CHIMPP 304
  CIRAN 14
  Forest, Trees and People Network
    125
  Foundation for Ethnobiology
    (FEB) 14
  IATSIS (Institute for Aboriginal
    and Torres Strait Islander
    Studies) 13
  Indigenous Knowledge Resource
    Centres 125
  National Cancer Institute (NCI)
    237
  OXFAM 138
  Shaman Pharmaceuticals Inc. 16
  TRAMIL 329–330
  UNESCO 14
  United Nations 88, 365
  WHO 234–235
  Working Group on Traditional
    Intellectual, Cultural and
    Scientific Resources 368, 369
  WWF 14, 84
Oryza barthii 148
Oryza spp 148
overharvesting 213–214, 359–361

Pacific Islanders (*see also* kava) 128
palaeobotany 12 (*see also*
    palaeoethnobotany)
palaeoethnobotany (*see also* pollen
    analysis, plant fossils,
    archaeobotanical methods,
    applied palaeoethnobotany,
    domestication)
  as a field of ethnobotany 15, 17
  definition of 11, 278
  history of 12
  interpreting the evidence 300–304

methods of study 119–122,
    280–284
  plant–human relationships
    304–311
  Pueblo Indians 6
  sources of evidence 279–300
    art and literature 294–300
    living plants 289–294
    the archaeobotanical record
    280–289
palaeopathology (*see*
    archaeobotanical methods)
Palmer, Edward 6
palms
  as food 56, 151
  economic significance 331
  in material culture 46, 51, 200,
    206, 214
  in soil management 184
  management of 214
  symbolic role 67, 68, 75, 250
palynology (*see* pollen analysis)
Panama hat palm (*Carludovica
    palmata*) 46
Pandanus spp 141, 151
Panicum australiense 154
Panicum decompositum (wild millet)
    135
Panicum laetum 135
Panicum turgidum 149
papaya 250
paper 49
paper stitch (*Betula papyrifera*) 201,
    211
Paraguay 131
participatory ethnobotanical
    research (PER) 80
participatory research 79–81, 87–88
  projects (*see also* Plants and People
    initiative)
    Australia 84
    Solomon Islands 84
pastoralists
  extant group 128–129, 137,
    144–145
  livestock management 146
  resource knowledge
    availability 137, 147
    fodder 137
    food 144, 145, 147–149

resource management 144–149,
155–157
pasture
ethnotaxonomy 147
management 153, 156–157
quality 144, 145, 147
peasant communities 128
Pennsylvania 6
People and Plants initiative 14,
82–83, 329–330
peppermint 230
peppers (*Capsicum* spp) 168
bell pepper 253
chili peppers 229, 230
Peru 163 (*see also mestizo*,
Aguaruna, Amuesha, Bora)
pest control (*see* biological control)
Petrohassa smithii (partridge pigeon)
157
pharmacognosy 11, 124
pharmacology 11
pharmacopoeia (*see also* herbals)
Caribbean 330
traditional 238, 300
US Pharmacopoeia 63
Phaseolus spp (beans) 174, 175
phenolics
effects on plant growth 39
nature and functions 38, 39, 40, 41
synthesis 38
phenotype 42
Philippines (*see also* Hanunóo) 69,
253
phonological variation (*see*
linguistics)
photosynthesis (*see* plant
biochemistry)
phytochemistry (*see also* botanical
methods, secondary products)
exploiting variation 55–57
phytogeography (*see* domestication)
phytoliths
definition of 50
in palaeoethnobotany 282, 286, 287
in plant identification 116
Picea spp 47
Pieris brassicae (cabbage white
butterfly) 42
pigments (*see also* plant–animal
relationships)

extraction and processing 233
sources of 51, 220, 224–225
pile sorting (*see* anthropological
methods)
Pimenta dioica (allspice) 4
pine 47
Pinus spp 39, 47
Piper jaquemontianum 101
Piper methysticum (kava) 108–111
Piper nigrum (black pepper) 8
Piper wichmanni 110
Pitjantjatjara (*see also* Aborigines)
255
pituri (*Dubiosia* spp) 13
PLANIMAL (Plant–Animal
Interactions) 125
plant–animal relationships (see also
plant–human relationships)
chemical mediation of 38, 53–55
in cabbage 42
in cacti 55
manipulation of 129, 137, 141–143,
154, 167, 168
mutualisms 54, 55
plant behaviour (*see also* defence
mechanisms in plants)
and classification 21
life cycles 24
plant biochemistry (*see also* plant
structures and functions)
biosynthetic pathways 43–44
enzymes 42–44
in plant identification 116, 118
in stress adaptation 52
nutrient assimilation 36
photosynthesis 22, 36, 45
primary and secondary pathways
34, 38
respiration 36
plant cells (*see also* plant tissues)
major components of 23–26
plant fossils (*see also*
archaeobotanical methods,
palaeoethnobotany, pollen
analysis)
macrofossils and microfossils 12,
280–283
methods of study 119–122
nature and formation of
280–284

sources of
  from animal remains   281,
    285–286
  from tools   122, 281, 283–284,
    287–289
plant growth   31–35
plant–human relationships (*see also*
    agriculture, anthropogenic
    effects, domestication)
  nature and evolution of   304–311
  humans in plant evolution
    308–311
  plants in cultural evolution
    307–308
  plants in human evolution   305,
    307
plant-lore studies   13, 67–68
plant names (*see also* ethnotaxonomic
    nomenclature)
  analysis of   7, 8, 112
  in Linnaean taxonomy   118
plant processing (*see also* extracts
    and exudates,food processing)
  bark technology   211–212
  experiments in   213–215
  fibre processing   212–213
  medicinal plants   242–243
  wood technology   210–211
plant structures and functions (*see
    also* material culture,
    reproductive structures)
  cotyledons   21
  emergences   50
  leaves   10, 12, 22, 23, 44–46
  pollen   11–12, 50, 51, 116–117,
    120
  roots   22, 47–49
  seeds   12, 23
  stems   22, 46–47
plant tissue (*see also* plant cells, cork,
    plant structures and functions,
    wood)
  distribution in plants   28–33
  epidermal   27, 31–33
  meristematic   22, 27
  parenchymatous   23, 27
  secretory   27, 30–31
  supporting   23, 27
  vascular   21, 23, 27, 28, 31, 33
plantain (*Musa* spp)   170, 182, 185

Pokomo
  material culture   191, 193, 205, 206,
    208
  medicinal plants   239
pollen (*see* plant structure and
    functions)
pollen analysis (*see also*
    archaeobotanical methods,
    plant fossils)
  as a dating method   123
  in vegetation change   12, 123, 282,
    283–284
    as an indicator of environmental
      change   11, 306
    as an indicator of human activity
      13, 305, 307
  methods   120
  palynology   10–13
Polysphaeria multiflora   197
potato (*Solanum tuberosum*)   4, 22,
    78
Powers, Stephen   6
preference ranking (*see*
    anthropological field
    methods)
privet (*Ligustrum ovafolium*)   67
Projeto Nordeste   83
Prostratin   65, 237, 243 (*see also*
    HIV)
Protium spp   138, 186
Provir   231
Prunus africanus   83, 359–361
Pseudomyrmex   54
psychiatric disorders   10
psychoactive plants   4, 11, 31, 216,
    108–11
  in art   295, 296–298, 301
  in shamanic healing   240–242
  ritual use of   5, 108–109
psychoactive substances
  and avoidance behaviour
  nature and activity   241–242
  sources   111, 241
publications   3, 14, 17, 125 (*see also*
    individual entries)
Pueblo Indians   6, 51

Quercus suber (cork oak)   47
quinine   217
quinones   40

radish (*Raphanus sativus*) 42
rangelands (*see also* resource
    management)
  conservation of 156, 324–325
  in Africa 155, 324–325
  traditional management of 13,
    154
rattan (*Calamus* spp)
  material culture 46, 201
  resource management 213–214
reproductive behaviour (in plants)
  20, 22, 23
reproductive structures (in plants)
  22, 23 (*see also* material
  culture, plant structures and
  functions)
research methods (*see also*
    anthropological field methods,
    archaeobotanical methods, art,
    botanical methods, ecological
    methods, environmental
    economies, ethnobotany,
    linguistics)
  criticisms of 66
  general ethnobotanical techniques
    91–118
  general palaeoethnobotanical
    techniques 119–122
  specialist techniques 122–123,
    124, 160, 236–237, 254–277
resin (*see also* extracts and exudates)
  classification 267–268
  nature and sources 39, 220, 231,
    233
resource management (*see also*
    farming practices, fire
    technology, forests)
  by agriculturalists 160–189
  by hunter-gatherers 152–157
  by pastoralists 144–149, 152–157
  influence of sociocultural factors
    68, 72, 75, 153
  influenced by environmental
    perception 246
Rhizophora spp 48, 49
rice (*Oryza* spp) 165, 253, 327
Richards, Paul 13, 85
ritual and ceremony (*see also* cultural
    ecology)
  and floral symbolism 68

behaviour 5, 8, 72, 250–251, 255
  repercussions of 72, 73, 75, 155
roots (*see* material culture, plant
    structures and functions)
Rosa canina 118
rose 39, 230
Royal Botanic Garden, Kew (*see also*
    People and Plants initiative)
  14, 125
rubber (*Hevea brasiliensis*) (*see also*
    extracts and exudates) 4, 39
Rudgley, Richard 14
Runa (*see* shifting cultivation)
Rungruw 287–289
rural peoples' knowledge (RPK) (*see
    also* traditional knowledge)
  60
Ryania spp 137

Saccharum officinarum (sugarcane)
  9
safflower 224
sago palm 46
Sagotia racemosa 186
Salicornia europa 52
Samoa 64, 65, 237, 243
San Francisco 16
San Salvadsor 4
SAPIR (South American Prehistory
    Inference Resource) 301–302
saponins (*see also* fish poisons)
  nature and activity 40
  sources and uses of 221, 227–228
Sarothamnus scoparius (broom) 68
scanning electron microscopy (*see*
    microscopy)
Scheelea zonensis (corozo palm)
schottenol 55
secondary products (*see also*
    alkaloids, essential oils,
    extracts and exudates, latexes,
    lignin, phenolics, plant
    biochemistry, resins, saponins,
    sulphur compounds, tannins,
    terpenoids, toxins)
  ecological roles 51–52, 53, 54
  nature of 31, 33, 39, 40
  uses of 56, 220–221
seeds (*see* material culture, plant
    structures and functions)

SEPASAL (Survey of Economic Plants for Arid and Semi-arid Lands) 125
Serengeti 324–325 (*see also* rangelands)
Seri
    material culture 205, 208, 212–213
        basketry 212–213
        ironwood sculptures 214–215
    medicinal and powerful plants 239, 251
    wild plant resources 130, 131, 133, 138
shade plants 51
Shaman Pharmaceuticals Inc. 16, 321–322
shamanism (*see* ritual and ceremony, traditional medicine)
shifting cultivation (*see also* Kayapó, low-input agriculture, swidden agriculture)
    definition of 161, 178, 180
    in Amazonia 178–188
        extant societies 180
        crop management 181–188
        fallow management 184–186, 187, 209–210
        soil management 162, 182–184
Sicily 10
silica
    in plants 50, 51, 54
    uses of 51
Sinapis alba 116
sinigrin 41, 42
sisal 46
social factors (*see* traditional knowledge)
Societies 8, 13
socio-cultural analysis 68
Socrates 41
soil (*see also* ethnotaxonomic classification, ethnotaxonomies)
    management 166, 167, 181–184
        fertility 167
        stability 19, 51, 156
    study of
        classification 162, 163, 259
        ecological maps 138
Solanum chippendalei 136

Solanum tuberosum 4
Solomon Islands 84
Somali pastoralists 137
Sorghum 290–291
South America 4, 128
Spanish 4, 6
Spartina pectinata 45
species diversity 122
spores 12. 50
sporopollenin 12
spruce 47
Spruce, Richard 4, 5, 6
stems (*see* material culture, plant structures and functions)
Steward, Julian 73, 74
stinging nettle (*Urtica dioica*) 50
strychnine 41
Strychnos nux-vomica 41
suberin 33, 35
subsistence
    and environmental perception 252
    and ethnotaxonomy 270–271
    and knowledge 77, 130
    strategies 127–129
sugar 9
sugarcane (*Sacharum officinarum*) 9, 52, 127
sulphur compounds
    functions 54
    nature and activity 39, 40, 41
sumpweed 293
sunflower 293
sustainable development 331–342 (*see also* biocultural reserves, environmental economics, forests, NTPPs, sustainable plant use)
    commercialisation and conservation 340–342
    developing markets 335–336
    economic value of NTPPs 331–335
    problems 347–362
    protecting local resources 331–332, 336–340
    solutions 362–373
    cultural impact assessments 338–340
    versus other development initiatives 340–341

sustainable plant use 82–83
sweet potato (*Ipomoaea* spp) 182, 183, 253, 289
swidden agriculture (*see also* shifting cultivation)
    changes in species composition 180–181, 209–210
    swidden cycle 180–181, 184, 185, 187, 210
symbolic analysis (*see* art, kava, myth)
Syzygium spp 151

taboos
    and plant use 194
    cultural ecology 74, 75
Tacca leontopetaloides (Polynesian arrowroot) 151
Tannins
    economic importance 233
    functions 54
    nature and activity 40
    sources 220, 224, 226
taro 288–289
tarragon (*Artemisia dracunculus*) 30, 31, 32, 33, 39, 229
taxonomy 256–257 (*see also* botanical methods, ethnotaxonomic systems, Linnaean classification)
tea (*Camellia sinensis*) 9
tea tree (*Melaleuca alternifolia*) 9
technology (*see* art and technology)
Terminalia ivorensis 200
terminalia spp 151
terpenoids (*see also* cardiac glycosides)
    functions 39, 54
    nature and activity 39, 40, 41
textiles (*see* fibres)
Themedra triandra 156
Theobroma cacao (cocoa) 4
timber (*see also* forests, wood)
    management of 209, 210
    sources of 193–195
        factors affecting use 195–199
    uses of 192–193, 194
Tobacco (*Nicotiana* spp) 3, 4, 5, 242
tools (*see* art and technology)

toxins (*see also* alkaloids, arrow poisons, cardiac glycosides, cyanogenic glycosides, cycads, fish poisons) 219–223
    in fodder and fuels 137, 138, 147
    in plant defence 53
    in seeds 39
    in Solanaceae 56
traditional knowledge (*see also* knowledge systems, traditional scientific and cultural property, Western scientific knowledge)
    and Western scientific knowledge 79–88
        integrated studies 63
        learning from 79–80
        Western view of 61, 84–85
    characteristics of 85, 86
    dissemination 76, 77
    documentation and interpretation 61–79
    dynamics and distribution 73, 76–78
    indigenous agricultural knowledge (IAK) 60
    indigenous technical knowledge (ITK) 60
    protection of 85–88
    rural peoples' knowledge (RPK) 60
    sociocultural influences 68, 76–78, 247, 249–253
    traditional agricultural knowledge (TAK) 17, 51, 159–189
        and conservation 160–171, 327–329
        and rural development 2
        research into 8, 160–161
    traditional botanical knowledge (TBK)
        definition of 60
        of the Hanunóo 71
        research into 13, 15, 62–73
    traditional ecological/environmental knowledge (TEK)
        and wild resources 129, 137–157
        definition of 60
        in agricultural practice 160–189

in managing national parks 156–157, 323–326
of hunter-gatherers 80–81, 149–157
traditional scientific knowledge (TSK) 82, 84
ethnoscientific knowledge 59, 60
traditional medicine (*see also* efficacy, ethnopharmacology, medicinal plants)
definition of 234
dependence on 234
plant remedies 237–243
study of 10, 236–237
traditional medical practitioners (TMPs) 236, 237
traditional medical systems 6, 13, 76, 234–237
herbalism 6, 11, 13, 235
shamanism 235
traditional peoples
changing attitudes towards 15
cultural diversity 3, 83, 88
in:
Africa 128, 137, 144, 147
Asia 128, 131, 133, 246
Central America 68, 128, 257
Europe 128
North America 4, 6, 7, 88, 128
Oceania 8, 128, 150
South America 128, 187, 238
USSR 128
traditional phytochemistry (*see* extracts and exudates)
traditional plant management (*see* domestication, low-input agriculture, rangelands, wild plant resources)
traditional scientific and cultural property 88, 91, 92
training 83 (*see also* education)
TRAMIL 329–330
Trema micantha 184
Tichocereus pachanoi 301
trichomes (*see also* fibres)
definition of 50
nature 31
types 31–33, 49, 50
Triticum aestivum (wheat) 21

domestication of 291–292, 293–294
yields 170
Triticum dicoccoides 289
Triticum timopheevii 289
tubocurarine 221

United Nations (*see also* legislation) 14
United States (*see* North America, traditional peoples)
unlucky plants
in Amazonia 765
in UK 67, 68
Upper Volta 138
Urtica dioica 50
use-values (*see* anthropological field methods)
use-wear analysis 287 (*see also* archaeobotanical methods)
uses of plants (*see also* art, art and technology, environmental functions, food plants, material culture, medicinal plants, ritual and ceremony, wild plant resources)
and overharvesting 213–215
reflected in plant names 269–277
research into 62–66
utensils (*see* art and technology)
utilitarian ethnobotany (*see also* efficacy)
history of study 62–63, 64–66
nature and limitations 63

vegetable oils 56
vegetation change (see anthropogenic effects, pollen analysis)
vegetation management (*see* anthropogenic effects, fire, technology, shifting cultivation, wild plant resources)
Vigna spp 152
viral diseases (*see also* HIV] 65, 317, 321–322
Virend 231
Vitex gaumeri 101
Vityus vinifera (grape) 20
volatile oils (*see* essential oils)
von Post, Lennart 12

voucher specimens (see botanical methods)

Wadi Kubbaniya 287
Waimiri Atroari
  extracts and exudates 230
  food and fuel 127, 130, 131, 133, 135, 137, 138
  material culture 49, 113, 191, 193, 194, 205, 208
  medicinal plants 238, 239
  wood technology 210–211
water
  and disease 254–255
  in traditional agricultural systems 161
  management of 167
waxes 226–227
Western Desert Aborigines (see also Aborigines) 75, 130, 131, 133, 136
Western scientific knowledge (WSK) 79–80 (see also knowledge systems, traditional knowledge)
  characteristics of 85, 86
  Green Revolution 13
  interest in traditional knowledge 3
  rediscovering the wheel 79
wheat (see also Triticum aestivum) 305, 327
white birch 201
white mangrove 204
wild plant foods (see also food plants, food processing, food storage) 8, 65
  and mode of subsistence 147–152
  dependence on 127, 129
  exploitation and management 149, 151–152, 154
  organs exploited 132, 133, 135, 148
  range and quality of 130–132
wild plant resources (see also wild plant foods) 127–158
  dependence on 127–129
  exploitation and management 137–141, 141–157, 209–210
  fodder and fuel 128, 137
  knowledge of 127–158

material culture 209–210
  resource units 138–139
wildebeest 155
witchcraft (see also folklore, myth) 343–4
Withering, William 10, 11
woad (see also dyes) 224
women
  and environmental perception 253
  and specialist knowledge 77
  as cultivators 77
wood (see also art and technology, material culture, plant processing, plant tissues, timber)
  fossilised 12
  specific gravity 195–196
  tissues 27
Working Group on Traditional Intellectual, Cultural and Scientific Resources 368, 369
World Conservation Strategy 80
World Directory of Ethnobotanists 125
world history, generalised scheme of 306
WWF 14, 84
Wyrie Swamp 156

yams
  domestication of 294
  ethnotaxonomy of 264
  growth requirements 183
  management of 75, 135, 182
  toxic strains 304
Yanomami 128, 180–181
yellow cedar 206
yellow mustard 116
Yolngu (see also Aborigines)
  land allocation and resource management 68
  numbers of 150
Yucca spp 33
Yucca Mountain, Nevada 339–340

Zambia 61, 162
Zea mays (see also corn) 4
Zetzsche, Fritz 12
zoopharmacognosy 111, 303–304